Progress in Scientific Computing
Volume 7

Large Scale Scientific Computing

P. Deuflhard,
B. Engquist,
Editors

1987

Birkhäuser
Boston · Basel · Stuttgart

P. Deuflhard
Konrad-Zuse-Zentrum
für Informationstechnik
D-1000 Berlin 31
Federal Republic of Germany

B. Engquist
Department of Mathematics
University of California at Los Angeles
Los Angeles, CA 90024
U.S.A.

Library of Congress Cataloging in Publication Data
Large scale scientific computing.
(Progress in scientific computing ; v. 7)
Proceedings of a meeting on large scale scientific computing held at the Oberwolfach Mathematical Institute, July 14–19, 1985, under the auspices of the Sonderforschungsbereich 123 of the Univ. of Heidelberg.
Bibliography: p.
1. Science—Data processing—Congresses.
2. Engineering—Data processing—Congresses.
3. Science—Mathematical models—Congresses.
4. Engineering—Mathematical models—Congresses.
5. Computers—Congresses. I. Deuflhard, P. (Peter)
II. Engquist, Björn, 1945– . III. Sonderforschungs-bereich 123—"Stochastische Mathematische Modelle."
IV. Series.
Q183.9.L37 1987 502.8'5 87-10340

CIP-Kurztitelaufnahme der Deutschen Bibliothek
Large scale scientific computing / P. Deuflhard ;
B. Engquist.—Boston ; Basel ; Stuttgart :
Birkhäuser, 1987.
(Progress in scientific computing ; Vol. 7)
ISBN 3-7643-3355-3 (Basel . . .)
ISBN 0-8176-3355-3 (Boston)
NE: Deuflhard, Peter [Hrsg.] ; GT

ISBN 0-8176-3355-3
ISBN 3-7643-3355-3

Printed and bound by R.R. Donnelley & Sons, Harrisonburg, Virginia.
Printed in the U.S.A.

9 8 7 6 5 4 3 2 1

PREFACE

In this book, the new and rapidly expanding field of *scientific computing* is understood in a double sense: as computing for scientific and engineering problems and as the science of doing such computations. Thus scientific computing touches at one side mathematical modelling (in the various fields of applications) and at the other side computer science. As soon as the mathematical models describe the features of real life processes in sufficient detail, the associated computations tend to be *large scale*. As a consequence, interest more and more focusses on such numerical methods that can be expected to cope with large scale computational problems. Moreover, given the algorithms which are known to be efficient on a traditional computer, the question of implementation on modern supercomputers may get crucial.

The present book is the proceedings of a meeting on "Large Scale Scientific Computing", that was held at the Oberwolfach Mathematical Institute (July 14-19, 1985) under the auspices of the Sonderforschungsbereich 123 of the University of Heidelberg. Participants included applied scientists with computational interests, numerical analysts, and experts on modern parallel computers. The purpose of the meeting was to establish a common understanding of recent issues in scientific computing, especially in view of large scale problems. Fields of applications, which have been covered, included

semi-conductor design,
chemical combustion,
flow through porous media,
climatology,
seismology,
fluid dynamics,
tomography,
rheology,
hydro power plant optimization,
subway control,
space technology.

The associated mathematical models included ordinary (ODE) and partial (PDE) differential equations, integral equations (IE) and mixed integer-real optimization problems.

The present book comprises 6 parts to be described here from an introductory point of view. For all chapters, *nonlinearity* plays a dominant role. Moreover, the book reflects an interesting issue of present research: ODE techniques of a rather high degree of sophistication gradually intrude into the PDE region. For this reason, the first two parts jointly cover ODE and PDE problems and techniques.

Part I covers initial value problems (IVP) for ODE's and initial boundary value problems (IBVP) for parabolic PDE's - recall that a method of lines approach to such IBVP's anyway leads to large scale stiff IVP's of a special structure. Chapter 1 (Bank, Fichtner, Rose, Smith) deals with computational techniques for semiconductor device modelling, which plays an important role in chip fabrication. Mathematically speaking, the models represent 3-D nonlinear PDE's. At present, the 2-D case is regarded as tractable (multigrid techniques, moderate size models), whereas the 3-D case is still in its infancy. In chapter 2 (Yserentant) the new and fascinating "hierarchical" finite element (FE) approach, which has been developed for 2-D elliptic PDE's, is transferred to the time-dependent case. The new approach seems to be rather promising especially for essentially parabolic PDE systems that are strongly coupled. After spatial discretization, where typically nonlinearity requires moving grid points, implicit ODE's or differential algebraic equations (DAE's) arise - which are treated in chapter 3 (Deuflhard, Nowak). In this chapter rather recent, highly sophisticated extrapolation techniques are described including order and stepsize control and index monitoring (recall that index >1 DAE's require further analytical preprocessing). The only illustrative example of this chapter stems from combustion. The next chapter (Warnatz), written by a combustion expert, gives a general survey of the demands arising from computational combustion. These include the numerical solution of auto-ignition problems (0-D), stationary flame problems (1-D), and instationary ignition/quenching problems (1-D). Both the above extrapolation techniques and recent multistep techniques play a role in that field. In chapter 5 (Fu, Chen) physical considerations led to the suggestion of a numerical scheme valid for large time scales. The method represents an asymptotic expansion around the equilibrium points of the physical systems under consideration. Finally, the last chapter (Knabner) could as well be subsumed into part II of the book: this chapter deals with both saturated and unsaturated see page flow through porous media, which mathematically leads to a mixed parabolic-elliptic free boundary value problem. As the paper focusses slightly on the question of time discretization and time step control, the editors decided to put it into part I.

Part II covers boundary value problems (BVP's) for both ODE's and elliptic PDE's. Unless the elliptic problems are embedded into parabolic problems (see part I), the typical nonlinearities of real life models strongly suggest the use of *numerical continuation techniques* - which are described in the first 3 chapters of this part. In chapter 7 (Deuflhard, Fiedler, Kunkel) a rather efficient numerical pathfollowing technique, known from bifurcation diagram computations in algebraic equations, is transferred to the case of ODE-BVP's. The method described carries over to PDE problems, wherein the arising large linear systems can be solved by a direct sparse solver (compare e.g. chapter 20). Illustrative numerical comparisons with competing techniques are included. In chapter 8 (Jarausch, Mackens) 2-D nonlinear elliptic PDE problems are treated. The treatment applies the hierarchical basis techniques (cf. chapter 2). The main idea of the chapter is to condense the effect of the nonlinearity numerically in a small system - which is then treated using a variant of the continuation techniques of chapter 7. The following chapter 9 (Giovangigli, Smooke) suggests a special ODE/PDE version of the so-called pseudo-arclength continuation method. In order to preserve the block structure of the arising linear systems, the embedding parameter is added as a trivial ODE - thus leading to continuation only in a selected boundary value. The method is illustrated by impressive combustion problems including extensive chemical mechanisms. The last two chapters of this part deal with *hybrid algorithms*. In chapter 10 (Ascher, Spudich), which is joint work of a numerical analyst and a seismologist, roughly 10^4 BVP's are to be solved in order to evaluate seismograms. Analytical preprocessing techniques include Hankel and Fourier-Bessel transforms. For actual computation a well-designed mixture of the known approaches is applied: collocation, orthogonal function expansion, multiple shooting (via transfer matrices). The last chapter (Hebeker) of this part aims at special 3-D Stokes problems. The hybrid nature of the advocated approach shows up in the mixed use of fundamental solutions, boundary element method and spectral method. One of the illustrative examples is parabolic, so that part I might also have been a place for presentation. At the same time, this chapter also touches the next part of the book.

Part III deals with hyperbolic fluid dynamics, a traditionally central part of scientific computing that is well represented in the literature. For this reason, only two state-of-the-art papers are included here. The first chapter (Engquist, Harten, Osher) derives new, rather general shock-capturing schemes of high order. The schemes automatically adjust the locally used computational stencil without making use of limiters - a distinguishing feature of this approach. The given treatment covers 1-D shocks and uniform grids. The efficiency of the smoothing around shocks is demonstrated by the illustrative example of Euler equations for the polytropic gas. In the next chapter (Rizzi) sophisticated 3-D computations for the compressible or incompressible Euler equations are presented using a finite volume method with up to 6.10^5 grid cells! The question studied is whether vorticity in the Euler equations is created by discretization errors or is a phenomenon depending on nonlinearity - a question, which seems to be far beyond elucidation by analytical techniques. In the examples included (traverse circular cylinder, two delta wings, prolate spheroid) both situations occur. In addition, one of the examples seems to exhibit a spiral singularity - an observation that will certainly stimulate further investigations (both numerical and analytical).

The following *part IV* shortly surveys some recent results for inverse problems. The first two chapters deal with two alternative ways of treating the inverse Radon transform problem arising in *computer tomography*. In chapter 14 (Louis, Lewitt) certain optimal filtering techniques are proposed and worked out to solve the 3-D discrete inversion problem on a minimal set of data. In chapter 15 (Kruse, Natterer) the ill-posed first-kind Fredholm problem (2-D case) is directly attacked leading to some Toeplitz matrix problem that can be solved in a fast way including Tikhonov regularization. Finally, chapter 16 (Friedrich, Hofmann) presents numerical techniques for identification and optimal control in parabolic PDE's in a common theoretical framework - thus also leading over to the next part of the book. The paper advocates to regularize the inverse problem in close connection with discretization of the PDE. An illustrative example from rheology is sketched.

Part V concentrates on large scale optimization and optimal control problems. In chapter 17 (Spielberg, Suhl), production software for mixed integer-real optimization is described. The mixing in the problem induces a mixing of methods: a fast linear programming solver (utilizing deep hardware structure of a special computer) is combined with branch and bound techniques. As it turns out, a crucial part in the whole computational speed-up is played by preprocessing of the model, e.g. separation of reducible graphs to irreducible subgraphs or increase of constraints, which are then more easily handled. In chapter 18 (Krämer-Eis, Bock) feedback control techniques for control and state constrained problems are worked out in the setting of the multiple shooting method for ODE-BVP's. Two interesting applications are given: energy minimization on a subway car ride and a space shuttle re-entry flight. Techniques of the preceding two chapters should, in principle, also apply to the problem of chapter 19 (Wacker). This chapter describes both the modelling and the computations necessary for optimizing the production scheme of a real life hydro power plant system in Austria. The results of these computations helped to drastically improve the actual performance of this plant system.

Last not least, adaptations of various algorithms to supercomputers are arranged within *part VI*. In chapter 20 (Duff), the performance of well-known sparse linear equation solvers on different supercomputers is compared. The insight coming from the detailed studies is: frontal methods are highly efficient on vector machines, but cannot take advantage of parallelism, whereas multi-frontal methods can, but are not as efficient on pure vector machines; general sparse codes, however, are generally not very efficient on present day supercomputers. The next chapter (Gropp) presents a modification of adaptive gridding that was especially developed in view of supercomputers: the basic idea is to work with a global uniform coarse grid and a local uniform fine grid simultaneously. Experience from the actual implementation of this concept is reported. Finally, chapter 22 (Hayes) presents a study of a special iterative method, which has been developed for FE methods on rather irregular 3-D domains. The field of application of this method includes cryopreservation of tissues, design of suits for deep space or deep sea exploration, or whole body cooling for premature infants (a basis for surgery in extreme cases). As is typical also for other scientific computing applications, the simulations have a pilot function before the explicit real life test.

Summarizing, the book presents the main recent issues of large scale scientific computing, such as adaptive time stepping and space gridding, treatment of strong nonlinearities, algorithm implementation for large scale problems subject to computing time and storage restrictions. As many of the described problems are near the border of tractability, all of the different issues are typically connected in the treatment of a particular problem. A basic knowledge of all of the different aspects described herein seems to be extremely helpful when attacking any realistic computational problem.

As the organizers of the associated meeting, the editors of this proceedings would like to thank the SFB 123 of the University of Heidelberg (the former place of one of us) for gracious financial support. In addition, they wish to thank S. Wacker for her extreme care and invaluable help in the final preparation of the manuscripts for this book.

December 1986

P. Deuflhard, Berlin
B. Engquist, Uppsala/Los Angeles

CONTENTS

(* indicates speaker among several authors)

PART I

INITIAL VALUE PROBLEMS
FOR
ODE'S AND PARABOLIC PDE'S

ALGORITHMS FOR SEMICONDUCTOR DEVICE SIMULATION

R.E. Bank, W. Fichtner, D.J. Rose, R.K. Smith

Abstract: Semiconductor device simulation is a very challanging problem for the numerical analyst. Here we give a brief survey of the the problem, and describe those algorithms which we have found to be effective in its solution.

Introduction

In this work, we present a brief survey of our research over the past several years in the area of semiconductor device modeling. The solution of three nonlinear partial differential equations that govern the behavior of electrons and holes in semiconductor devices is a very challenging problem in scientific computation. It is our goal here to provide a brief introduction to the numerical analysis aspects of the problem with a view towards the non specialist in VLSI.

Within the VLSI design process, device simulation is logically situated between process modelling and circuit simulation. In process modelling, various steps in the chip fabrication process are studied. The numerical problem here is the solultion of diffusion dominated, time dependent partial differential equations [7]. The output of such a simulation is a description of the impurity concentration within the device (doping profile) as well as some information on device geometry (such problems often involve moving boundaries). The doping profile serves as the forcing function for the device simulator.

In device simulation, the electrical characteristics of individual devices simulated under various operating conditions [9, 16]. The output of such a study may be a relation between applied voltages at the contacts and the current in the device (see below). These relations serve as data for circuit simulation. Here one is attempting to model the electrical interaction among many devices, for example, the components of an adder. The equations governing this situation are Kirchhoff's current and voltage laws and constitutive relations. This yields a system of (ordinary) differential/algebraic equations [5, 15].

Overall, the goal of using such simulations as a design tool is to reduce the need of actually fabricating prototype chips in order to study their behavior, an expensive and time consuming process. As such, one expects simulators to be run routinely and often, mainly by engineers who may be non experts on the underlying numerical techniques. It is thus important to have efficient and robust algorithms which can be implemented as "black boxes" with respect to the user.

The remainder of this paper is organized as follows: in section 1, we describe the partial differential equations used in (time independent) device simulation. In section 2, we describe the box method discretization scheme of the type commonly used today in device models. Finally, in section 3, we briefly describe our own experience in the use of multigrid techniques to solve the discretized equations.

1 EQUATIONS AND BOUNDARY CONDITIONS

In dimensionless variables, the equations used in device modeling are

(1.1a) $$-\Delta u + n - p - k = 0,$$

(1.1b) $$-\Delta \cdot J_n - G + R = 0, \qquad J_n = -\mu_n n \nabla u + D_n \nabla n,$$

(1.1c) $$\nabla J_p - G + R = 0, \qquad J_p = -\mu_p p \nabla u - D_p \nabla p$$

where

u is the electrostatic potential

$n(p)$ is the electron (hole) density

J_n (J_p) is the electron (hole) current

k is the doping profile

$\mu_n(\mu_p)$ is the electron (hole) mobility coefficient

D_n (D_p) is the electron (hole) diffusion coefficient

G (R) is the generation (recombination) term.

Assuming the Einstein relation [16] $\mu_n = D_n, \mu_p = D_p$ is valid, it is traditional to write (1.1) in terms of the quasifermi potentials for electrons and holes, v and w, respectively, given by

(1.2) $$n = e^{u-v}, \quad p = e^{w-u}$$

Then equations (1.1) become

(1.3a) $$-\Delta u + e^{u-v} - e^{w-u} - k = 0$$

(1.3b) $$\nabla (\mu_n e^{u-v} \nabla v) - G + R = 0$$

(1.3c) $$-\nabla (\mu_p e^{w-u} \nabla w) - G + R = 0$$

The quasifermi levels are generally smoother functions than the corresponding densities, and thus are in some respects more suitable for numerical approximation.

Problems can be posed in one, two, or three space dimensions. Today, 1-D models are of somewhat limited interest; 2-D models are computationally tractable and are adaquate for many purposes. They have reached a fairly sophisticated level of development and are the most widely used today. Some effects are inherently 3-dimensional and hence require a 3-D model. Obviously the computational costs and hardware requirements of large scale 3-D simulation are enormous, and compared with 2-D simulation, the development of 3-D models is still in its infancy.

The computational domain for a typical 2-D problem for a MOSFET (metal oxide semiconductor field effect transistor) is shown in figure 1.1

This region represents a vertical cross section of the device. Dirichlet boundary are specified on the contacts (source and drain) and at the base. These boundary conditions may be derived as follows: Assuming space-charge neutrality at these contacts we obtain from (1.1)

(1.4a) $$n - p - k = 0$$

Carrier equilibrium defines the condition

(1.4b) $$n \cdot p = 1 \qquad \text{(or } v = w\text{)}$$

at the contacts. If $V = v = w$ the (normalized) applied voltage at a contact, then (1.2) and (1.4) imply

(1.5) $$n = \frac{k}{2} + \sqrt{\frac{k^2}{4} + 1}$$

$$p = -\frac{k}{2} + \sqrt{\frac{k^2}{4} + 1}$$

$$u = v + \log n = V - \log p$$

the vertical edges of the silicon region, and also for n and p at the oxide-silicon interface.

In the oxide region the potential u satisfies

(1.6) $$-\varepsilon \nabla u = 0$$

(n and p are not defined in the oxide region). The potential satisfies a Dirichlet boundary condition $u = V$ at the gate, homogeneous Neumann boundary conditions on the vertical edges of the oxide region, and natural continunity conditions for u and its normal derivative (times the diffusion coefficient) at the oxide-silicon interface.

The internal interfaces within the silicon region represent (near) discontinuities in the doping profile (k in (1.1)). The doping profile may vary by many orders of magnitude near such p-n junctions.

A typical coarse mesh with 467 vertices for a mosfet is shown in figure 1.2. With the doping profile $k = 10^{10}$ in the n-regions, $k = -10^5$ in the p-region, $\varepsilon = 1/3$ in (1.6), and applied voltages $V_s = V_b = 0$ at the source and base, $V_g = V_d = 40$ at the gate and drain, we solved the problem in the quasifermi variables; normalized plots of the solution (u, v, and w) are shown in figures 1.3-1.5.

In problems of this type, an important goal is to calulate the current flow through the source and/or drain as a function of varying voltages at the contacts. It is also possible to vary the doping and some aspects of the device geometry in order to optimize a design. In either event, it is the solution of a sequence of related problems, rather than a single problem which is often required of the device simulator. Computational efficiency is then very important in this context.

2. Discretization via the Box Scheme

As one might infer from figures 1.2-1.5, the "action" in most problems usually occurs in a relatively small fraction of the computational domain. As a result, it is now almost universally agreed that computational meshes should be appropriately non-uniform, and adapted to the internal geometrical structure of the device. Because of the geometric flexibility of triangles, the discretization schemes we have favored are based on triangulations of the computational domain. Others [10, 13] have proposed schemes using rectangles as the basic element, but our brief discussion here will be limited to the triangular case.

For any triangulation of the domain, we construct a dual mesh of elements, called boxes, which can be placed in 1-to-1 correspondence with vertices in the triangulation. The box corresponding to a given vertex v_i is found by connecting the perpendicular bisectors of the triangle edges incident to the vertex as illustrated in figure 2.1

This construction of the dual (box) mesh supposes all angles in the triangular mesh are bounded by π. Several modifications have been proposed to deal with obtuse angles (see [11, 14], for example) but we do not consider this problem in detail here. For simplicity, we assume $\mu_n = \mu_p = 1$, $R = G = 0$ in (1.3).

For the nonlinear Poisson equation, we begin by integrating (1.3a) over the box b_i associated with vertex ν_i in figure 2.1.

$$(2.1)\quad \iint_{b_i} -\Delta u + n-p-k \, dx = \int_{\partial b_i} -\nabla u\cdot n \, ds + \iint_{b_i} e^{u-v} - e^{w-u} - k \, dx \; 0$$

where for the boundary integral n is the outward pointing normal for the boundary of b_i (not to be confused with the electron density). The line integral is broken up into a sum of integrals along each straight line segment. The integral along the perpendicular bisector of the triangle edge connecting ν_i and ν_j in figure 2.1 is approximated by

$$\left(\frac{u_i-u_j}{h_{ij}}\right) \ell_{ij}$$

where h_{ij} is the length of the edge connecting ν_i and ν_j

ℓ_{ij} is the length of the perpendicular bisector, and u_i and u_j are the point values of u at the vertices. The term $\iint e^{u-v}\,dx$ is approximated by $e^{u_i-v_i} a_i$ where a_i is the area of box b_i. Thus we obtain

$$(2.2)\quad \sum_{k\in S_i} \left(\frac{u_i-u_k}{h_{ik}}\right) \ell_{ik} + a_i \left(e^{u_i-v_i} - e^{w_i-u_i} - k_i\right) = 0$$

where S_i is the index set of vertices adjacent to vertex ν_i. Points on the boundary are treated by appropriate and obvious extensions of this scheme. The continity equations are quite similar with respect to discretization, so we consider only (1.3b) in detail. By applying the divergence theorem, we obtain

$$(2.3)\quad -\int_{\partial b_i} e^{u} \, \nabla e^{-v}\cdot n \, ds = 0$$

As in the nonlinear Poisson equation, this integral is approximated on an edge by edge basis. For the perpendicular bisector of the edge connection ν_i and ν_j we approximate the gradient term by

$$\frac{e^{-v_i}-e^{-v_j}}{h_{ij}} \quad \ell_{ij}$$

The approximation of e^u is less clear. Since the equation is quite like a singular perturbation problem, one would expect a harmonic average $\left(\frac{e^{-u_i}+e^{-u_j}}{2}\right)^{-1}$ to be appropriate. This leads to

$$(2.4) \quad \sum_{k \in S_i} \left(e^{u_i - v_i} \, C(u_i - u_k) - e^{u_k - v_k} \, C(u_k - u_i) \right) \frac{\ell_{ik}}{h_{ik}} = 0$$

$$C(x) = 2/(e^x+1)$$

A second, and very widely used alternative, which may be derived via a more complicated approximation of e^u or by the Sharfetter-Gummel method [9, 13, 14] leads to

$$(2.5) \quad \sum_{k \in S_i} \quad e^{u_i - v_i} \, B(u_i - u_k) - e^{u_k - v_k} \, B(u_k - u_i) \quad \frac{\ell_{ik}}{h_{ik}} = 0$$

$$B(x) = x/(e^x-1)$$

Both these discretizations have the property that they are linear in the discrete density $n_i = e^{u_i - v_i}$.

It is important to note that all these discretizations may be derived starting form the standard Galerkin finite element formulation using continuous piecewise linear polynomials for the triangular mesh. Using the standard Lagrange (nodal) basis to represent both trial and test spaces, and a careful choice of (nonstandard) quadrature formulae one is lead to (2.3)-(2.5). See [1,4] for details. Thus, in assembling the system of equations, one may use either an equation by equation process, as suggested in (2.3)-(2.5) or a triangle by triangle assembly process more common in finite elements. For each triangular element, which contains parts of 3 boxes, one can compute contributions to equations corresponding to the 3 triangle vertices in the form of element stiffness (Jacobian) matrices and right hand sides (nonlinear residuals). These are then added up in the usual finite element fashion to obtain global quantities. These two possibilities allow a great deal of algorithmic flexibility in formulating the assembly process, which is of great practical value in a computing environment with substantial vector or parallel processing capabilities.

3 Multigrid Solution

In this section, we discuss the solution of the discrete systems of nonlinear equations by approximate Newton algorithms and multigrid techniques. Multigrid algorithms enter the computation in two ways. First, they enter in the overall strategy of nested iterations (also called full multigrid). In this scheme, one has a sequence of discretizations of increasing fineness $\{\mathcal{D}_j\}$, corresponding to problems of geometrically increasing dimension N_j. Typically in 2-dimensional computation $N_{j+1} \sim 4N_j$. For the classes of problems considered here, one can expect to use 2-4 discretization levels. The nested iteration algorithm is summarized below:

Nested Iteration Algorithm:

(i) Solve the problem for $\mathcal{D}_1$ using some intial guess, obtaining an approximate solution S_1

(ii) for $j > 1$, solve the problem. for $\mathcal{D}_j$ using the solution S_{j-1} as initial guess, obtaining an approximate solution S_j.

Using this algorithm, most of the iterations will be carried out on the smallest problem, where such iterations are least costly. As j increases, one expects S_{j-1} to become an increasingly good estimate of S_j (since both are increasingly good approximations of the same continuous solution), and hence fewer iterations should be required on finer levels. A rigorous analysis showing that asymptotically only one approximate Newton iteration per level is required is given in [2].

Within reason, one may expect to receive these benefits of computation reduction through iteration reduction more or less independently of the algorithm chosen to solve the problem on each individual level (obviously in some cases the benefits will be greater than others, depending on the algorithm). However, since the machinery is already available, it seems natural to use a multigrid iterative method using J levels to solve the problems for discretization $\mathcal{D}_j$.

Here we will consider an approximate Newton method for solving the entire set of equations. There is an alternate approach where the 3 pdes are solve individually in cyclic fashion. This latter scheme is called Gummel's method in the field, and many of the ideas described below can be reformulated in nonlinear terms and applied to this scheme as well.

In analyzing the solution of linear systems to be solved in each iteration of the approximate Newton process, we consider two (classes of) orderings for the equations and unknowns. In the first, we group all equations and unknowns corresponding to each pde together yielding a block 3x3 Jacobian with NxN blocks (dropping the subscript j) of the form

(3.1)
$$J = \begin{pmatrix} A_1 & D_2 & D_3 \\ B_2 & A_2 & 0 \\ B_3 & 0 & A_3 \end{pmatrix}$$

Each of the diagonal blocks A_i corresponds to a linear, 2nd order, scalar elliptic pde. A_1, which corresponds to the nonlinear Poisson equation, is symmetric and positive definite; A_2 and A_3, corresponding to the continunity equations, are diagonally similar to symmetric positive definite matrices. The D_i are diagonal, while the B_i have the same sparsity structure as the A_i. The continuity equations are not coupled to each other if there is no generation / recombination.

The second alternative is to group the 3 unknowns corresponding to a given triangle vertex together. In this scheme one ends up with a block NxN jacobian with 3x3 blocks of the form

(3.2)
$$\tilde{J} = [c_{ij}] \qquad 1 \leq i, j \leq N$$

The matrix $\tilde{J}$ (which is just a permutation of J of the form $\tilde{J} = P J P^T$ for an appropriately defined permutation matrix P) has a block nonzero structure corresponding to the graph of the underlying triangular mesh. The nonzero blocks of $\tilde{J}$ have the generic form

$$C_{ii} = \begin{bmatrix} a_1 & d_2 & d_3 \\ b_2 & a_2 & 0 \\ b_3 & 0 & a_3 \end{bmatrix} ,$$

$$C_{ij} = \begin{bmatrix} a_1 & 0 & 0 \\ b_2 & a_2 & 0 \\ b_3 & 0 & a_3 \end{bmatrix} , \quad i \neq j$$

We have not yet found a multigrid scheme which we believe to be "the best." We have tried several successful schemes but for various reasons remain somewhat dissatisfied with the results. Below we briefly summarize our experience.

A. Block Symmetric Gauss Seidel

Let J of (3.1) be written as $J = \mathcal{D} - \mathcal{L} - \mathcal{U}$ where $\mathcal{D}$ is block diagonal and $\mathcal{L}$ and $\mathcal{U}$ are block lower and upper triangular, respectively.

We can apply a block symmetric Gauss Seidel (SGS) iteration [17] to $Jx = b$, or, equivalently, use block incomplete factorization of the form

$$J = (\mathcal{D} - \mathcal{L}) \, \mathcal{D}^{-1} \, (\mathcal{D} - \mathcal{U}) + R$$

where R is the matrix of ignored fill in. The iteration can be accelerated by orthomin [8]. This scheme is somewhat like a linear version of Gummel's method. In this scheme, we are required to solve systems involving the diagonal blocks A_i of (3.1). These systems can be solved using standard multigrid techniques for scalar linear elliptic pdes.

The main disadvantage of this scheme is that, like Gummel's method, it is slowly convergent for many problems of practical interest (e.g. for high currents, when the D_i in (3.1) have large diagonal entries). See [12] for an analysis of Gummel's method.

B. Alternating Block Iteration.

In an attempt to better capture coupling between the pdes, we have tried several alternating block iterations (ABI), in which the iteration scheme described in A above is alternated with, say, a block Jacobi scheme using the Jacobian in the form (3.2). Such an iteration involves solving N 3x3 linear systems, which is relatively inexpensive and easy to implement. Although an improvement over using just the SGS scheme, the improved rate of convergence is not always as impressive as one might hope. We are presently exploring other ABI schemes, in which the alternate blocks are 3x3 for most of the region, but with a few larger blocks, corresponding to the vertices along a p-n junction for example. These larger blocks basically correspond to 1-dimensional problems and can be handled efficiently by band elimination.

C. Incomplete $\mathcal{L}\mathcal{U}$ factorization for $\tilde{J}$

We have tried incomplete $\mathcal{L}\mathcal{U}$ factorizations for the jacobian in the form (3.2). In this scheme we have allowed additional block fillin. We can define a sequence of $\mathcal{I}\mathcal{L}\mathcal{U}$ factorizations $\mathcal{L}_k\ \mathcal{U}_k$ as follows:
$\mathcal{L}_0 + \mathcal{U}_0$ has the same (block) nonzero structure as $\tilde{J}$; for $k > 0$ $\mathcal{L}_k\ \mathcal{U}_k$ allows (block) fillin in the nonzero blocks of R_{k-1}, given by

$$\tilde{J} - \mathcal{L}_{k-1}\ \mathcal{U}_{k-1} = R_{k-1}.$$

Obviously, for k sufficiently large $R_k \equiv 0$ and the $\mathcal{I}\mathcal{L}\mathcal{U}$ become a complete factorization. For some $k \geq 0$, the $\mathcal{I}\mathcal{L}\mathcal{U}$ factorization, accelerated by orthomin, can be used as a smoother for a multigrid iteration applied to the entire coupled system. This is an effective scheme in terms of convergence, but it can be very expensive in time and space.

This is mainly due to the fact that, despite the nice properties of the diagonal blocks of (3.1), the overall jacobian matrix can be highly indefinite. Or the coarser meshes, one must therefore allow a lot of fillin in order to obtain a convergent iteration matrix. On the finer meshes it is not as necessary for the smoothing iteration to be convergent, since the coarse grid correction phase of the multigrid iteration will periodically annihilate those error components which are diverging. [3].

The $\mathcal{I}\mathcal{L}\mathcal{U}$ factorization is quite sensitive to the ordering of the grid points, not unlike sparse Gaussian elimination in general. We have found that a local minimum fillin ordering, taking into account the magnitude of the ignored fillin elements produces good orderings for this scheme. Obtaining an effective procedure for determining a priori how much fillin is required on each grid level over a wide class of problems appears to be difficult. In view of the high cost of the procedure in our experience, we have not yet made a serious attempt at finding a good heuristic for this problem.

D. Block Incomplete $\mathcal{L}\mathcal{U}$ factorization for J.

We are presently working on block $\mathcal{I}\mathcal{L}\mathcal{U}$ schemes for (3.1) of the form

$$J = (\tilde{\mathcal{D}} - \tilde{\mathcal{L}})\tilde{\mathcal{D}}^{-1}(\tilde{\mathcal{D}} - \tilde{\mathcal{U}}) + R$$

where $\tilde{\mathcal{D}}$ is block diagonal the same nonzero structure as A_i in (3.1), and $\mathcal{L}$ and $\tilde{\mathcal{U}}$ are block lower and upper triangular, respectively, with blocks having (at worst) sparsity comparable to the A_i. This scheme is similar in spirit to that proposed by Concus, Golub and Meurant [6]. The three diagonal blocks of $\tilde{\mathcal{D}}$ will still correspond to linear elliptic equations, and standard or slightly modified multigrid iterations for linear elliptic problems could be used, as in the SGS scheme.

The orthomin and ABI enhancements are also possible here. We expect to report on our numerical experiences with this scheme in the near future.

Acknowledgement:
The work of the first and the third author was supported in part by the Office of Naval Research.

References

(1) R. E. Bank and D. J. Rose "Some error estimates for the box method", submitted, SIAM J. Numer. Anal.

(2) R.E. Bank and D. J. Rose "Analysis of a mult-level iterative method for nonlinear finite element equations," Math. of Comp 39 (1982) pp. 453-465.

(3) R. E. Bank, "A comparison of two multilevel iterative methods for nonsymmetric and indefinite finite element equations", SIAM J. Numer. Anal 18 (1981) pp.724-743.

(4) R. E. Bank, D. J. Rose, and W. Fichtner "Numerical methods for semiconductor device simulation," IEEE Trans on Elect. Dev. ED-30 (1983) pp.1031-1041.

(5) R. E. Bank, W. M. Coughran Jr., W. Fichtner, E. H. Grosse, D. J. Rose, and R. K. Smith, "Transient simulation of silicon devices and circuits," IEEE Trans on Elect. Dev. to appear.

(6) P. Concus, G. H. Golub and G. Meurant, Block Preconditioning for the Conjugate Gradient Method, Technical Report LBL-4856, Laurence Berkeley Laboratory, 1982.

(7) R. Dutton, "Modeling of the silicon intergrated circuit design and manufacturing process" IEEE Trans. on Elect. Dev. ED-30 (1983) pp.968-985.

(8) H. C. Ellman, Iterative Methods for Large, Sparse, Nonsymmetric Systems of Linear Equations, Yale University Computer Science Department Research Report 229, 1982.

(9) W. Fichtner, D. J. Rose, and R. E. Bank, "Semiconductor Device Simulation", IEEE Trans on Elect. Dev. ED30 (1983) pp.1018-1030.

(10) A. F. Franz, G. A. Franz, S. Selberherr, C. Ringhofer, P. Markowich, "Finite boxes a generalization of finite differences suitable for semiconductor device simulation," IEEE Trans on Elect Dev ED30, (1983) pp.1070-1082.

(11) M. D. Huang, "Constant flow patch test - a unique guideline for the evaluation of discretization schemes for the current continunity equations," IEEE Trans on Elect. Dev. , to appear.

(12) T. Kerkhoven, On the Dependence of the Convergence of Gummel's Algorithm on the Regularity of the Solution, YaleUniversity Computer Science Department Technical Report RR-366, 1985.

(13) M. Kurata, Numerical Analysis for Semiconductor Devics. Heath and Co, Lexington Mass. 1982.

(14) M. S. Mock, "Analysis of a discretization algorithm for stationary continunity equations in semiconductor devices," COMPEL 2 (1983) pp.117-139.

(15) A. R. Newton and A. L. Sangiovanni-Vincentelli, " Relaxation based electrical sinulation", IEEE Trans on Elect. Dev ED 30 (1983) pp.968-985.

(16) S. Sze Physics of Semiconductor Devices, 2nd ed. Wiley, New York, 1981.

(17) R. S. Varga, Matrix Iterative Analysis, Prentice Hall, Englewood Cliffs, New Jersey, 1962.

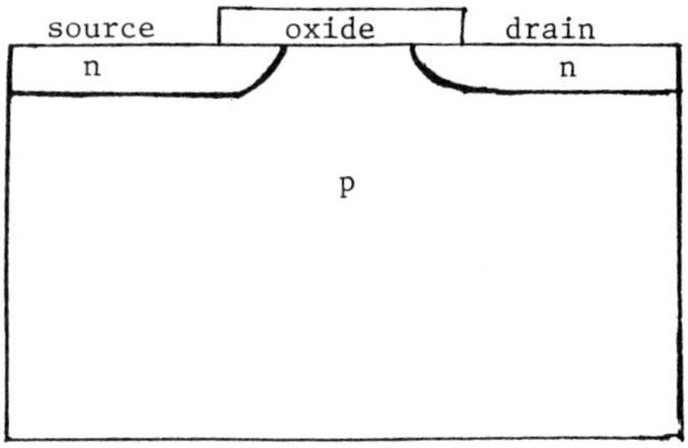

figure 1.1
The geometry of a MOSFET
(not drawn to scale)

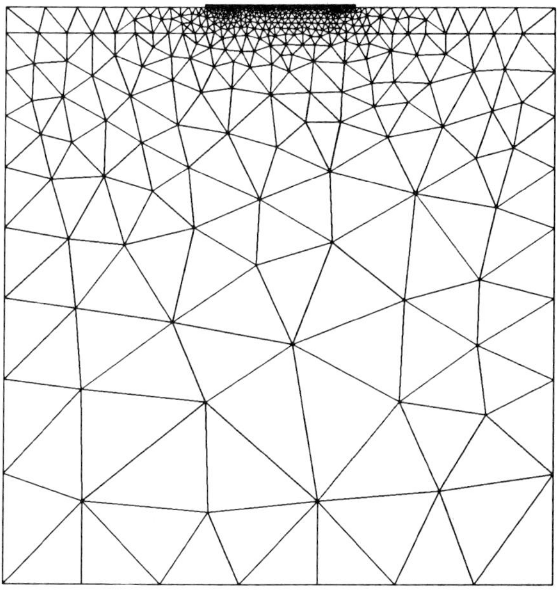

figure 1.2
The coarse triangulation for the MOSFET example.

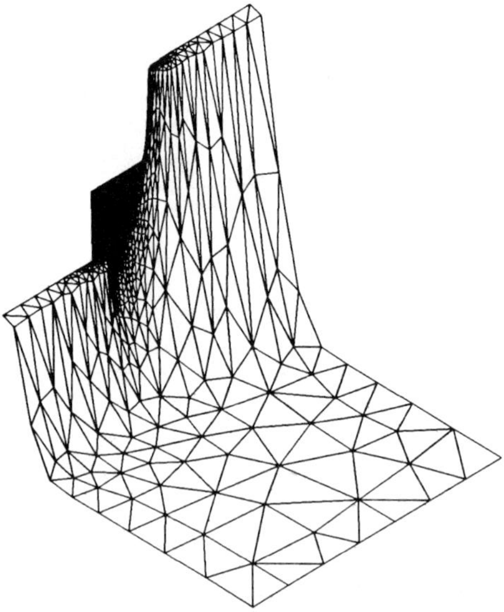

figure 1.3

The electrostatic potential u.

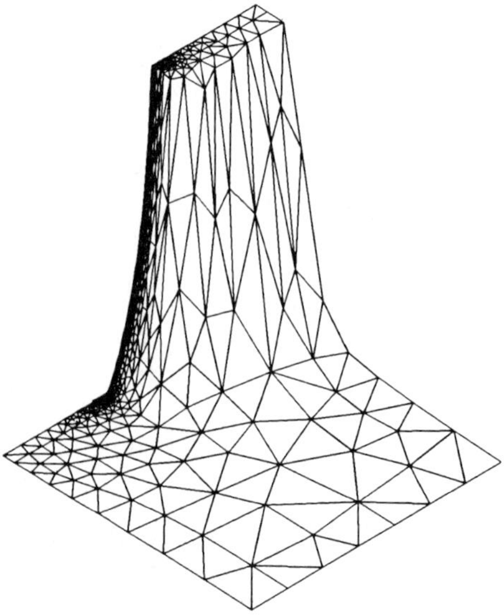

figure 1.4

The quasi fermi level for electrons v.

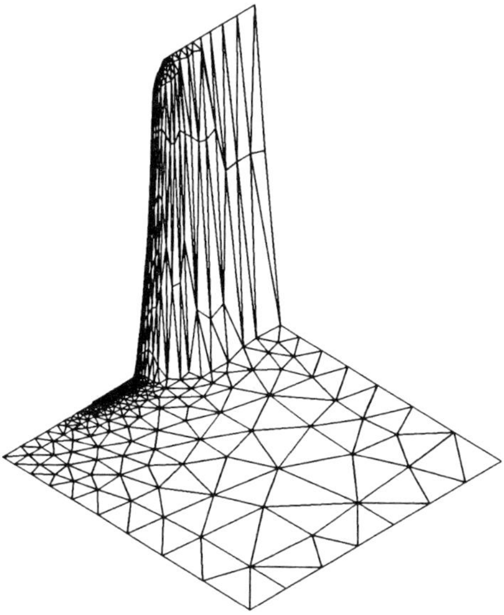

figure 1.5

The quasi-fermi level for holes w.

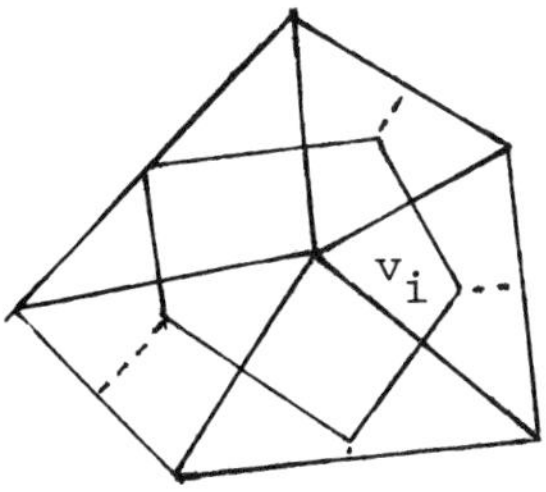

figure 2.1

The box b_i corresponding to vertex ν_i

HIERARCHICAL BASES IN THE NUMERICAL SOLUTION OF PARABOLIC PROBLEMS

Harry Yserentant

Summary:

We continue our work concerning the use of hierarchical bases in finite element computations. Here we consider parabolic initial boundary value problems. We show that the linear systems arising at each time step can be solved at least as efficiently as in the elliptic case provided that the multi-level splitting of the finite element space is stopped on a certain level depending on the stepsize in time.

AMS-Subject Classification: 65F10, 65F35, 65M99, 65N20, 65N30

Key words: fast solvers, parabolic equations, finite element methods

1. Introduction

Hierarchical bases of finite element spaces originate in a very natural way if one constructs the computational grid by successively refining a coarse initial triangulation of the domain under consideration. As it has been shown in [5], [6] and [7] they provide a very useful tool for the solution of the large systems of linear equations arising in the numerical solution of plane second order elliptic boundary value problems by finite element methods. If one formulates these linear systems with regard to such hierarchical bases and solves the systems by conjugate gradient type methods, one gets, at least in the symmetric case, algorithms of the nearly optimal computational complexity $O(n \log n)$. Therefore this method is nearly as fast as conventional multigrid methods. But it has a simpler structure, it is easily programmable, and, unlike multigrid methods, it does not make use of the regularity properties of the problem to be solved.

In the present paper it is demonstrated how this approach has to be modified for the linear systems arising in the solution of parabolic initial boundary value problems at each time step, if one uses an unconditionally stable time discretization. The main result is that the multi-level splitting of the finite element space has to be stopped at a certain refinement level depending on the stepsize in time. If this is done, the linear systems can be solved as or even more efficiently as in the elliptic case and with only a slightly larger amount of work than in the case of unlumped explicit time integration schemes which are unreasonable due to their stability properties.

2. The structure of the linear systems to be solved at each time step

Let Ω be a bounded polygonal region in the plane. Let

$$\|u\|_{0,2;\Omega} = \Big(\int_\Omega |u(x)|^2\, dx \Big)^{1/2} \tag{2.1}$$

be the L_2-norm on the space $L_2(\Omega)$, and equip the Hilbert-space $W^{1,2}(\Omega)$ with the usual seminorm

$$|u|_{1,2;\Omega} = (\|D_1 u\|^2_{0,2;\Omega} + \|D_2 u\|^2_{0,2;\Omega})^{1/2} \tag{2.2}$$

and the norm

$$\|u\|_{1,2;\Omega} = (|u|^2_{1,2;\Omega} + \|u\|^2_{0,2;\Omega})^{1/2} \quad . \tag{2.3}$$

By $H(\Omega)$ we denote either $W^{1,2}(\Omega)$ itself or a linear subspace of $W^{1,2}(\Omega)$ consisting of all functions in $W^{1,2}(\Omega)$ vanishing on a given boundary piece Γ of Ω in the sense of the trace operator. $H(\Omega)$ is the solution space of the parabolic initial boundary value problem

$$(\dot u(t),v) + B(t;u(t),v) = (f(t),v) \quad , \quad v \in H(\Omega) \quad , \tag{2.4}$$

to be solved approximately. The brackets denote the L_2-inner product inducing the norm (2.1),and the point the time derivative of the solution $u(t) \in H(\Omega)$, $t \geq 0$, the initial state $u(0)$ of which is given. $f(t) \in L_2(\Omega)$ is a given right-hand side, and the possibly time-dependent and not necessarily symmetric bilinear form B is assumed to satisfy the estimate

$$B(t;u,v) \leq M \|u\|_{1,2;\Omega} \|v\|_{1,2;\Omega} \tag{2.5}$$

for all $u,v \in H(\Omega)$ and the Gårding-type inequality

$$B(t;u,u) \geq \delta |u|^2_{1,2;\Omega} - \omega^2 \|u\|^2_{0,2;\Omega} \tag{2.6}$$

for all $u \in H(\Omega)$ where $M, \delta > 0$ and ω^2 are constants independent of u,v and $t \geq 0$.

As an example consider the heat equation

$$u_t = \Delta u + f \quad \text{on} \quad \Omega \tag{2.7a}$$

with the boundary conditions

$$u = 0 \qquad \text{on} \quad \partial\Omega . \tag{2.7b}$$

Here the solution space $H(\Omega)$ is the subspace $W_0^{1,2}(\Omega)$, and the bilinear form B is given by

$$B(u,v) = \sum_{i=1}^{2} \int_\Omega D_i u \; D_i v \; dx \tag{2.8}$$

and satisfies (2.5) and (2.6) with $M = 1$, $\delta = 1$ and $\omega^2 = 0$.

We want to solve the initial boundary value problem (2.4) by the finite element method using piecewise linear functions; other finite element functions could be treated in a similar way. For this purpose we assume that a coarse initial triangulation $\mathcal{T}_0$ of the domain Ω is given. Beginning with this triangulation we construct a nested family $\mathcal{T}_0, \mathcal{T}_1, \mathcal{T}_2, \ldots$ of further triangulations of Ω. $\mathcal{T}_{k+1}$ is obtained from $\mathcal{T}_k$ by subdividing any triangle of $\mathcal{T}_k$ into four congruent subtriangles. By S_k we denote the finite element space consisting of all functions which are continuous on $\overline{\Omega}$ and piecewise linear on the triangles $T \in \mathcal{T}_k$. These functions are determined by their values at the nodes $x \in \mathcal{N}_k$, which are the vertices of the triangles $T \in \mathcal{T}_k$. We assume that the boundary piece Γ in the definition of the solution space $H(\Omega)$ is the union of certain sides of the triangles in the initial triangulation $\mathcal{T}_0$. This means that the discrete solution spaces $S_j \cap H(\Omega)$ consist of all functions $u \in S_j$ vanishing at the nodes $x \in \mathcal{N}_j \cap \Gamma$.

If one wants to determine an approximate solution $u(t) \in S_j \cap H(\Omega)$ of the initial boundary value problem (2.4) one is led, in the first step, to a large system of ordinary differential equations. This system has the same weak form

$$(\dot{u}(t),v) + B(t;u(t),v) = (f(t),v) \quad , \quad v \in S_j \cap H(\Omega) \quad , \tag{2.9}$$

on $S_j \cap H(\Omega)$ as the continuous problem on the whole solution space $H(\Omega)$. As the mass matrix induced by the L_2-inner product is nonsingular independent of the choice of the basis of the discrete solution space, (2.9) is equivalent to an explicit linear system of ordinary differential equations. Therefore it has a unique solution $u(t) \in S_j \cap H(\Omega)$, $t \geq 0$, if one fixes the initial state $u(0)$.

In the second step the system (2.9) of ordinary differential equations is solved approximately by an appropriate numerical method for initial value problems. A simple example is the implicit Euler method, where the approximate solution u_{i+1} of (2.9) for the time $t_{i+1} = t_i + \tau_i$ is determined by the approximate solution for time t_i via the equations

$$\left(\frac{u_{i+1}-u_i}{\tau_i} \, , \, v\right) + B(t_{i+1};u_{i+1},v) = (f(t_{i+1}),v) \, , \, v \in S_j \cap H(\Omega) \, . \tag{2.10}$$

Another example is the second order implicit midpoint rule

$$\left(\frac{u_{i+1}-u_i}{\tau_i} \, , \, v\right) + B\left(t_i + \frac{\tau_i}{2} ; \frac{u_i+u_{i+1}}{2} \, , \, v\right) = \left(f(t_i + \frac{\tau_i}{2}) \, , \, v\right) \quad , \tag{2.11}$$

$v \in S_j \cap H(\Omega)$.

If one has $\omega^2 = 0$ in (2.6), both schemes are unconditionally stable independent of the time stepsizes τ_i. (2.10) leads to the system

$$\tau_i \, B(t_{i+1};u_{i+1},v) + (u_{i+1},v) = f^*_{i+1}(v) \quad , \quad v \in S_j \cap H(\Omega) \quad , \tag{2.12}$$

of linear equations for the unknown function $u_{i+1} \in S_j \cap H(\Omega)$ where the known right hand side is given by

$$f^*_{i+1}(v) = (u_i,v) + \tau_i(f(t_{i+1}),v) \quad . \tag{2.13}$$

One gets analogous systems of linear equations in the case of the second order scheme (2.11) and for any other linear multistep, one-leg or semiimplicit Runge-Kutta method which is unconditionally stable for $\omega^2 = 0$.

So our problem can be stated as follows: Solve systems

$$\tau \, B(t;u,v) + (u,v) = f^*(v) \quad , \quad v \in S_j \cap H(\Omega) \quad , \tag{2.14}$$

of linear equations for $u \in S_j \cap H(\Omega)$, where the bilinear form B is assumed to satisfy (2.5) and (2.6) and $\tau > 0$ is a constant multiple of the actual time stepsize. Our method should perform uniformly well in τ.

To study the spectral properties of the coefficient matrices of these linear systems we introduce the τ-dependent norm

$$\|u\|_\tau^2 = \tau\,|u|_{1,2;\Omega}^2 + \|u\|_{0,2;\Omega}^2 \tag{2.15}$$

on $W^{1,2}(\Omega)$. Using (2.5) and (2.6) one can easily prove the estimates

$$\tau\,B(t;u,v) + (u,v) \le \max(M,1+\tau M)\,\|u\|_\tau\,\|v\|_\tau \tag{2.16}$$

$$\tau\,B(t;u,u) + (u,u) \ge \min(\delta,1-\tau\omega^2)\,\|u\|_\tau^2 \tag{2.17}$$

for the functions $u,v \in H(\Omega)$ and therefore also for the finite element functions satisfying the boundary conditions. Because of (2.17) the linear system (2.14) is uniquely solvable, provided the time stepsize is chosen such that

$$1 - \tau\,\omega^2 \ge \delta_0 > 0 \tag{2.18}$$

is satisfied. If one has $\omega^2 = 0$, (2.18) holds for all $\tau > 0$, whereas in the case $\omega^2 > 0$ (2.18) restricts the time stepsize to a value depending on the properties of the continuous problem but not on the discrete solution spaces. By (2.17) and (2.18) there exists a constant $\delta^* > 0$ with

$$\tau\,B(t;u,u) + (u,u) \ge \delta^*\,\|u\|_\tau^2 \tag{2.19}$$

for all functions $u \in H(\Omega)$. In addition to (2.18) we assume that the time stepsize is bounded by some quantity related to the diameter of the domain Ω. Then one has in addition to (2.19)

$$\tau\,B(t;u,v) + (u,v) \le M^*\,\|u\|_\tau\,\|v\|_\tau \tag{2.20}$$

for $u,v \in H(\Omega)$, and (2.15) is the appropriate norm for describing the properties of the linear system (2.14).

3. The condition number of the linear systems when using hierarchical bases

To solve the linear systems (2.14) one has to specify a basis of the finite element space $S_j \cap H(\Omega)$. The natural choice seems to be the usual nodal basis which consists of basis functions which take the value one at an associated node $x \in N_j$ and vanish at all other nodes. But this choice has the big disadvantage that the resulting matrices are rather ill-conditioned which makes the iterative treatment of the linear systems difficult; the spectral condition number grows exponentially when the time stepsize is fixed and the number j of refinement levels increases. Therefore we propose to use hierarchical bases of the finite element spaces. To build up these hierarchical bases we start with a given refinement level m depending on the actual stepsize in time. The hierarchical basis of $S_m \cap H(\Omega)$ is the nodal basis of this space. To get the hierarchical basis of $S_{k+1} \cap H(\Omega)$, $k \geq m$, we take the hierarchical basis of $S_k \cap H(\Omega)$ and add the nodal basis functions of $S_{k+1} \cap H(\Omega)$ vanishing at the nodes $x \in N_k$.

To describe the mathematical properties of hierarchical bases we need the finite element interpolation operators I_k . By $I_k u$ we denote the function of S_k interpolating u at the nodes from N_k . We have

$$I_k u \in S_k \quad ; \quad (I_k u)(x) = u(x) \quad , \quad x \in N_k \quad . \tag{3.1}$$

For $j \geq k$ the function $u \in S_k$ is reproduced by the interpolation operator I_j . Therefore one has for $m = 0,1,\ldots,j-1$ the representation

$$u = I_m u + \sum_{k=m+1}^{j} (I_k u - I_{k-1} u) \tag{3.2}$$

of the functions $u \in S_j$. This is a decomposition of u into fast oscillating functions corresponding to the different refinement levels; $I_k u - I_{k-1} u$ vanishes at all nodes of level $k-1$. With the multi-level splitting (3.2) of the finite element space S_j we associate the mesh-dependent seminorm

$$|u|_m^2 = \sum_{k=m+1}^{j} \sum_{x \in N_k \setminus N_{k-1}} |(I_k u - I_{k-1} u)(x)|^2 \quad . \tag{3.3}$$

This seminorm has a very simple representation when using a hierarchical basis with level m as initial refinement level: it is the Euclidean length of the vector of its coefficients with exception of those corresponding to level m.

To fix the level m depending on the time stepsize let the diameters $d(T)$ of the triangles from the coarse initial triangulations $\mathcal{T}_0$ of the domain Ω satisfy

$$c_1 H \leq d(T) \leq c_2 H \quad , \quad T \in \mathcal{T}_0 \quad , \tag{3.4}$$

with positive constants c_1, c_2 and H. Because of the construction of the following triangulations (3.4) implies

$$c_1 \frac{H}{2^k} \leq d(T) \leq c_2 \frac{H}{2^k} \quad , \quad T \in \mathcal{T}_k \quad , \tag{3.5}$$

for all refinement levels k. We assume that the quantity $\tau > 0$ in the linear system (2.14) satisfies

$$\left(\frac{H}{2^j}\right)^2 \leq \tau \leq H^2 \quad . \tag{3.6}$$

This is a very reasonable restriction: $\tau \geq H^2$ would correspond to an unusually large time stepsize. In this case the linear system to be solved totally behaves like a discrete elliptic boundary value problem so that we can refer to [5], [6]. On the other hand, if one has $0 < \tau \leq (H/2^j)^2$, the time stepsize satisfies a Courant-Friedrichs-Lewy type condition on the computational grid of level j which is very seldom necessary when using an unconditionally stable time integration scheme. In this case the coefficient matrix behaves like the mass matrix and a scaled nodal basis is totally appropriate; we refer to [4]. So we can fix a level m, $0 \leq m \leq j-1$, such that

$$\left(\frac{H}{2^{m+1}}\right)^2 \leq \tau \leq \left(\frac{H}{2^m}\right)^2 \tag{3.7}$$

holds. For a large time stepsize m is small and the number of refinement levels is large, whereas for a small time stepsize m comes near to j and the number of refinement levels is small.

With the choice (3.7) of m we define the discrete norm

$$|||u|||_\tau^2 = \|I_m u\|_{0,2;\Omega}^2 + \tau\, |u|_m^2 \tag{3.8}$$

on the finite element space S_j . The main mathematical result of this paper is that this norm essentially behaves like the norm (2.15).

<u>Theorem</u>

There are positive constants K_1 and K_2 with

$$\frac{K_1}{(j-m+1)^2}\, |||u|||_\tau^2 \le \|u\|_\tau^2 \le K_2\, |||u|||_\tau^2$$

for all functions $u \in S_j$ and all τ's satisfying (3.6). These constants depend only on a lower bound for the interior angles of the triangles in the initial triangulation T_0 and on the constants c_1 and c_2 in (3.4) describing the quasiuniformity of this triangulation.

Because of (2.19) and (2.20) one has

$$\tau\, B(t;u,v) + (u,v) \le M^*\, K_2\, |||u|||_\tau\, |||v|||_\tau\ , \tag{3.9}$$

$$\tau\, B(t;u,u) + (u,u) \ge \delta^* \frac{K_1}{(j-m+1)^2}\, |||u|||_\tau^2\ . \tag{3.10}$$

Because of (3.5) and (3.7) the discrete norm (3.8) is equivalent to the Euclidean norm of the coefficient vectors of the functions in S_j with regard to the given hierarchical basis of this space. Therefore (3.9) and (3.10) mean that the matrix of the linear system (2.14) with regard to this hierarchical basis has a spectral condition number behaving like $O((j-m+1)^2)$. In terms of the Courant-number τ/h^2 the condition number behaves like $O((\log(\tau/h^2))^2)$ instead of $O(\tau/h^2)$ for a nodal basis; $h = H/2^j$ is the average gridsize on the final level.

The proof of our theorem is quite simple, if one makes use of the results obtained in [5]. We summarize these results in the following lemma which is equivalent to Lemmas 2.5, 2.6, 2.8 and 2.9 in [5].

Lemma:

There are positive constants C_1 , C_2 , C_3 and C_4 , which depend only on a lower bound for the interior angles of the triangles in the initial triangulation of level zero, with

(i) $|I_m u|^2_{1,2;\Omega} + |u|^2_m \leq C_1(j-m+1)^2 \; |u|^2_{1,2;\Omega}$

(ii) $\|I_m u\|^2_{0,2;\Omega} \leq C_2(j-m+1)\left\{\|u\|^2_{0,2;\Omega} + H_m^2 \; |u|^2_{1,2;\Omega}\right\}$

and

(iii) $|u|^2_{1,2;\Omega} \leq C_3 \left\{|I_m u|^2_{1,2;\Omega} + |u|^2_m\right\}$

(iv) $\|u\|^2_{0,2;\Omega} \leq C_4\left\{\|I_m u\|^2_{0,2;\Omega} + H_m^2 \; |u|^2_m\right\}$

for all functions $u \in S_j$. Here H_m denotes an upper bound for the diameters of the triangles of level m , $0 \leq m \leq j-1$.

Parts (i) and (ii) imply that one has

$$|||u|||^2_\tau \leq C_5(j-m+1)^2\left\{(\tau + H_m^2) \; |u|^2_{1,2;\Omega} + \|u\|^2_{0,2;\Omega}\right\} \tag{3.11}$$

for the functions $u \in S_j$, and by parts (iii) and (iv) one gets

$$\|u\|^2_\tau \leq C_6\left\{\|I_m u\|^2_\tau + (\tau + H_m^2) \; |u|^2_m\right\} \tag{3.12}$$

where $C_5 = \max(C_1,C_2)$ and $C_6 = \max(C_3,C_4)$. To replace the norm (2.15) of $I_m u$ in (3.12) by the L_2-norm of this function we need the inverse inequality

$$|v|^2_{1,2;\Omega} \leq C_7 \; h_m^{-2} \; \|v\|^2_{0,2;\Omega} \tag{3.13}$$

for the functions $v \in S_m$ where h_m is a lower bound for the diameters of the triangles of level m. It implies

$$\|I_m u\|^2_\tau \leq C_8(1 + \tau \, h_m^{-2}) \; \|I_m u\|^2_{0,2;\Omega} \quad . \tag{3.14}$$

As we have by (3.5) and (3.7)

$$\tau \, h_m^{-2} \leq \left(\frac{1}{c_1}\right)^2 \quad , \quad H_m^2 \leq 4c_2^2 \; \tau \tag{3.15}$$

the estimates (3.11), (3.12) and (3.14) prove the theorem.

4. The numerical solution of the linear systems by conjugate gradient type methods

Because of the small condition number of the coefficient matrix the hierarchical basis formulation of the linear system (2.14) is very well suited to conjugate gradient type methods [3]. In the symmetric case one can show that one needs at most

$$\left[\frac{1}{2}\sqrt{\kappa}\,\left|\log\left(\frac{\varepsilon}{2}\right)\right|\right] + 1 \tag{4.1}$$

steps of the conjugate gradient method to reduce the energy norm of the initial error by the factor $\varepsilon \in (0,1)$; see [1] or [3]. This energy norm is induced by the coefficient matrix of the linear system, and κ is the spectral condition number of the matrix. In our case we get the upper bound

$$C(j-m+1)\,\left|\log\left(\frac{\varepsilon}{2}\right)\right| \quad . \tag{4.2}$$

In terms of the Courant-number τ/h^2

$$\tilde{C}\left(\left|\log\left(\frac{\tau}{h^2}\right)\right| + 1\right)\left|\log\left(\frac{\varepsilon}{2}\right)\right| \tag{4.3}$$

is the maximal number of conjugate gradient steps needed to reduce the norm (2.15) of the initial error by the factor ε .

On the other hand the matrix A arising from the hierarchical basis formulation is relatively dense and complicated compared with the corresponding nodal basis matrix $\hat{A}$. This fact seems to compensate many of the mathematical advantages of the hierarchical basis formulation. But one has to realize that conjugate gradient type methods do not need the matrix A in an explicit form. It is sufficient to provide an efficient algorithm for computing Ax . Therefore we can make use of the factorization

$$A = S^T\hat{A}S \tag{4.4}$$

of the hierarchical basis matrix A where $\hat{A}$ is the nodal basis matrix introduced above and S is the matrix transforming the representation of a function with regard to the hierarchical basis into its representation with regard to the nodal basis. S and S^T depend only on the choice of the initial level m in the construction of the hierarchical basis and the refinement structure but not on the bilinear form B defining the initial boundary value problem. The factorization (4.4) makes it easy to compute Ax if fast algorithms for computing Sx and S^Tx are available. Such algorithms have been developed in [5]. They have a very simple structure. In the case of linear elements they use less than $2n$ additions and n divisions by 2 where n is the dimension of the finite element space. Computing Sx means evaluating a finite element function given by its coefficients with regard to the hierarchical basis in the nodal points. This is done recursively beginning with the nodes of the given initial level m where the values of the function are explicitly known. If the values of the function at a level below the final level are given, the values at the new nodes of the next level are computed by interpolation and by adding the corresponding hierarchical basis coefficients. The algorithm for computing S^Tx works similarly but in the reverse direction.

Thus we can conclude that the computation of the product of the hierarchical basis matrix with a given coefficient vector is nearly as cheap as the computation of the product of the corresponding nodal basis matrix with this vector. Using (4.2) we get an upper bound

$$C(j-m+1)\ n\ |\log(\tfrac{\varepsilon}{2})| \tag{4.5}$$

for the number of computer operations necessary for reducing the norm (2.15) of the initial error by the factor ε , provided that the bilinear form B is symmetric. As above n is the dimension of the finite element space which increases exponentially in j . Therefore the bound (4.5) behaves like $O(n \log n)$ for large time stepsizes and like $O(n)$ for small time stepsizes, which means that our method is of nearly optimal or optimal computational complexity independent of the choice of the stepsize in time.

To check these estimates we have considered the boundary value problem

$$-\tau \Delta u + u = f \tag{4.6}$$

on the unit square $\overline{\Omega} = [0,1]^2$ with zero Dirichlet boundary conditions and the right hand side $f = 1$. Its solution $u \in H(\Omega) = W_0^{1,2}(\Omega)$ satisfies

$$\tau\, B(u,v) + (u,v) = (f,v) \quad , \quad v \in H(\Omega) \quad , \tag{4.7}$$

where B is the bilinear form (2.8). We used bilinear finite elements and a regular grid of gridsize $h = 1/64$. The initial "triangulation" of level 0 consists here of the square itself. $H = 1$ is the constant in (3.4), (3.6) and (3.7), and at most $j := 6$ refinement levels are possible. Because of (3.7) we have chosen 4^{-k} , $k = 0,1,\ldots,j$, as values for τ . To justify our choice (3.7) of the initial level m the linear systems corresponding to these boundary value problems have been formulated with regard to every level $m = 0,1,\ldots,j$ as initial refinement level. Because of the zero boundary conditions there is no nontrivial function of level 0 , and the bases for $m = 0$ and $m = 1$ coincide. For $m = j$ the hierarchical basis is the nodal basis of the finite element space. The linear systems were solved by the conjugate gradient method. The iteration has been stopped when the norm (2.15) of the error became smaller than or equal to ε times the norm of the initial error; $\varepsilon \in (0,1)$ is a given accuracy. As initial approximation we took $u = 0$. Note that the energy norm (2.15) of the error cannot be computed without knowing the exact solution of the linear system but that by our theory for the correct choice of m the Euclidean norm of the residual is a good measure for the energy norm of the error.

The number of iteration steps needed to satisfy the low accuracy requirement $\varepsilon = 10^{-2}$ are given in Table 4.1. One can see that the "hierarchical" bases always work quite well and that our choice (3.7) of the initial level m leads to good results. For such low accuracies unpredictable initial effects determine the iteration essentially, mainly in favour of hierarchical bases with a large number of refinement levels. To fade out such effects we also solved the systems with the high accuracy $\varepsilon = 10^{-6}$; these results are shown in Table 4.2. For all values of τ our choice (3.7) of the initial level m nearly minimizes the number of iteration steps. For this choice of m the necessary number of iterations turned out to be very stable against changes of the right hand side f whereas for extremely "false" choices of m the number of iteration steps sometimes increased considerably. As an example the results for a randomly chosen right hand side f are given in Table 4.3. Note that in the elliptic case $\tau = 1$ with the correct choice of m the number of iteration steps is only twice as large as for the very small value $\tau = 1/4096$. This indicates

$\varepsilon = 10^{-2}$	m = 0	m = 1	m = 2	m = 3	m = 4	m = 5	m = 6
$\tau = 1$	7	7	10	11	14	21	34
$\tau = 1/4$	7	7	10	11	13	20	34
$\tau = 1/16$	7	7	9	10	13	19	31
$\tau = 1/64$	7	7	8	9	10	16	27
$\tau = 1/256$	7	7	9	9	9	10	16
$\tau = 1/1024$	8	8	9	10	8	6	8
$\tau = 1/4096$	11	11	10	14	9	5	4

Table 4.1: The number of iteration steps for the right hand side $f = 1$ and the accuracy $\varepsilon = 10^{-2}$

$\varepsilon = 10^{-6}$	m = 0	m = 1	m = 2	m = 3	m = 4	m = 5	m = 6
$\tau = 1$	27	27	26	26	30	40	65
$\tau = 1/4$	27	27	26	26	29	40	65
$\tau = 1/16$	27	27	26	25	28	39	63
$\tau = 1/64$	29	29	26	25	26	36	58
$\tau = 1/256$	34	34	30	25	23	28	44
$\tau = 1/1024$	43	43	40	33	23	19	26
$\tau = 1/4096$	67	67	63	51	32	18	13

Table 4.2: The number of iteration steps for the right hand side $f = 1$ and the accuracy $\varepsilon = 10^{-6}$

$\varepsilon = 10^{-6}$	m = 0	m = 1	m = 2	m = 3	m = 4	m = 5	m = 6
$\tau = 1$	28	28	27	27	35	60	101
$\tau = 1/4$	28	28	27	27	34	58	99
$\tau = 1/16$	29	29	27	26	32	53	90
$\tau = 1/64$	30	30	28	26	29	42	71
$\tau = 1/256$	37	37	33	28	24	29	47
$\tau = 1/1024$	53	53	51	40	25	20	27
$\tau = 1/4096$	87	87	82	64	37	20	13

Table 4.3: The number of iteration steps for a randomly chosen right hand side f and the accuracy $\varepsilon = 10^{-6}$

that the amount of work does not change very much for a large range of values for τ .

Without doubt such simple computations can give only a limited impression of the performance of our method. On the one hand in the solution of parabolic initial boundary value problems its success is not only dependent on its asymptotic speed of convergence, from the last time levels one has good starting values for the iteration which differ mainly in the higher frequency parts from the exact solution. On the other hand we believe that problems as in our example are not able to demonstrate the main advantages of our method, its inherent robustness and flexibility. Probably the most promising area of application are the adaptively refined families of grids which are used in Bank's finite element package PLTMG [2]. As it is shown in [8] the finite element spaces corresponding to these triangulations can be imbedded into our spaces. Therefore our theory can be applied and appropriately scaled hierarchical bases lead to a nearly optimal formulation of the linear systems to be solved there.

References

[1] Axelsson, O., Barker, V. A.:
Finite element solution of boundary value problems: Theory and computation
Academic Press, 1984

[2] Bank, R. E.:
PLTMG user's guide, Edition 4.0
Technical report, Department of Mathematics, University of California at San Diego, March 1985

[3] Stoer, J.:
Solution of large linear systems of equations by conjugate gradient type methods
In: Mathematical Programming, the State of Art
Edited by A. Bachem, M. Grötschel, B. Korte
Springer-Verlag, 1983

[4] Wathen, A. J.:
Attainable eigenvalue bounds for the Galerkin mass matrix
Submitted to IMA J. Numer. Anal.

[5] Yserentant, H.:
On the multi-level splitting of finite element spaces
Submitted to Numerische Mathematik

[6] Yserentant, H.:
Hierarchical bases of finite element spaces in the discretization of nonsymmetric elliptic boundary value problems
To appear in Computing

[7] Yserentant, H.:
On the multi-level splitting of finite element spaces for indefinite elliptic boundary value problems
To appear in SIAM J. Numer. Anal.

[8] Yserentant, H.:
Hierarchical bases give conjugate gradient type methods a multigrid speed of convergence
Submitted to Applied Mathematics and Computation

EXTRAPOLATION INTEGRATORS FOR QUASILINEAR IMPLICIT ODEs

P. Deuflhard, U. Nowak

0. Introduction

This paper deals with quasilinear implicit ODEs of the form

$$B(y)\, y' = f(y) \; . \qquad (0.1)$$

The main emphasis of the paper will be on problems, where B is nonsingular (index = 0). Extensions to problems with index = 1 , where B is singular, are also included. In large scale scientific computing, problems of the type (0.1) may arise e.g. in chemical reaction kinetics, when thermodynamic equations are added, or in method of lines treatment for time-dependent PDEs with moving spatial grids (compare Miller [8,9] , Hyman [6]).

At present, the numerical solution of (0.1) is typically attacked by BDF-type codes such as LSODI due to Hindmarsh [5] and DASSL due to Petzold [13]. The latter code is also designed for the treatment of problems with index > 0 . Recently, associated extrapolation techniques have been developed for:(a) problems (0.1) with index = 0 - see Deuflhard [1], (b) problems (0.1) with constant B and index = 1 - see Deuflhard/Hairer/Zugck [2]. It is the purpose of the present paper to report about recent progress made beyond [1,2].

In section 1 , the theoretical background of (0.1) with index 0 and 1 will be reviewed, leading to some matrix pencil that can be assumed to be nonsingular. In section 2 , several variants of the semi-implicit Euler discretization are discussed and worked out in detail. Numerical comparisons among these variants and with the BDF-codes LSODI and DASSL are given in section 3 .

1. Theoretical Background

Consider system (0.1) of dimension n with B(y) possibly singular, for the time being. Let (0.1) have a C^1-solution $\bar{y}$ and let $\delta y \in C^1$ denote a variation around $\bar{y}$. In first order, the following *variational equation* is obtained

$$B(\bar{y})\ \delta y' = [f_y(\bar{y}) - \Gamma(\bar{y},\bar{y}')]\delta y \tag{1.1}$$

where

$$\Gamma(\bar{y},\bar{y}')\delta y := (B_y(\bar{y})\delta y)\bar{y}'$$

Componentwise, $\Gamma = (\Gamma_{ik})$ is defined with $B = (B_{ij})$, $y = (y_k)$ as

$$\Gamma_{ik} := \sum_{j=1}^{n} \frac{\partial B_{ij}(y)}{\partial y_k} \cdot y_j' \ .$$

Local uniqueness of $\bar{y}$ means that (1.1) has locally only the trivial solution $\delta y \equiv 0$, whenever $\delta y(0) = 0$. Hence, pointwise application of the theory of Gantmacher [3] leads to the *necessary* condition that the matrix pencil

$$\{B(y) - \lambda[f_y(y) - \Gamma(y,y')]\} \tag{1.2}$$

must be nonsingular in a neighborhood of $(\bar{y},\bar{y}')$. For constant B , Γ vanishes and (1.2) reduces to the standard matrix pencil; in this case, the regularity of (1.2) is also sufficient for uniqueness.

For variable B(y) and index = 0 , (0.1) is equivalent to

$$y' = B(y)^{-1}\ f(y) =: g(y)\ . \tag{1.3}$$

In this case, a sufficient uniqueness condition is that the Lipschitz constant of g is bounded. This means that

$$g_y(y) = B(y)^{-1}[f_y(y) - \Gamma(y,y')] \tag{1.4}$$

y' from (1.3)

is bounded.

For index > 0 , B is singular. For convenience, one may introduce the *orthogonal projectors*

$$P := B\,B^+ \,, \qquad P^\perp := I - P \,, \qquad (1.5)$$
$$\bar{P} := B^+B \,, \qquad \bar{P}^\perp := I - \bar{P} \,,$$

where B^+ is the Moore-Penrose pseudoinverse. With this notation, the derivative y' can be naturally split according to

$$y' = u' + v' \qquad (1.6)$$
$$u' = \bar{P}y' = B^+f \,, \qquad v' = \bar{P}^\perp y'$$

Application of $P^\perp$ to (0.1) yields the *algebraic conditions* (cf. März [7])

$$F(y) := P(y)^\perp f(y) = 0 \,, \qquad (1.7)$$

which define a manifold M . The (given) initial values y_0 are said to be *consistent*, iff $y_0 \in M$. In this case, uniqueness can be studied following the lines of the theory of Rheinboldt [14].

Lemma.
Let $\mathrm{rank}(B(y)) = m < n$, m fixed in a neighborhood of a solution $\bar{y}$ of (0.1) subject to (1.7). Then

$$\mathrm{rank}\{P^\perp\,[f_y(\bar{y}) - \Gamma(\bar{y},\bar{y}')]\,\bar{P}^\perp\} = n\text{-}m \qquad (1.8)$$

implies the local uniqueness of $\bar{y}$.

Proof. Following [14], the vector field must satisfy the overdetermined system

$$\begin{bmatrix} B(y) \\ F_y(y) \end{bmatrix} y' = \begin{bmatrix} f(y) \\ 0 \end{bmatrix} \,. \qquad (1.9)$$

Upon applying the differentiation of pseudoinverses (cf. [4]) to (1.7), one obtains:

$$F_y(y) = P^{\perp} f_y(y) - \frac{\partial P}{\partial y} f = P^{\perp} f_y - [P^{\perp} B_y B^+ + (B_y B^+)^T P^{\perp}] f .$$

For $y \in M$, the third term vanishes, which yields

$$F_y(\bar{y}) = P(\bar{y})^{\perp} [f_y(\bar{y}) - \Gamma(\bar{y},\bar{u}')]$$

The second term may be rewritten according to

$$P(y)^{\perp} \Gamma(y,u') = P(y)^{\perp} \Gamma(y,y') , \tag{1.10}$$

for any y' satisfying (1.6). To verify this result, one may calculate

$$B_y y' = \frac{\partial}{\partial y} (B(y)y') = \frac{\partial}{\partial y} (B \bar{P} y') =$$

$$= B_y \bar{P} y' + B[B^+ B_y \bar{P}^{\perp} + \bar{P}^{\perp} (B^+ B_y)^T] y' = B_y u' + P B_y v' ,$$

since $B \bar{P}^{\perp} \equiv 0$. From this one obtains

$$P^{\perp} B_y u' = P^{\perp} [B_y y' - P B_y v'] = P^{\perp} B_y y' ,$$

which confirms (1.10). Upon partitioning (1.9) using the projectors (1.5) one obtains

$$\begin{bmatrix} P B \bar{P} & 0 \\ P^{\perp} F_y \bar{P} & P^{\perp} F_y \bar{P}^{\perp} \end{bmatrix} \begin{bmatrix} u' \\ v' \end{bmatrix} = \begin{bmatrix} Pf \\ 0 \end{bmatrix} \tag{1.9'}$$

Since $\text{rank}(PB\bar{P}) = \text{rank}(B) = m$, this system can be directly solved for u' . Solvability for v' then leads to (1.8) , when (1.10) has been inserted. ☒

The above Lemma gives a sufficient uniqueness criterion for problems (0.1) with index = 1 . For the purpose of the present paper, the essential item is that the extension of constant B to variable B(y) here also just induces the replacement of f_y by $f_y - \Gamma$.

2. Semi-Implicit Euler Discretizations

In this section, the algorithmic realization of an extrapolation method based on the semi-implicit Euler discretization (see [1], code EULSIM) will be discussed in detail. To start the derivation, assume index = 0 in (0.1), which is then equivalent to the explicit ODE (1.3). Formally, the basic discretization of (1.3) reads:

$$\hat{A} := g_y(y_0)\ ,\ \{n_j\} \in F\ , \qquad H : \text{basic stepsize} \tag{2.1}$$

$$j = 1,2,\ \dots\ , j_{max} : h := \frac{H}{n_j}$$

$$k = 0,\ \dots\ , n_j - 1 : y_{k+1} := y_k + h(I - h\hat{A})^{-1}\ g(y_k)\ .$$

This form, however, is not well-suited for implementation. Rather, the expressions from (0.1) will be inserted:

$$f(y) = B(y)\ g(y)\ , \qquad A := B(y_0)\hat{A} = f_y(y_0) - r(y_0, y_0')$$

This leads to *variant I* of the discretization:

$$\begin{aligned} y_{k+1} &:= y_k + h(B_0 - hA)^{-1}\ B_0\ B_k^{-1}\ f(y_k) \\ B_k &:= B(y_k) \end{aligned} \tag{2.2}$$

For constant B, one has $B_0\ B_k^{-1} = I$, which yields the standard case already treated in [1].
For variable $B(y)$, (2.2) requires (DEC = decomposition, SUBST = substitution, MV = matrix/vector multiplication):

1	DEC : $(B_0 - hA)$, $(n_j - 1)$ DEC : B_k,	
n_j	SUBST : $y_{k+1} - y_k$	
(n_j-1)	SUBST + MV : $B_0\ B_k^{-1}\ f(y_k)$	(*)

in each discretization loop (k-loop). In general, this implementation requires too much storage and too much computing time (because of (*)). It is therefore only advised in special cases - see Nowak/Walkowiak [11] for a class of chemical kinetics problems. To derive further discretizations, (2.2) may be rewritten as

$$y_{k+1} := y_k + h(B_k - h\, B_k\, B_0^{-1}\, A)^{-1}\, f(y_k) \tag{2.2'}$$

In this form, the replacement

$$B_k\, B_0^{-1} \rightarrow I$$

may be interpreted as just introducing an approximation error into the Jacobian matrix. This leads to *variant II*

$$y_{k+1} := y_k + h(B_k - hA)^{-1}\, f(y_k) \quad . \tag{2.3}$$

Compared with (2.2) , this discretization requires

$$\begin{array}{lll} n_j & \text{DEC} : B_k - hA & . \\ n_j & \text{SUBST} : y_{k+1} - y_k & . \end{array}$$

Obviously, the above amount (*) is avoided and storage is reduced. In order to reduce the number of decompositions, one may implement an *iterative realization* of (2.3):

$$\begin{aligned} & y_{k+1}^{0} := y_k \; , \quad \Delta_{k+1}^{i} := y_{k+1}^{i+1} - y_{k+1}^{i} \; , \\ & \Delta_{k+1}^{0} := h(B_0 - hA)^{-1}\, f(y_k) \\ & i = 1,2, \ldots : \\ & \Delta_{k+1}^{i} := -(B_0 - hA)^{-1}\, \Delta B_k \cdot \Delta_{k+1}^{i-1} \\ & \Delta B_k := B(y_k) - B_o \end{aligned} \tag{2.4}$$

This fixed point iteration is a modification of the suggestion in [1] , where, however, the term Γ in A was missing. For a careful examination of the convergence behavior, one needs the following mild assumptions:

$$\begin{aligned} & \| \Delta_{k+1}^{0} \| \leq \alpha \cdot h \\ & \| (B_0 - hA)^{-1}\, (B(y_k) - B_0) \| \leq \gamma \cdot k \cdot h \end{aligned} \tag{2.5}$$

This leads to the estimates:

$$\text{a)}\quad \|\Delta_{k+1}^{2}\| \le \alpha\gamma\ k\cdot h^2 = \alpha\gamma \cdot \frac{k}{n_j^2} \cdot H^2 \tag{2.6}$$

$$\text{b)}\quad \frac{\|\Delta_{k+1}^{i}\|}{\|\Delta_{k+1}^{i-1}\|} \le \gamma\cdot k\cdot h = \gamma \cdot \frac{k}{n_j} \cdot H$$

$$\text{c)}\quad \|\Delta_{k+1}^{m}\| \le \alpha\gamma^m \cdot \left(\frac{kH}{n_j}\right)^{m+1} \cdot \frac{1}{k} < \frac{1}{n_j}\ \alpha\gamma^m \cdot H^{m+1}$$

For a reasonable basic stepsize H one will require sufficient accuracy after m iterations, which means:

$$\|\Delta_{k+1}^{m}\| \le \frac{\varepsilon}{n_j} \quad , \qquad \varepsilon : \text{ prescribed error tolerance} \tag{2.7}$$

(The factor n_j takes care of possible error accumulation). Combining (2.6.c) and (2.7) leads to

$$\alpha\gamma^m \cdot H^{m+1} \ \dot{\le}\ \varepsilon \quad , \tag{2.8}$$

which is an upper bound for the basic stepsize. The quantities α, γ can be computationally estimated from (2.6. a,b) - for $k = 1$, $n_j = 2$, say

$$\alpha\gamma \doteq \frac{4}{H^2}\|\Delta_2^1\| \tag{2.9}$$

$$\gamma \doteq \frac{2}{H} \cdot \hat{\Theta}_2 \ , \ \hat{\Theta}_2 := \|\Delta_2^1\| \ / \ \|\Delta_2^0\|$$

Under the assumption that $\hat{\Theta}_2 < \frac{1}{2}$, the necessary number $\bar{m}$ of iterations can be directly estimated as

$$\bar{m} \doteq |\log(2\ \hat{\Theta}_2)| \ / \ \left|\log\left(\frac{\varepsilon}{2\|\Delta_2^0\|}\right)\right| \tag{2.10}$$

This represents the maximum number of iterations, since $k = n_j$ was inserted in (2.6.c). The amount of work in (2.4) for m iterations is

m (SUBST + MV) , 1 SUBST ,

to be compared with

1 DEC , 1 SUBST

in the direct solution of (2.3). In this situation, *variant III* of the discretization realizes (2.4) directly, whenever

$$\bar{m}\ (\mathrm{SUBST} + \mathrm{MV}) \ \leq\ \mathrm{DEC} \quad \text{and} \quad \hat{\Theta}_2 \ \leq\ \frac{1}{4}\ , \tag{2.11}$$

otherwise (2.3). This strategy requires one to start each basic step with (2.4) until the quantities $\|\Delta_2^0\|$, $\hat{\Theta}_2$ are available. For the illustrating combustion problem presented in section 3 , internal time monitoring led to $\bar{m} \leq 4$ from (2.11).

Remark 1. The above formulas (2.8) and (2.9) also lead to a reasonable maximum stepsize

$$\bar{H}_m \doteq \frac{H}{2} \cdot \left(\frac{\varepsilon}{\|\Delta_2^1\|\ \hat{\Theta}_2^{m-1}} \right)^{\frac{1}{m+1}} \tag{2.12}$$

for given maximum iteration number m . An alternative strategy would be to impose the restriction $H \leq \bar{H}_m$ throughout the integration and apply only (2.4). Numerical experiments showed that such a strategy is less efficient.

Remark 2. A more detailed study of the convergence pattern leads to estimates $m_{k,j}$ of iterations (2.4) with k,n_j given. On this basis, one might refine (2.11) accordingly and split between (2.3) and (2.4) *within* each k-loop. Such a strategy would just complicate the code without leading to significant improvements.

Independent of the selected discretization variant, the matrix

$$\begin{aligned} A &= f_y(y_0) - \Gamma(y_0, y_0') = \\ &= \frac{\partial}{\partial y}\ [f(y) - B(y)\ y_0']_{y_0} \end{aligned} \tag{2.13}$$

needs to be computed - typically by using finite difference approximations. Obviously, this procedure requires the value y_0' , which is not available in the usual course of integration. To compute y_0' for *index = 0*, one may either solve the linear system (typically sparse)

$$B(y_0)\ y_0' = f(y_0)$$

or apply h-extrapolation to the terms

$$\{ \frac{y_k - y_{k-1}}{h} \Big|_{k=n_j} \} . \qquad (2.14)$$

For problems with *index = 1* , both variants II and III based on the discretization (2.3) are promising - in combination with (2.14). For such problems, however, the existence of a perturbed asymptotic h-expansion is only guaranteed (up to now), if (0.1) can be transformed into a Kronecker canonical form - for details see [2].

3. Numerical Comparisons

The above developed *variant III* led to the implementation of a refined version of the original code LIMEX due to [1,15]. This code is compared with the two BDF codes DASSL due to [13] and LSODI due to [5]. All experiments were run in FORTRAN double precision on the IBM 3081 D of the Computing Center of the University of Heidelberg. Scaling and error criteria of the codes were made comparable.

Numerical comparisons with *index = 1* have been presented in [2] for nearly constant B . In this test set (from chemical mass action kinetics) the code LSODI turned out to be significantly less robust in critical examples, but efficient in non-critical examples. The performance of DASSL and LIMEX appeared to be roughly comparable.

In the present paper, the emphasis is on problems with *index = 0* and "sufficiently nonlinear" B(y). The main interest is concentrated on solving combustion problems by method of lines with *moving* spatial grids. In realistic sensitive combustion problems, re-gridding techniques will be needed in addition: in this situation, one-step methods like LIMEX will have a natural advantage over the multistep methods (because of the many re-starts). In order to obtain an unbiased performance of the IVP solvers, only a less sensitive combustion problem is selected, which does not require re-gridding.

Example. Combustion problem (Peters/Warnatz [12]).
The selected problem is test problem A , case 2 , from [12], p. 3-6.The PDE system reads (T : temperature, Y : chem. species):

a) $T_t = T_{xx} + R(T,Y)$ (3.1)

$$Y_t = \frac{1}{Le} Y_{xx} - R(T,Y)$$

$$R(T,Y) := \frac{\beta^2}{2\ Le} Y \exp\left[- \frac{\beta(1-T)}{1-\alpha(1-T)}\right]$$

$$\alpha = 0.8\ , \quad \beta = 10\ , \quad Le = 1$$

b) $T_x(\infty) = 1\ , \quad Y_x(\infty) = 0$

$T(-\infty) = 0\ , \quad Y(-\infty) = 1$

c) $t = 0\ ,\ x \leq 0 : T = \exp(x)\ , \quad Y = 1 - \exp(Le \cdot x)\ ,$

$t = 0\ ,\ x > 0 : T = 1\ , \quad Y = 0\ .$

Method of lines treatment with a slightly non-uniform initial grid of 94 nodes on [-25,10] is used. Linear finite differences in combination with a moving grid technique similar to Miller [8,9] and Hyman [6] are applied. Details are left to [10]. This treatment leads to a problem (0.1) of dimension n = 282, index = 0, where B, f_y, Γ are banded with a maximum total bandwidth b = 11. The subsequent tables use the following specifications:

TIME	-	total computing time [sec]
NSTEP	-	number of integration steps
NFCN	-	number of (B,f)-evaluations without Jacobian finite difference approximation
NJAC	-	number of Jacobian evaluations
NFJAC	-	number of (B,f)-evaluations for Jacobian approximations
NDEC	-	number of Gaussian decompositions (banded mode)
NSUBST	-	number of forward - backward substitutions

Table 1. Comparison results for the solution of (3.1) with $x \in [-25,10]$ fixed from $t = 0$ to $t = 12$.

		LIMEX	DASSL	LSODI
$TOL = 10^{-2}$:	TIME	2.8	2.8	1.7
	NSTEP	7	18	15
	NFCN	18	38	18
	NJAC	7	13	9
	NFJAC	77	143	99
	NDEC	20	13	9
	NSUBST	39	38	21
$TOL = 10^{-3}$:	TIME	2.7	3.5	2.7
	NSTEP	5	32	27
	NFCN	25	68	40
	NJAC	5	12	13
	NFJAC	55	132	143
	NDEC	14	12	13
	NSUBST	70	68	39
$TOL = 10^{-4}$:	TIME	5.2	5.2	3.2
	NSTEP	7	58	45
	NFCN	44	121	57
	NJAC	7	14	13
	NFJAC	77	154	143
	NDEC	26	14	13
	NSUBST	164	121	56

Table 2. Comparison results for the solution of (3.1) with $x(0) \in [-25,10]$, free boundaries, from t = 0 to t = 120.

		LIMEX	DASSL	LSODI
TOL = 10^{-2} :	TIME	2.0	3.5	1.9
	NSTEP	6	20	17
	NFCN	12	41	19
	NJAC	6	17	10
	NFJAC	66	187	110
	NDEC	12	17	10
	NSUBST	28	41	21
TOL = 10^{-3} :	TIME	2.4	4.1	3.3
	NSTEP	6	33	29
	NFCN	16	67	45
	NJAC	6	17	16
	NFJAC	66	187	176
	NDEC	14	17	16
	NSUBST	45	67	41
TOL = 10^{-4} :	TIME	5.7	6.1	3.7
	NSTEP	8	62	50
	NFCN	51	128	63
	NJAC	8	20	16
	NFJAC	88	220	176
	NDEC	30	20	16
	NSUBST	170	128	59

Roughly speaking, LSODI comes out to be 20 % faster (in total) than LIMEX , while DASSL is 20 % slower - in this test problem. Even though the percentages of work spent in the linear algebra part, the f-evaluations, and the overhead are rather different, the overall performance is comparable. The differences in these percentages are typical for the codes and may lead to a rather different performance in other examples.

A drastic improvement, however, can be seen by comparing the refined version of LIMEX (this paper) with the older version of [1,15]. To model the older version, $\Gamma := 0$ was set in the present code.

Table 3. LIMEX comparisons (TIME).

		new	old
Table 1 ,	TOL = 10^{-2}	2.8	3.4
	TOL = 10^{-4}	5.2	15.4
Table 2 ,	TOL = 10^{-2}	2.0	5.8
	TOL = 10^{-4}	5.7	24.0

With $\Gamma \neq 0$, the variants I and II (described in section 2 above) would yield only slightly slower runs in this special test problem.

Conclusion

The present version of the extrapolation integrator LIMEX is competitive with the BDF codes DASSL and LSODI for quasilinear implicit ODEs with index = 0 and index = 1 . For sufficiently sensitive nonlinear PDEs , that require re-gridding in method of lines treatment, the one-step method of LIMEX has a natural advantage over the multistep methods.

References

[1] P. Deuflhard: Recent Progress in Extrapolation Methods for ODEs. SIAM Rev., 27, 505-535 (1985)

[2] P. Deuflhard, E. Hairer, J. Zugck: One-Step and Extrapolation Methods for Differential-Algebraic Systems. Univ. Heidelberg, SFB 123: Tech. Rep. 318

[3] F. Gantmacher: The Theory of Matrices, Vol. 2, New York: Chelsea (1959).

[4] G.H. Golub, V. Pereyra: The Differentiation of Pseudoinverses and Nonlinear Least Squares Problems whose Variables Separate, SIAM J. Numer.Anal. 10, 413-432 (1973).

[5] A.C. Hindmarsh: LSODE and LSODI, Two New Initial Value Ordinary Differential Equation Solvers, ACM-SIGNUM Newsletters 15, 10-11 (1980).

[6] J.M. Hyman: Moving Mesh Methods for Initial Boundary Value Problems, Los Alamos National Laboratory, Tech.Rep. (1985).

[7] R. März: Correctness and Isolatedness in the Case of Initial and Boundary Value Problems in Differential Algebraic Equations, Humboldt-Univ. Berlin, Sektion Mathematik: Preprint Nr. 73 (1984).

[8] K. Miller: Moving Finite Elements II , SIAM J.Numer.Anal. 18, 1033-1057 (1981).

[9] K. Miller, R.N. Miller: Moving Finite Elements I , SIAM J.Numer.Anal. 18, 1019-1032 (1981).

[10] U. Nowak: Extrapolation Methods for Instationary Reaction-Diffusion Systems, work done in preparation of a dissertation (1985).

[11] U. Nowak, D. Walkowiak: Numerical Simulation of Large Chemical Reaction Kinetics Including Thermodynamics, in preparation.

[12] N. Peters, J. Warnatz (Eds.): Numerical Methods in Laminar Flame Propagation, Notes on Numerical Fluid Dynamics 6 , Vieweg (1982).

[13] L. Petzold: A Description of DASSL: A Differential/Algebraic System Solver, Proc. IMACS World Congress 1982, to appear.

[14] W.C. Rheinboldt: Differential-Algebraic Systems as Differential Equations on Manifolds, Math.Comp. 43, 473-482 (1985).

[15] J. Zugck: Numerische Behandlung linear-impliziter Differentialgleichungen mit Hilfe von Extrapolationsmethoden, Univ. Heidelberg, Inst.Angew.Math.: Diplomarbeit (1984).

NUMERICAL PROBLEMS ARISING FROM THE SIMULATION OF COMBUSTION PHENOMENA

J.Warnatz

1. Introduction

Combustion processes result from a complex interaction of convection, transport, and chemical reaction. Thus, simulation needs much computational effort and must (at present) be restricted to relatively simple cases. Combustion calculations on three different levels of complexity are considered:

(1) Chemistry of auto-ignition processes e.g. in shock tubes and Otto engines: Simulation of this zero-dimensional problem leads to large stiff and non-linear differential/algebraic systems which are solved by implicit and semi-implicit methods.

(2) Flame propagation and flame front structure: This stationary one-dimensional problem is solved with an implicit finite-difference method, using a simplified linearization considering diagonal elements of the Jacobian only.

(3) Ignition/quenching problems: This instationary one-dimensional problem, after discretization by finite differences, leads to large differential/algebraic systems which are solved both by BDF and extrapolation methods.

2. Simulation of the Chemistry of Auto-Ignition

2.1 Background

Engine knock is a decades old problem connected with the operation of Otto engines, in special since the development of high performance versions (e.g. for airplanes). Very early it became apparent, that knocking is kinetically controlled, and must be closely connected with the chemistry of combustion, as can be seen for instance from the strong effect of addition of traces of inhibitors like lead tetraethyl (for reference see [1]).

Extension of reaction schemes of the high temperature oxidation of hydrocarbons [2-6] to lower temperatures now allows the explanation of ignition delay up to octane using sophisticated mechanisms for the alkyl radical decomposition of higher hydrocarbons [7-9], as will be demonstrated in the following.

2.2 Calculation Method

Calculation of ignition delay times is done by solving the conservation equations of the reaction system, considering the respective experimental conditions of pressure and volume (temperature T is specified by the experiment). Treating a homogeneous mixture, this is a zero-dimensional system with time t as only independent variable, consisting of total mass conservation and species conservation:

$$dm/dt = 0 \quad \text{or} \quad \varrho V = \text{const.} \tag{1}$$

$$dc_i/dt = R_i \; , \quad i = 1,\ldots,n \tag{2}$$

Here c_i = concentration of species i, R_i = (molar scale) chemical rate of formation of species i, m = total mass of the mixture, n = number of species, ϱ = density of the mixture, V = volume of the mixture.

These equations represent a differential/algebraic system containing one algebraic (1) and n ordinary differential equations (2), with typically n = 50.

The dependent variables in the equations are P (pressure calculated from the ideal gas law) and c_i, if the reaction volume is given (shock tube conditions), and V and c_i, if the pressure is given (engine conditions). The solution of this initial value problem is obtained by use of the programs DASSL [10] or LIMEX [11], both of them designed to solve stiff differential/algebraic systems. Typical calculation times are 10 s on an IBM 3081D.

2.3 Reaction Mechanism

The reaction mechanism used for the description of the oxidation of molecules up to C_8-species is very similar to that discussed in [2-6] (see Section 3). The attack on higher alkanes by H, O, OH, HO_2, etc. can be described by summing up data for the attack of single primary, secondary, and tertiary C-H bonds [3].

The decomposition of the alkyl radical produced can be predicted to be β-decomposition, as originally descibed by Rice in the work on alkane pyrolysis [8] (for reference see [3,7]). This procedure leads to the smaller alkyl radicals CH_3, C_2H_5, C_3H_7, or C_4H_9, whose oxidation then is considered in detail. Besides, reaction of the alkyl radicals with O_2 to form HO_2 and the corresponding alkene must be considered at lower temperatures.

The oxidation of the alkenes formed is so fast, that it is not rate limiting (see e.g. [4,9]). For this reason, a global reaction to directly form C_2H_4 is used to describe its decomposition (detailed mechanisms of higher

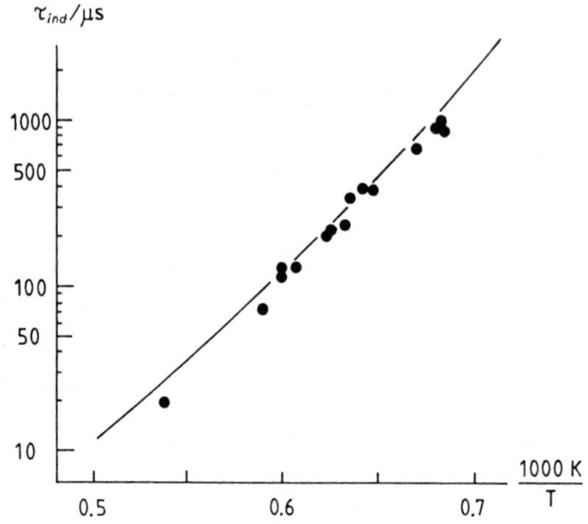

Fig.1: Experimental and calculated ignition delay times in a mixture of 2% CH_4 with artificial air (Ar/O_2 = 80/20), $\varrho = 2.5 \cdot 10^{-5}$ mol/cm $^{-3}$ [12]

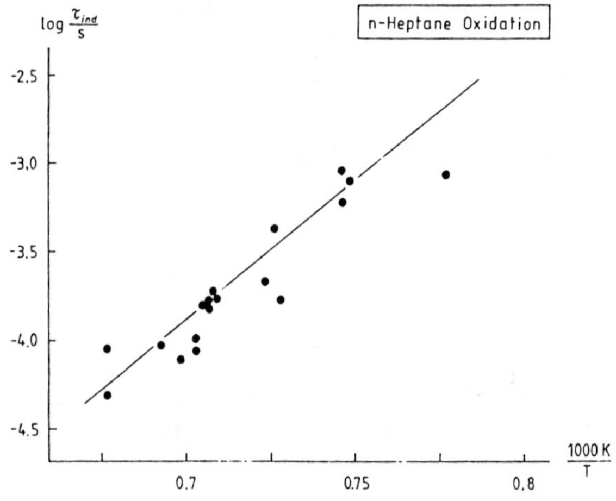

Fig.2: Experimental and calculated induction times in n-heptane-oxygen-argon (0.17/1.83/98.0) mixtures [13], P = 1.5 bar.

Table 1: Global Formulations of Alkane and Alkyl Radical Decomposition in Octane and Heptane Combustion

n-octane

$n\text{-}C_8H_{18}$	$\longrightarrow$	$n\text{-}C_4H_9 + n\text{-}C_4H_9$
$1\text{-}n\text{-}C_8H_{17}$	$\longrightarrow$	$3\ C_2H_4 + C_2H_5$
$2\text{-}n\text{-}C_8H_{17}$	$\longrightarrow$	$C_3H_6 + 2\ C_2H_4 + CH_3$
$3\text{-}n\text{-}C_8H_{17}$	$\longrightarrow$	$3\ C_2H_4 + C_2H_5$
$4\text{-}n\text{-}C_8H_{17}$	$\longrightarrow$	$C_3H_6 + 2\ C_2H_4 + CH_3$
$O_2 + C_8H_{17}$	$\longrightarrow$	$HO_2 + 4\ C_2H_4$

i-octane (2,2,4-trimethyl-pentane)

$i\text{-}C_8H_{18}$	$\longrightarrow$	$2\ C_3H_6 + 2\ CH_3$
$1\text{-}i\text{-}C_8H_{17}$	$\longrightarrow$	$C_3H_6 + 2\ C_2H_4 + CH_3$
$2\text{-}i\text{-}C_8H_{17}$	$\longrightarrow$	$C_3H_6 + 2\ C_2H_4 + CH_3$
$3\text{-}i\text{-}C_8H_{17}$	$\longrightarrow$	$C_3H_6 + 2\ C_2H_4 + CH_3$
$5\text{-}i\text{-}C_8H_{17}$	$\longrightarrow$	$C_3H_6 + 2\ C_2H_4 + CH_3$
$O_2 + C_8H_{17}$	$\longrightarrow$	$HO_2 + 4\ C_2H_4$

n-heptane

$n\text{-}C_7H_{16}$	$\longrightarrow$	$2\ C_2H_4 + C_2H_5 + CH_3$
$1\text{-}n\text{-}C_7H_{15}$	$\longrightarrow$	$3\ C_2H_4 + CH_3$
$2\text{-}n\text{-}C_7H_{15}$	$\longrightarrow$	$C_3H_6 + C_2H_4 + C_2H_5$
$3\text{-}n\text{-}C_7H_{15}$	$\longrightarrow$	$3\ C_2H_4 + CH_3$
$4\text{-}n\text{-}C_7H_{15}$	$\longrightarrow$	$C_3H_6 + C_2H_4 + C_2H_5$
$O_2 + C_7H_{15}$	$\longrightarrow$	$HO_2 + 2\ C_2H_4 + C_3H_6$

alkene oxidation are not available). The resulting mechanisms of alkyl radical decomposition are given in Table 1 for octane.

2.4 Results at Isothermal Conditions

Fig.1 shows measured ignition delay times and corresponding simulations for methane-oxygen-argon mixtures [12]. The ignition delay time both in the experiments and in the calculations has been determined from the fuel consumption profiles. Data for the highest hydrocarbon available in the literature (n-heptane) are presented in Fig.2, using OH^* and CO profiles to define the ignition delay time [13]. In both cases, the simulations are in agreement with the experimental material, if the error limits of the measurements are taken into consideration.

Interesting is a comparison of data for n-octane with iso-octane and for n-heptane with iso-heptane (iso-octane, ROZ = 100, and n-heptane, ROZ = 0, are used to define the research octane number scale characterizing knock tendency in Otto engines). As expected, in the temperature range of interest for knocking ($T < 1100$ K) ignition delay times for the iso-alkanes are larger than that for the n-alkanes by one order of magnitude (see Fig.3), explaining the completely different knocking tendency (e.g. ROZ($i\text{-}C_7H_{16}$)= 112, ROZ($n\text{-}C_7H_{16}$)= 0 [14]).

2.5 Results at Changing Temperature and Pressure

To simulate end-gas autoignition in engines, the computer code is modified to calculate ignition delay times for changing temperature and pressure, which are known from experiments in engines.

The example considered here (Fig.4) is knock in an engine driven with n-butane [15]. Both knocking and

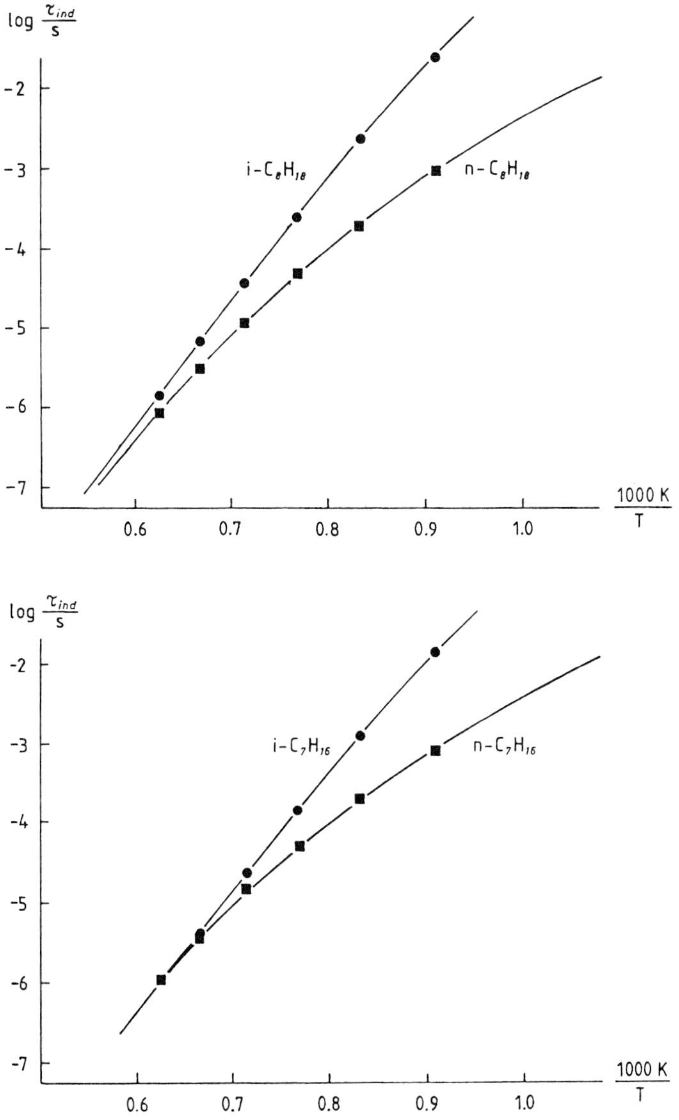

Fig.3: Calculated ignition delay times at atmospheric pressure in stoichiometric mixtures of air with different octanes and heptanes.

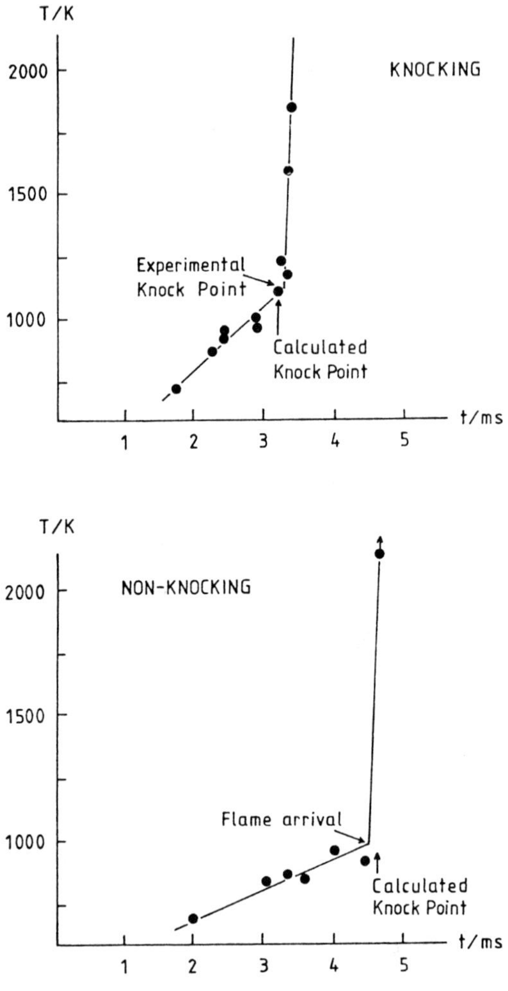

Fig.4: Measured and calculated knocking behaviour in an engine fueled with n-butane. Experimental data are taken from [15]

non-knocking case can be predicted with the mechanism taken from [9]. No global steps have to be included in the reaction mechanism for this relatively simple fuel.

3. Simulation of Stationary Flame Propagation

3.1 Background

If one is particularly interested in a detailed knowledge of the reactions in such a complex system it is advisable to choose simple flow conditions and a simple geometry of the flame. Therefore, the calculation of a laminar flat flame which allows one-dimensional treatment shall be considered here. Calculated temperature and concentration profiles can then be compared directly with profiles determined experimentally in a flat flame. Moreover, flame velocities are obtained which can be directly compared with experimental values.

3.2 Calculation Method

For a quantitative treatment of the measurements, the corresponding conservation equations for a premixed laminar flat flame must be solved. Conservation of enthalpy and of mass of species i leads to the time-dependent equations [2,16,17]

$$\varrho \frac{\partial T}{\partial t} = - \varrho v \frac{\partial T}{\partial z} - j_H \frac{\partial T}{\partial z} + \frac{1}{A c_p} \frac{\partial}{\partial z}\left(A \lambda \frac{\partial T}{\partial z}\right) - \frac{\Sigma r_i h_i}{c_p} \qquad (3)$$

$$\varrho \frac{\partial w_i}{\partial t} = - \varrho v \frac{\partial w_i}{\partial z} - \frac{1}{A} \frac{\partial (A \cdot j_i)}{\partial z} + r_i \qquad (4)$$

where the diffusion fluxes j_i and the mean diffusion flux j_H are given by

$$j_H = \frac{\sum c_{p,i} j_i}{c_p} \quad ; \quad j_i = - D_{i,M} \varrho \frac{\partial w_i}{\partial z} - D_{i,T} \frac{\partial \ln T}{\partial z} \qquad (5)$$

(A = area ratio, c_p = specific heat capacity, $D_{i,T}$ = thermal diffusion coefficient, h = specific enthalpy, r = mass scale chemical rate of formation, t = time, T =

temperature, v = flow velocity, w = mass fraction, z = cartesian space coordinate, λ = mixture heat conductivity, ϱ = density.)

A simplified transport model

$$\lambda_M = 0.5 \left[\sum x_i \lambda_i + \left(\sum x_i / \lambda_i \right)^{-1} \right] \qquad D_{i,M} = \frac{1 - w_i}{\sum_{j \neq i} x_j / \mathcal{D}_{ij}} \quad ; \quad \sum_i j_i = 0 \quad (6)$$

($\mathcal{D}_{ij}$ = binary diffusion coefficients, x = mole fraction, λ_i = species heat conductivity) is used because comparison with multicomponent transport models results in relatively small errors [16,18-20].

Due to the stiffness of the system of differential equations (3) and (4), an implicit finite difference method is chosen for solution [2,16]. This method starts with arbitrary profiles of temperature T and mass fractions w_i at time zero. With the aid of an adaptive (grid point density proportional to temperature gradient) non-uniform grid point system (see Fig.5) the derivatives are replaced by finite difference expressions assuming a parabolic

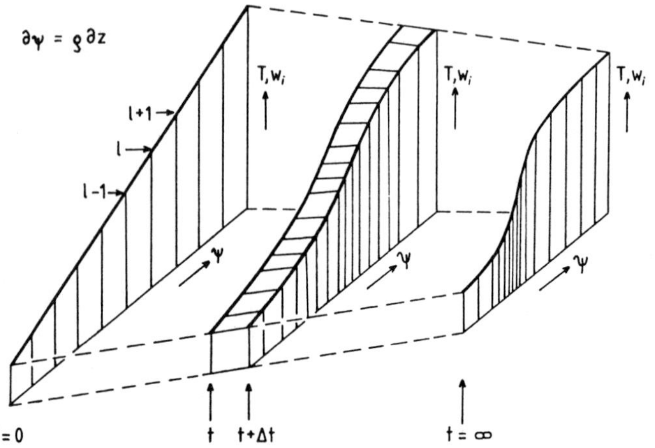

Fig.5: Development of solutions of the time-dependent conservation equations (see text)

approach between three neighbouring grid points in each case.

This procedure reduces the given problem to the solution of a tridiagonal linear equation system, if at the edges of the grid point system the values of temperature T and mass fractions w_i are specified by means of proper boundary conditions [2,16,21].

Typical calculation times are of the order of 3o min on an IBM 3081D for 50 species, 41 grid points, i.e. about 2000 ordinary differential equations.

3.3 High temperature reaction mechanism

The reaction mechanism used in this paper is based on a critical review of 200 elementary reactions which may occur in hydrocarbon combustion [3]. Unimportant reactions are eliminated by comparison of the rates of reaction in the flames considered here. Details are explained elsewhere [2,6].

The extensive literature on the H_2-O_2-CO reaction system shall not be discussed here in detail, since there are comprehensive reviews on the elementary reaction and rate coefficients in this system [23-25].

The mechanism of the oxidation of C_1- and C_2-hydrocarbons is shown in Fig.6. In the same way, combustion of C_3- and C_4-hydrocarbons can be described by detailed reaction schemes consisting of elementary steps.

Apart from the initial attack on the fuel, oxidation of higher aliphatic hydrocarbons (see Fig.7) can be treated by simplified global reactions, because the rate-limiting steps are contained in the H_2-O_2-CO-C_1-C_2 system (see [2,26]).

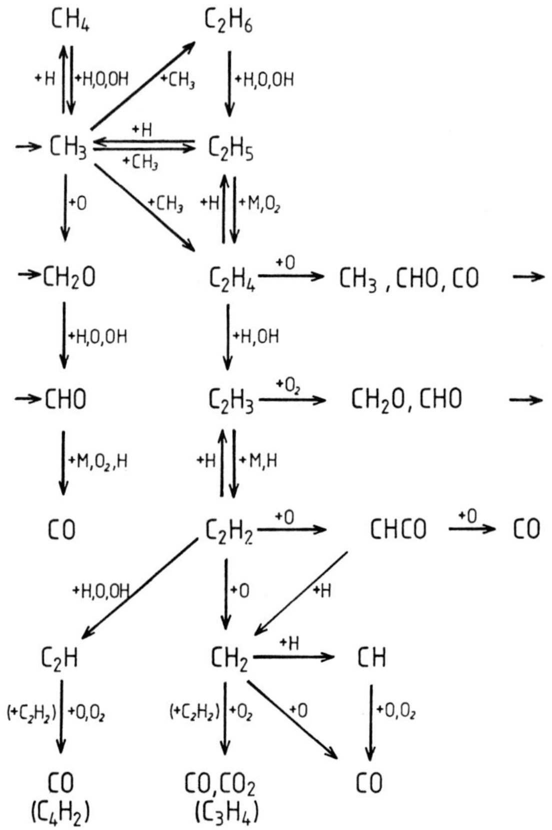

Fig.6: Schematic mechanism of oxidation of C_1- and C_2-hydrocarbons [2,22]

3.4 Results on Stationarily Propagating Flame Fronts

This mechanism for the oxidation of C_1- to C_8-hydrocarbons presented here should predict experimental results in lean and moderately rich flames of alkanes, alkenes, and alkynes.

In fact, within the limits of experimental errors, there is agreement between experiments and calculations for

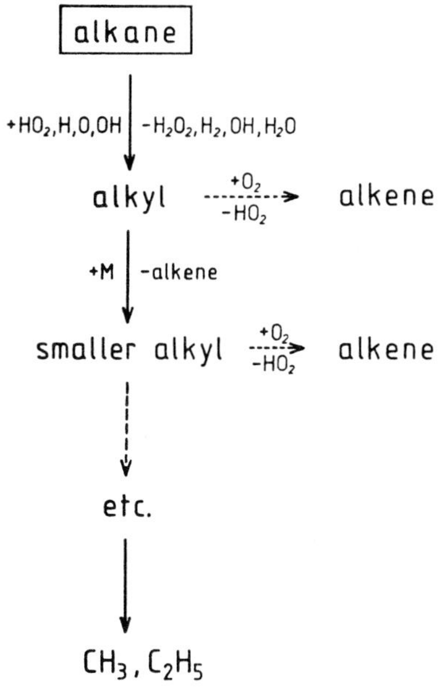

Fig.7: General mechanism of radical pyrolysis of alkanes leading to CH_3 and C_2H_5 radicals

free flame velocities of these fuels up to octane (example in Fig.8) and for concentration and temperature profiles in burner-stabilized lean hydrocarbon-air (or O_2) flames (see Fig.9; details given in [2,6]).

More insight into the reaction mechanism is provided by a sensitivity test carrying out a systematic variation of rate coefficients in calculations of the flame velocity of e.g. a stoichiometric C_4H_{10}-air flame at atmospheric pressure (see Fig.10). For each reaction the rate coefficient is reduced individually by a factor of five. The ratio of the original and the new flame velocity is given in the illustration. For the large majority of reactions the flame velocity changes by less than 2% (hatched area). For this reason, these reactions are not listed.

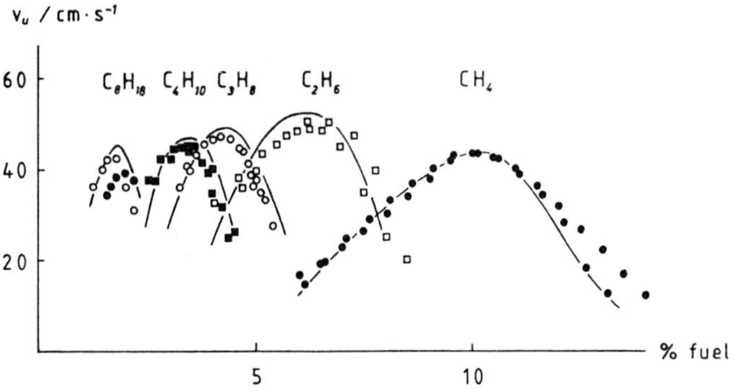

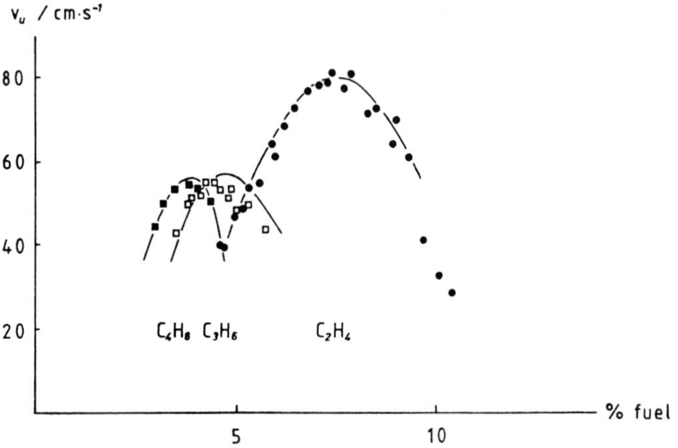

Fig.8: Free flame velocities in alkane- and alkene-air mixtures, P=1bar, T_u=298 K. Points: measurements of different workers (for reference see [2,27], lines: calculations

It is clearly shown that the flame velocities are relatively insensitive to reactions specific for the hydrocarbon considered. However, there is a strong influence of the two unspecific reactions of H atoms with O_2 and of CO with OH radicals. This means that lean and nearly stoichiometric hydrocarbon flames, on the whole, can be characterized as H_2- O_2- CO flames fed by the hydrocarbon.

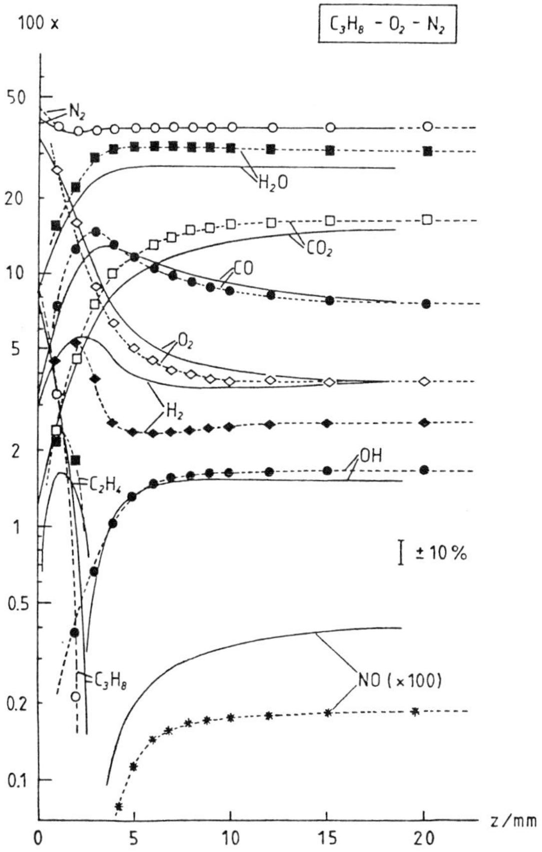

Fig.9: Experimental and calculated mole fraction and temperature profiles in a propane-oxygen-nitrogen low pressure flame (for reference see [28])

4. Simulation of Instationary Flame Propagation

4.1 Background

The understanding of instationary processes is very important, in special if ignition or quenching phenomena are considered (Otto engines, safety considerations).

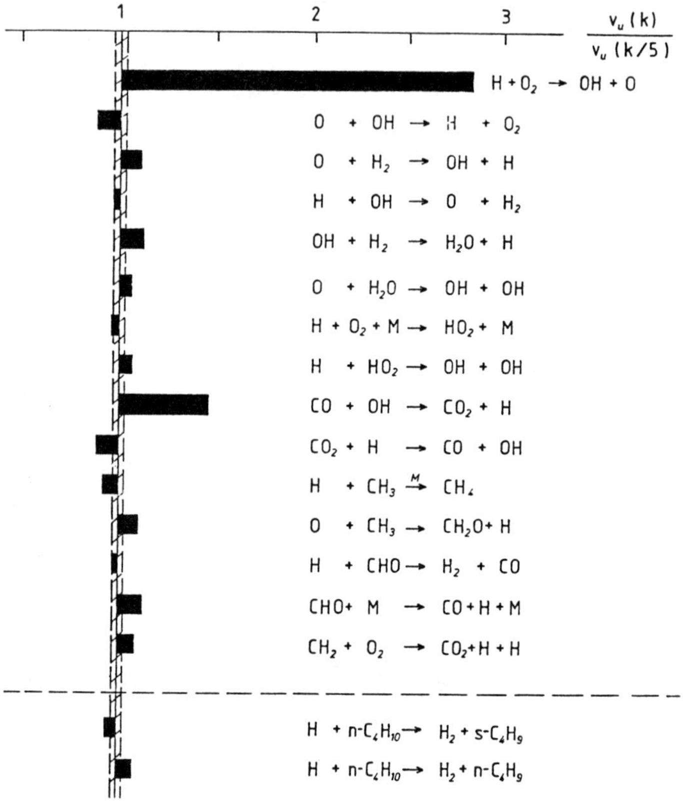

Fig.10: Influence of a reduction of rate coefficients by a factor of 5 on the free flame velocity in a stoichiometric flame of $n\text{-}C_4H_{10}$ with air ($P=1$ bar, $T_u=298$ K)

To study the influence of chemistry on instationary flame propagation, thermal ignition in a one-dimensional configuration is studied by computer simulation for O_2-O_3 mixtures to enable comparison with corresponding experiments [29,30]. To interpret these measurements, the corresponding time-dependent one-dimensional conservation equations (assuming uniform pressure) are integrated by various differential/algebraic solvers.

As shown below, experimental results on thermal minimum ignition energies in O_2-O_3 mixtures can be simulated by the calculations within the limits of experimental error. On the basis of this "calibration", further simulations are carried out to study the dependence of minimum igition energy on ignition source diameter and ignition time to get more information on the nature of the ignition process.

4.2 Calculation Method

After transformation to Lagrangian coordinates $(r,t) \rightarrow (\psi,t)$

$$(\partial\psi/\partial r)_t = \varrho r^n \quad ; \quad (\partial\psi/\partial t)_r = -\varrho v r^n \tag{7}$$

the system of equations under consideration is given by

$$\varrho \frac{\partial w_i}{\partial t} = -\varrho v \frac{\partial w_i}{\partial r} - \frac{1}{r^n} \frac{\partial j_i}{\partial r} + R_i \tag{8}$$

$$\varrho c_p \frac{\partial T}{\partial t} - \frac{\partial p}{\partial t} = -\varrho v c_p \frac{\partial T}{\partial r} - j_H c_p \frac{\partial T}{\partial r} + \frac{1}{r^n} \frac{\partial}{\partial r}\left(r^n \lambda \frac{\partial T}{\partial r}\right) - \sum R_i h_i \tag{9}$$

where c_p = specific heat capacity, h = specific enthalpy, n = 0 for linear, = 1 for cylindrical, = 2 for spherical geometry, p = pressure, R = (molar scale) chemical rate of formation, r = cartesian coordinate, T = temperature, v = flow velocity, w = mass fraction, λ = mixture heat conductivity, ϱ = density.

Diffusion fluxes and heat conductivity, again, are given by (5) and (6).

Together with the uniform pressure assumption

$$\partial p/\partial\psi = 0 \tag{10}$$

and the boundary conditions (R = vessel radius)

$$\psi = 0: \quad r = 0 \quad , \quad \partial T/\partial\psi = 0 \quad , \quad \partial w_i/\partial\psi = 0 \qquad (11)$$

$$\psi = \psi_R: \quad r = R \quad , \quad \partial T/\partial\psi = 0 \quad , \quad \partial w_i/\partial\psi = 0 \qquad (12)$$

the problem can be reduced to a differential/algebraic system by discretization using simple finite difference methods (see [29]). This system then is solved by the programs DASSL (Sandia National Laboratories [10]) or LIMEX (Heidelberg University [11]).

Because of the large ratio of vessel diameter and flame front thickness, adaptive gridding has to be used to describe flame propagation following the ignition process. An example is given in Fig.11, using static regridding with the grid point density proportional to the temperature gradient, using local monotone piecewise cubic interpolation [31].

Typical calculation times amount to about 5 min for the O_2-O_3 system (IBM 3081 D).

4.3 Comparison of Experiments and Calculations

A typical simulation of ignition in oxygen-ozone mixtures is given in Fig.11. After a relatively long induction period (confirming the corresponding experimental observation) there is a thermal explosion in the ignited gas pocket. This explosion initiates flame propagation across the vessel with subsequent equilibration by transport processes.

Calculated minimum thermal ignition energies are compared with the experimental ones in Fig.12. The simulations clearly show a pressure dependence of the minimum ignition energy which cannot be resolved experimentally due to missing sensitivity of the energy

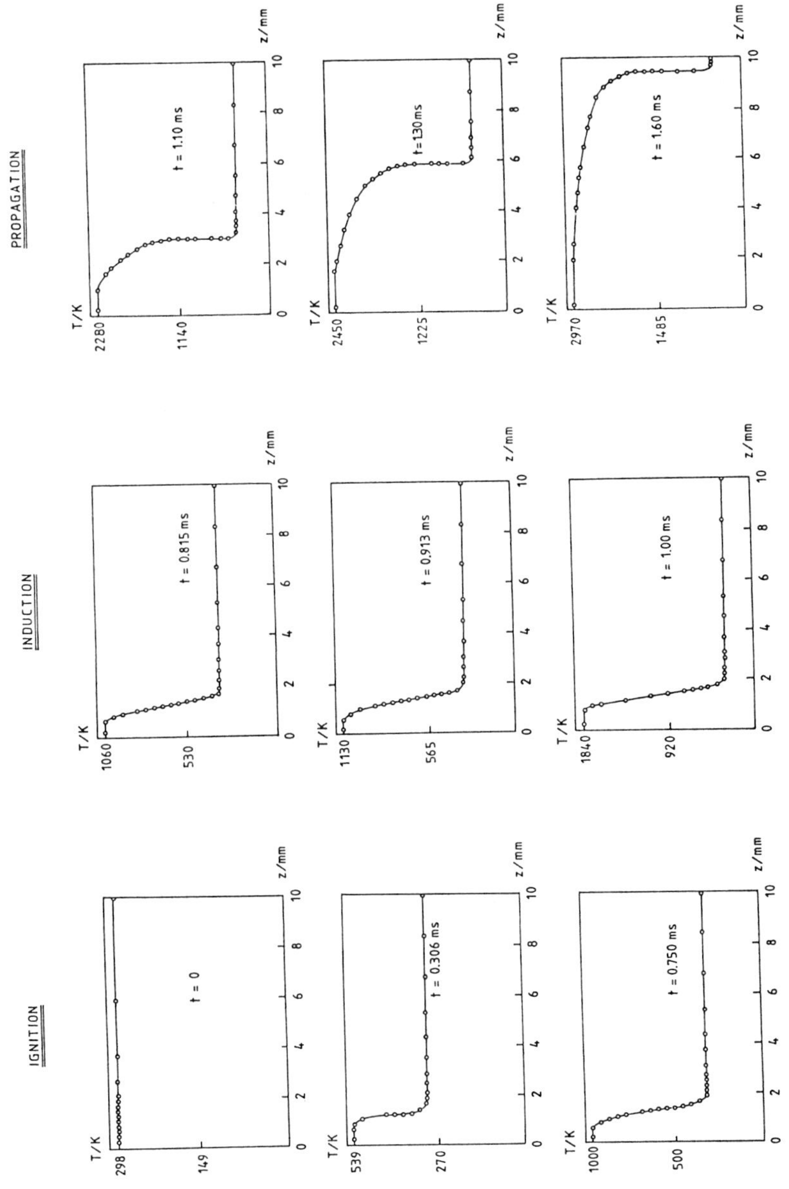

Fig.11: Progress of the ignition process in equimolar O_2/O_3 mixtures, represented by calculated temperature profiles; P = 0.34 bar, T_u=298 K, t_{ign} = 1 μs

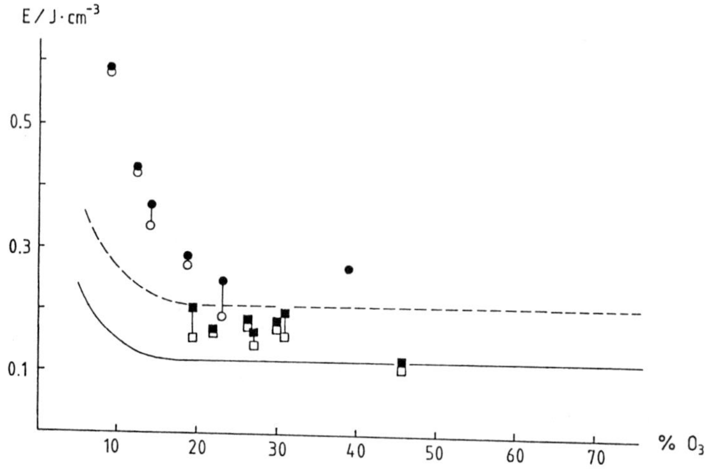

Fig.12: Measured (points) and calculated (lines) minimum thermal ignition energy in O_2/O_3 mixtures at P = 0.34 and 0.68 bar, T_u = 298 K, t_{ign} = 1 μs

detectors used in these measurements. Nevertheless, there is a rather good agreement of calculations and measurements confirming reasonable operation of the experiment. On the other hand, the experiment yields a verification of the calculation method, thus enabling reliable extrapolation to larger reaction systems (if a detailed reaction mechanism is available).

4.5 Results on Auto-Ignition in Hydrogen-Oxygen Mixtures

Since the detection of P-T explosion limits in hydrogen-oxygen mixtures in the Thirtieth and Fourtieth, many efforts have been made to explain this phenomenon quantitatively. But all of these attempts had to include some serious restrictions like truncation of the reaction mechanism, quasi-steady state assumptions, or reduction to zero-dimensional systems etc.

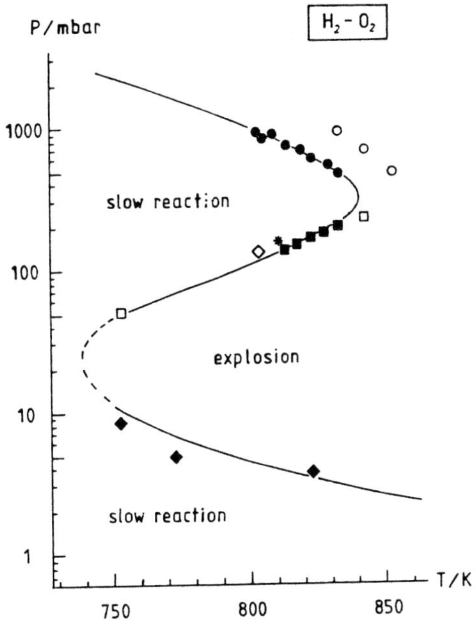

Fig.13: Calculated (line) and experimental (points) ignition limits in a stoichiometric H_2-O_2 mixture; (see [32] for further reference)

The present status of knowledge on reaction kinetics in the hydrogen-oxygen system and recently developed methods for the solution of time-dependent one-dimensional partial differential equation systems now allow the simulation of all of these explosion limits, using a common detailed reaction mechanism (consisting of 37 elementary reactions [3,32], a multi-species transport model [2,18-20], and realistic surface chemistry basing on surface collision numbers and experimentally determined surface destruction efficiencies [32]. None of the restrictions mentioned above has to be applied. Solution of the partial differential equation system, again, is done by spatial discretization by finite differences, leading to an ordinary differential/algebraic equation system.

Calculated P-T explosion limits in the hydrogen-oxygen system at various conditions are presented in Fig.13. As can be seen, these values are in quite good agreement with the experimental results. Sensitivity analysis identifies the rate-limiting processes and shows areas which should be object of further research.

References

[1] B.Lewis, G.von Elbe, Combustion, Flames, and Explosions in Gases, Academic Press, New York (1961)

[2] J.Warnatz, 18th Symposium (International) on Combustion, p.369. The Combustion Institute, Pittsburgh (1981)

[3] J.Warnatz, Survey of Rate Coefficients in the C/H/O System, in: W.C.Gardiner (ed.), Combustion Chemistry. Springer, New York (1984)

[4] J.Warnatz, 20th Symposium (International) on Combustion, p.845. The Combustion Institute, Pittsburgh (1985)

[5] J.Warnatz, H.Bockhorn, A.Möser, and H.W.Wenz, 19th Symposium (International) on Combustion, p.197, The Combustion Institute, Pittsburgh (1983)

[6] J.Warnatz, Ber. Bunsenges. Phys. Chem.87, 1008 (1983)

[7] F.Dryer, I.Glassman, in: C.T.Bowman, J.Birkeland (ed.), Alternative Hydrocarbon Fuels, Combustion and Kinetics. AIAA, New York (1979)

[8] F.O.Rice, JACS 55, 3035 (1933)

[9] C.Esser, U.Maas, J.Warnatz, Proc. Intl. Symp. on Diagnostics and Modelling of Combustion in Reciprocating Engines, p.335. The Japanese Society of Mechanical Engineers, Tokyo (1985)

[10] L.Petzold, "A Description of DASSL: A Differential/Algebraic System Solver," IMACS World Congress, Montreal (1982)

[11] P.Deuflhard, E.Hairer, J.Zugck, "One-Step and Extrapolation Methods for Differential-Algebraic Systems," Uni Heidelberg, SFB 123, Technical Report 318 (1985)

[12] W.C.Gardiner, J.M.Simmie, R.Zellner, K.Niemitz, J.Warnatz, in: Flames, Lasers, and Reactive Systems (J.R.Bowen, N.Manson, A.K.Oppenheim, and R.I.Soloukhin eds.), Progress in Astronautics and Aeronautics 88, 252 (1983)

[13] C.M.Coats, A.Williams, 17th Symposium (International) on Combustion, p.611. The Combustion Institute, Pittsburgh (1979)

[14] H.K.Kamptner, in: K.Winnacker, L.Küchler, Chemische Technologie, Vol.3, p.176. Hansen Verlag, München (1971)

[15] J.R.Smith, R.M.Green, C.K.Westbrook, W.J.Pitts, 20th Symposium (International) on Combustion (1985), p.91. The Combustion Institute, Pittsburgh (1985)

[16] J.Warnatz, Ber. Bunsenges. Phys. Chem. 82, 193 (1978)

[17] J.Warnatz, Ber. Bunsenges. Phys. Chem. 82, 643 (1978)

[18] J.M.Heimerl, T.P.Coffee, in: N.Peters, J.Warnatz (ed.), Numerical Methods in Laminar Flame Propagation, Vieweg, Wiesbaden (1982)

[19] J.Warnatz, in: N.Peters, J.Warnatz (ed.), Numerical Methods in Laminar Flame Propagation, Vieweg, Wiesbaden (1982)

[20] R.J.Kee, J.A.Miller, J.Warnatz, "A Fortran Computer Code Package for the Evaluation of Gas Phase Viscosities, Conductivities, and Diffusion Coefficients," Report SAND83-8209. Sandia National Laboratories, Livermore (1983)

[21] J.Warnatz, Ber. Bunsenges. Phys. Chem. 82, 834 (1978)

[22] J.Warnatz, in: K.Ebert, P.Deuflhard, W.Jäger (ed.), Modelling of Chemical Reaction Systems, p.162. Springer, Heidelberg (1981)

[23] D.L.Baulch, D.D.Drysdale, J.Duxbury, S.Grant, Evaluated Kinetic Data for High Temperature Reactions Vol.1: Homogeneous Gas Phase Reactions of the H_2-O_2 System. Butterworths, London (1972)

[24] D.L.Baulch, D.D.Drysdale, D.G.Horne, A.C.Lloyd, Evaluated Kinetic Data for High Temperature Reactions Vol.3: Homogeneous Gas Phase Reactions in the O_2-O_3 System, the CO-CO_2-H_2 System, and of the Sulfur-containing Species. Butterworths, London (1976)

[25] J.Warnatz, Comb. Sci. Technol. 34, 177 (1983)

[26] J.Warnatz, in: V.K.Baev (ed.), Proceedings of the Workshop on Gas Flame Structure, p.190. USSR Academy of Sciences, Novosibirsk (1983)

[27] J.Warnatz, in: R.Glowinski, B.Larrouturou, R.Temam, Numerical Simulation of Combustion Phenomena. Springer, Heidelberg (1985)

[28] H.Bockhorn, W.Weyrauch, J.Warnatz, publication in preparation

[29] B.Raffel, J.Warnatz, H.Wolff, J.Wolfrum, R.J.Kee, Progress in Aeronautics and Astronautics. AAIA (1986), in press

[30] B.Raffel, J.Warnatz, J.Wolfrum, Appl. Phys. B37, 189 (1985)

[31] F.N.Fritsch, J.Butland, SIAM J. Sci. Stat. Comput. 5, 300 (1984)

[32] U.Maas, J.Warnatz, submitted for publication in "Comb. Flame"

NUMERICAL COMPUTATION OF STIFF SYSTEMS FOR NONEQUILIBRIUM IONIZATION PROBLEMS

Fu Hongyuan and Chen Guannan

1. INTRODUCTION

In the study of physics, many problems require to investigate the nonequilibrium ionization processes. For example, some material is irradiated by laser, then the plasma with low density and higher temperature may be produced. The radiation and the material are in nonequilibrium states. If the target is made of high Z elements, the electrons are ionized generally not completly. In order to study the absorption of laser energy, plasma dynamics and specially the transformation from laser into x-ray, one consider the ionizability and electron distribution with different occupation states. To describe such physical phenomena, the fluid dynamics systems, multi-group energy equations of photons, electrons, ions and equations for electron occupation probabilities are considered. There are nonlinear coupled ordinary and partial differential equations, which have much different time scales. It is a large scale scientific computing problem.

It is difficult to calculate the whole process, which contains very rapid changes and slow nonequilibrium phenomena, if the time step of numerical computation is constraint by the very fast process. The so called stiff system is arising in many research field of physics. We want to reduce the computational work for some parts of the system. Here we discuss the numerical computation for nonequilibrium ionization problem with average ion model. The electron's states and the change of ionizabilities are described by the following equations [1][2]

$$dp_n/dt = A_n - B_n P_n + \beta_n + Q_n \quad , \quad n=1,2,\ldots,M \ ,$$
$$dNe/dt = (N_0/A) \sum_{n=1}^{M} C_n (A_n - B_n P_n) \ , \qquad (1)$$

where P_n denote the occupation probabilities in state n, Ne the number of free electrons per unit mass, N_0 a constant, A denotes atomic weight, $C_n = 2n^2$. And A_n, B_n, β_n, Q_n are nonlinear functions dependent on P_n and Ne.

$$A_n = N\acute{e}(\beta_n^{(\nu)} + Ne\, \beta_n^{(e)}) \ ,$$

$$B_n = A_n + N\acute{e}\, \alpha_n^{(e)} + \alpha_n^{(\nu)},$$

$$\begin{aligned}\beta_n = {} & N\acute{e}\, \beta_{n-1,n} P_{n-1}(1-P_n) C_{n-1}/C_n + N\acute{e}\, \alpha_{n+1,n} P_{n+1}(1-P_n) \\ & - N\acute{e}\, \alpha_{n,n-1} P_n (1-P_{n-1}) C_{n-1}/C_n - N\acute{e}\, \beta_{n,n+1} P_n (1-P_{n+1}) \\ & + H_{n,n+1} \left[\gamma^{(1)}_{n+1,n} P_{n+1}(1-P_n) - \gamma^{(2)}_{n,n+1} P_n (1-P_{n+1})\right] + \\ & H_{n-1,n}\left[\gamma^{(2)}_{n-1,n} P_{n-1}(1-P_n) - \gamma^{(1)}_{n,n-1} P_n (1-P_{n-1})\right] C_{n-1}/C_n,\end{aligned}$$

$$\begin{aligned}Q_n = {} & \sum_{\substack{n'>n \\ n' \neq n+1}} H_{n,n'} \left[\gamma^{(1)}_{n',n} P_{n'} (1-P_n) - \gamma^{(2)}_{n;n} P_n (1-P_{n'}) \right] + \\ & \sum_{\substack{n''<n \\ n'' \neq n-1}}' H_{n'',n} \left[\gamma^{(2)}_{n'',n} P_{n''} (1-P_n) - \gamma^{(1)}_{n,n''} P_n (1-P_{n''}) \right] C_{n''}/C_n \ ,\end{aligned}$$

where $N\acute{e} = \rho\, Ne$ is the number of free electrons per unit volume, $\alpha_n^{(\nu)}$, $\beta_n^{(\nu)}$, $\alpha_n^{(e)}$, $\beta_n^{(e)}$, $\alpha_{n+1,n}$, $\beta_{n,n+1}$, $H_{n,n'}$, $\gamma^{(1)}$, $\gamma^{(2)}$ are coefficients that dependent on electron temperature and radiation field corresponding certain physical process.

From physics we know that $A_n - B_n P_n$ describe the ionization, β_n the transitions to the energy levels of neighborhood (i.e. $n \to n+1$ or $n \to n-1$) and Q_n the transitions to other levels. In this ordinary differential system the scales of times are very different, so it is a very strong stiff system. In order to avoid the strict condition for time step, in general the A-stable,[3] A(α)-stable[4] or stiff stable schemes may be used. The presently effective algorithm for this type of problems seems to be the method of Gear with variable orders. That is an A(α)-stable scheme with higher accuracy.[5]

To solve implicit schemes the Newton's iterations are usually applied, for which inverse matrices are computed. Sometimes it cost much computational work for large systems. Here we suggest a reduced iteration method to solve stiff systems with special character. As a model system we consider the ordinary differential system in the following form

$$du/dt = \varepsilon f(u,v) + h(u,v) \;,$$
$$dv/dt = \varepsilon g(u,v) \;, \qquad (2)$$

where $u = (u_n)_{n=1}^{M}$, $f = (f_n)_{n=1}^{M}$, $h = (h_n)_{n=1}^{M}$ and $v = (v_j)_{j=1}^{J}$, $g = (g_j)_{j=1}^{J}$ are M-dimensional and J-dimensional vector valued functions respectively. Suppose $|\varepsilon| \ll 1$. This problem means that from the physical or mathematical analysis we know that the stiff system with slow and rapid processes may be separated as the above mentioned form (2). In other words, the system of equations may be partitioned into sets which change at very different speeds. The absolut value of eigenvalues for the Jacobian matrix $\partial h/\partial u$ are much more bigger than that of $\varepsilon\partial f/\partial u$. For the nonequilibrium ionization problem (1) we have $J=1$ and $h_n = \beta_n$ where β_n describe the transition to the energy levels of neighborhood.

2. INVERSE ITERATION

Now we calculate the problem (2). Suppose the numerical solution of vector valued functions $u(t)$ and $v(t)$ for time $t \le t_n$ are already obtained, then the solution u , v for time $t_{n+1} = t_n + \tau$ is evaluated by the following iteration algorithm.

At first we consider a inverse iteration for differential equation. Let u^0 , v^0 denote the initial approximation, then the next formal approximation are denoted by u^{i+1} , v^{i+1} which are obtained by the following equations

$$du^i/dt = \varepsilon f(u^i, v^i) + h(u^{i+1}, v^{i+1}) ,$$

$$dv^{i+1}/dt = \varepsilon g(u^{i+1}, v^{i+1}) , \ i=0,1,\dots . \tag{3}$$

In the system (3) the terms dependent on u^i, v^i are known and u^{i+1}, v^{i+1} unknown.If (u^i, v^i) convergent to (u,v) as $i \to \infty$,then the limit (u,v) satisfies the system (2).

In general,it is incorrect.But for some cases,one can prove that,the system (3) is an approximation of the original system (2) for long time,i.e. for large time.For simple example

$$du/dt = f(t) - Hu . \tag{4}$$

Let $u^{I} = 0$,then

$$u^1 = f(t)/H ,$$

$$u^i(t) = [f - f/H + \dots + (-1)^{i-1} f^{(i-1)}/H^{i-1}]/H, i=1,2,\dots .$$

The exact solution of (4) is

$$u(t) = u_0 \exp(-Ht) + \exp(-Ht) \int_0^t f(\tau)\exp(H\tau)d\tau .$$

Integrating by parts,it follows

$$u(t) = u^i(t) + [u_0 - u^i(0) + (-1)^i \int_0^t \exp(H\tau) f(\tau) d\tau/H^i]$$

$$\exp(-Ht) .$$

Under certain conditions for $f(t)$ and H ,if the series convergent,then $u^i(t)$ is an approximation for large time.

This is the background to construct iteration algorithm for large step lenth τ .If the transient region for the components with rapid speed is passed,then one can take large step length for some absolute stable schemes. The numerical iteration will be described in next.

Consider a multistep formula defined by the following

$$\sum_{j=0}^{k} a_j u_{n+j} = \tau \sum_{j=0}^{k} b_j L(u_{n+j}) \qquad (5)$$

for the differential system $du/dt = L(u)$,where the consistency conditions are satisfied

$$\sum_{j=0}^{k} a_j = 0 , \quad \sum_{j=0}^{k} b_j = 1 , \quad b_k > 0 , \quad a_k > 0 .$$

This is an implicit scheme.

In order to solve this implicit scheme,the following inverse iteration is suggested.

Suppose L is nonsingular matrix with constant entries,then we have the iteration formula

$$a_k u^{i}_{n+k} + \sum_{j=0}^{k-1} a_j u_{n+j} = \tau s b_k L u^{i+1}_{n+k} + \tau \sum_{j=0}^{k-1} b_j L u_{n+j} + (1-s)\tau b_k L u^{i}_{n+k} ,$$

where $s \geq 1$ .It may be proved that this iteration is convergent,if the eigenvalues $\lambda_m (m=1,\ldots,M)$ of L are negative and the inequalities

$$s \tau \lambda_m b_k / a_k < -1 , \quad m=1,\ldots,M$$

are satisfied.For fixed τ there is sufficiently large s, such that the convergent conditions hold.If the scheme (5) is absolute stable, $|\lambda_m|$ are rather big and the solution already approximates equilibrium states,i.e.the transient region is passed for large negative eigenvalues,then for big τ it may be let $s=1$.

For nonlinear case we assume that the vector valued functions $L(u)$ is strict negative monotone : i.e. the scalar product satisfies the inequality as this

$$\langle L(u) - L(v) , u - v \rangle \leq \mu |u - v|^2 , \quad \mu < 0$$

where μ is negative.The following iterations

$$a_k u^{i}_{n+k} + \sum_{j=0}^{k-1} a_j u_{n+j} = \tau \left[b_k L(u^{i+1}_{n+k}) + \sum_{j=0}^{k-1} b_j L(u_{n+j}) \right]$$

convergent also for

$$\tau \mu b_k / a_k < -1 \quad .$$

It is satisfied,if the step length τ sufficiently large or the absolute $|\mu|$ rather big.

We want to note that,if τ is fixed and we have to find s ,such that the convergent condition

$$\tau \mu s b_k / a_k < -1$$

is satisfied, for nonlinear case we can prove the conclusion only for one-leg formula[6] as following

$$a_k u^i_{n+k} + \sum_{j=0}^{k-1} a_j u_{n+j} = \tau L(s b_k u^{i+1}_{n+k} + \sum_{j=0}^{k-1} b_j u_{n+j} + (1-s) b_k u^i_{n+k}) \quad .$$

It doesn't hold for general multi-step linear methods. Of course,for Euler's backward method it holds also, because the one-leg form is the same.

So we have the iteration formula for the system (2) as following

$$a_k u^i_{n+k} + \sum_{j=0}^{k-1} a_j u_{n+j} = \tau \varepsilon f(b_k w^i_{n+k} + \sum_{j=0}^{k-1} b_j w_{n+j}) + $$
$$+ \tau h(s b_k w^{i+1}_{n+k} + \sum_{j=0}^{k-1} b_j w_{n+j} + (1-s) b_k w^i_{n+k}) \quad ,$$
$$a_k v^{i+1}_{n+k} + \sum a_j v_{n+j} = \tau \varepsilon g(b_k w^{i+1}_{n+k} + \sum_{j=0}^{k-1} b_j w_{n+j}) \quad ,$$

where $w = (u,v)$ .For strong stiff systems we take $s=1$, if the transient region with rapid speed is passed and the large step length τ may be choosen.

3. INITIAL APPROXIMATIONS

The first approximation is denoted by u^o , v^o which satisfies the following algebraical and differential equations

$$h(u^o, v^o) = 0 \quad , \tag{6}$$

$$dv^{0}/dt = \varepsilon\, g(u^{0},v^{0}) \quad . \tag{7}$$

The differential system (7) is calculated by some descrete schemes.For example,the simpler scheme,the Euler's backward formula may be applied.

To compute the next approximation u^{1}, v^{1} ,the following relation is introduced

$$du^{0}/dt = (\partial u/\partial v)^{0}\, dv^{0}/dt \ , \tag{8}$$

where $\partial u/\partial v$ denotes the Jacobian matrix with $M{\times}J$ entries.From the formula of differentiation for implicit function it follows

$$(\partial h/\partial u)^{0}(\partial u/\partial v)^{0} + (\partial h/\partial v)^{0} = 0 \ ,$$

in which $\partial h/\partial u$ and $\partial h/\partial v$ express the Jacobian matrices with $M{\times}M$ or $M{\times}J$ entries respectively.Suppose that the vector valued functions h are smooth and the Jacobian matrix $\partial h/\partial u$ nonsingular,then there exists an inverse matrix denoted by $(\partial h/\partial u)^{-1}$ .Hence we obtain the expression

$$\partial u/\partial v = -(\partial h/\partial u)^{-1}\, \partial h/\partial v \quad . \tag{9}$$

Using the above equations,the system for u^{1},v^{1} may be written in the following form

$$\begin{aligned} &\varepsilon(\partial u/\partial v)^{0}\, g(u^{0},v^{0}) = \varepsilon f(u^{0},v^{0}) + h(u^{1},v^{1}) \ , \\ &dv^{1}/dt = \varepsilon\, g(u^{1},v^{1}) \ , \end{aligned} \tag{10}$$

in which $\partial u/\partial v$ is evaluted by (9).This system (10) consists of algebraical and differential equations and is the the computing formula for u^{1},v^{1} .The differential equations of the system may be solved by Euler's backward difference formula.

The linearization formulae for nonlinear case of $h(u,v)$ are suggested.

Applying the inverse iteration method we have some numerical experiments.For the nonequilibrium ionization problem it is J=1 and $h_n = \beta_n$ .It has proved the eigenvalues of $\partial h/\partial u$ are negative[1]. Because β_n are only dependent on the levels of neibourhood,only respect to P_{n-1} , P_n , P_{n+1} and Ne ,then P_n^o may be solved explicitly.

The numerical results obtained by the inverse iteration are compared with solutions by Gear's method.The results are almost the same.We see that the advantage of such method is that: to find the inverse matrix of $\partial h/\partial u$ (a part of the whole system) is much more simpler than the inverse matrix of the whole system.Special $\partial \beta/\partial P$ is a matrix with only three diagonal entries.

For large difference of speeds,when the transient region for rapid speed is passed,or to say,it may be negilible,then only a few times of iterations are required. For example,the eigenvalues of $\varepsilon \partial f/\partial u \sim 1$ and eigenvalues of $\partial h/\partial u \sim 10^4 \sim 10^6$,then for $\tau = 10^{-2} \sim 10^{-1}$ we need only one time iteration,i.e. $u^1, v^1 \Rightarrow u_{n+1}, v_{n+1}$ .This iteration method save much time.It may be applid to other problems.

We would like to thank prof.P.Deuflhard for his proposal about the convergent condition to take a better form.

REFERENCES

[1] Chang,T.Acta Physica Sinica,34(1985) No.4,528-536.

[2] Lokke,W.A. andGrasberger,W.H.,UCRL-52276,(1977).

[3] Dahlquist,G.,BIT 3(1963),27-43.

[4] Widlund,O.B.,BIT 7(1967),65-70.

[5] Gear,C.W.,Numerical Initial Problems in Ordinary differential equations,1971.

[6] Dahlquist,G.,Notes in Math.No.506,Springer Verlag,1975.

FINITE ELEMENT SIMULATION OF SATURATED-UNSATURATED FLOW THROUGH POROUS MEDIA

Peter Knabner

An algorithm for the numerical simulation of saturated-unsaturated flow through porous media is described and some aspects of its performance are outlined.

1. Introduction

In this paper we describe an algorithm for the numerical simulation of two-dimensional unsteady saturated-unsaturated seepage flow through porous media and report about its performance. The implementation of this algorithm has a twofold aim: First, to provide an instrument for the simulation of actual field study situations. Therefore (at the present stage) the algorithm follows closely the lines, which have proved to be rather successful in this respect (compare Hauhs [2], Neuman [7]). The second aim is to test systematically specific features in order to improve the algorithm towards an optimal performance. This is highly desirable, because the need to follow a field experiment over several years of real time causes a tremendous complexity (several tenthousands of elliptic problems).

Let us point out briefly the *mathematical model* (see e. g. Bear [1] for a detailed exposition). We denote by t [sec] the *time*, bei $\underline{x} = (y,z)[m]$ a plane in the *physical space*, where the flow takes place, either *horizontal* ($\gamma = 0$) or *vertical* ($\gamma = 1$) with z directed downwards. If we use $\Theta(x,t)[m^3/m^3]$ for the *volumetric water content* and $\underline{q}(\underline{x},t)[m/sec]$ for the *specific water flux density*, then, assuming the incompressibility of water, the *conservation of mass* yields

$$\Theta_t = -\operatorname{div} \underline{q} + S, \tag{1.1}$$

where $S[m^3/m^3/sec]$ is the volumetric source/sink density. If we consider only isotropic (but in general inhomogeneous) porous media, we can use the *water potential* $\psi[m]$, which is the primary observed quantity, to represent $\underline{q}$ in the following form, known as *Darcy's law*:

$$\underline{q}(\underline{x},t) = -K(\underline{x},\Theta(\underline{x},t))\,\operatorname{grad}(\psi(\underline{x},t) - \gamma z), \tag{1.2}$$

where K is the *hydraulic conductivity*. K is a characteristic of the porous medium, as the functional relationship between ψ and Θ:

$$\Theta(\underline{x},t) = \Theta(\underline{x},\psi(\underline{x},t)). \tag{1.3}$$

These *water characteristic curves* always have the following shape:

FIGURE 1.1

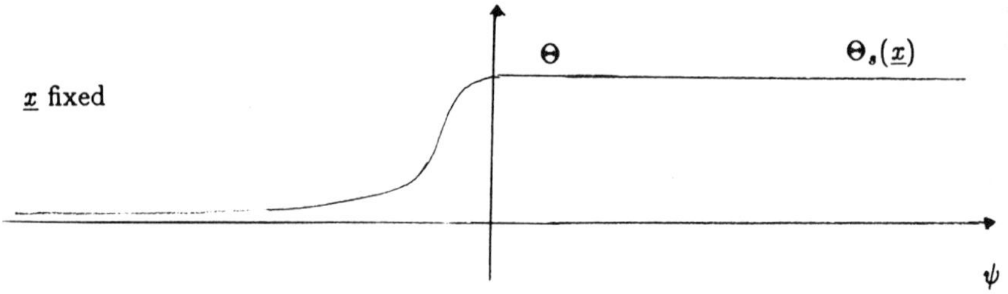

In particular, the region of saturation (where $\Theta(\underline{x}, t) = \Theta_s(\underline{x})$, the saturation values Θ_s being determined solely by the medium's internal geometry) corresponds to $\psi \geq 0$. The shape of $\psi \to K(\underline{x}, \psi)$ is similar. (1.1) – (1.3) now imply for the considered region $\Omega \subset \mathbb{R}^2$ and time interval $(0, T)$:

$$\Theta_t(\underline{x}, t) = \operatorname{div}(K(\underline{x}, \psi(\underline{x}, t))\operatorname{grad}(\psi(\underline{x}, t) - \gamma z)) + S(\underline{x}, t, \psi(\underline{x}, t)) \text{ in } \Omega \times (0, T). \quad (1.4)$$

Using $C(\underline{x}, \psi) := \frac{\partial \Theta}{\partial \psi}(\underline{x}, \psi)$, the *specific capacity*, we get the alternative formulation:

$$C(., \psi)\psi_t = \operatorname{div}(K(., \psi)\operatorname{grad}(\psi - \gamma z)) + S(., \psi). \quad (1.5)$$

From figure 1.1 we see that we do not have a uniformly parabolic equation, but rather an *elliptic-parabolic* one, where the subregions of different character are a priori unknown (a free boundary problem). Of course, we must supply an *initial condition*:

$$\psi(\underline{x}, 0) = \psi_0(\underline{x}), \ \underline{x} \in \Omega, \quad (1.6)$$

and *boundary conditions*, which take the form:

$$\begin{aligned} q_n(\underline{x}, t) + \alpha(\underline{x})\psi(\underline{x}, t) &= \overline{q}(\underline{x}, t, \psi(\underline{x}, t)) \quad \text{on } \Gamma_1, \\ \psi(\underline{x}, t) &= \overline{\psi}(\underline{x}, t) \quad \text{on } \Gamma_2, \end{aligned} \quad (1.7)$$

where $\underline{n}$ is the outer unit normal, $q_n := \underline{q} \cdot \underline{n}$, $\alpha \leq 0$, $\Gamma = \Gamma_1 \dot{\cup} \Gamma_2$.

2. The Numerical Algorithm

The approximation has the following structure: We use finite differences in time, treat the resulting nonlinear elliptic problems iteratively, which leads to linear elliptic problems in each iteration step. These are approximated by linear finite elements.

Let $\Omega \subset \mathbb{R}^2$ be polygonal (which is no restriction for the applications). Let $\mathcal{T}_h$ be a partition of Ω into disjoint triangles such that no vertex of any triangle lies in the interior of a side of another triangle, and $(P_j)_1^m$ the vertices of $\mathcal{T}_h$. We use

$$\begin{aligned} S_h &:= \{\varphi | \varphi \text{ continuous on } \overline{\Omega}, \text{ linear on every triangle of } \mathcal{T}\} \\ &= \operatorname{span}\{\varphi_1, ..., \varphi_m\}, \end{aligned} \tag{2.1}$$

where φ_i is defined by $\varphi_i(P_j) = \delta_{ij}$, $i, j = 1, ..., m$.

For the simplicity of exposition we do not consider Dirichlet conditions.

The *time stepping* is initialized by $t_0 := 0$ and

$$\begin{aligned} &\psi^0 \in S_h \text{ s. t. } \psi^0(P_i) = \psi_0(P_i), \\ &\Theta^0 \in S_h \text{ s. t. } \Theta^0(P_i) = \Theta(P_i, \psi^0(P_i)), \\ &C^0 \in S_h \text{ s. t. } C^0(P_i) = C(P_i, \psi^0(P_i)). \end{aligned} \tag{2.2}$$

(For the simulation of water flow in the field (ψ^0, Θ^0, C^0) is given as the result of an other appropriate simulation.)

Now let $k \in \mathbb{N}_0$, $t = t_k$, $\Delta t = \Delta t^{k+1}$ be computed by the step size control (see below) and ψ^k, Θ^k, $C^k \in S_h$ be given. The time stepping proceeds as follows:

We use a fully-implicit difference scheme in time, i. e. ψ^{k+1} is to solve approximately:

$$\begin{aligned} &(\Theta(., \psi^{k+1}) - \Theta^k)/\Delta t = \operatorname{div}(K(., \psi^{k+1})\operatorname{grad}(\psi^{k+1} - \gamma z)) + S(., t_{k+1}, \psi^{k+1}) \quad \text{in } \Omega, \\ &q_n(., t_{k+1}) + \alpha\psi^{k+1} = \overline{q}(., t_{k+1}, \psi^{k+1}) \quad \text{on } \Gamma. \end{aligned} \tag{2.3}$$

The Crank-Nicolson scheme is excluded here, because it would not treat appropriately at least the situation, where at $t = t_k + \Delta t$ Ω is fully saturated, i. e. ψ at t is independent of ψ at t_k.

To get an appropriate solution of this nonlinear elliptic boundary value problem we use the following *iteration procedure*: Assume that the starting values $\psi^{k+1,0}$, $\Theta^{k+1,0}$, $C^{k+1,0} \in S_h$ have been determined (see below). Then for $\nu \in \mathbb{N}_0$ the $(\nu + 1)$th step in the iteration is defined on the following way:

Let $\psi^{k+1,\nu}$, $\Theta^{k+1,\nu}$, $C^{k+1,\nu} \in S_h$ be given. Then $\psi^{k+1,\nu+1}$ is to solve approximately:

$$\begin{aligned} &(\Theta^{k+1,\nu} + C^{k+1,\nu}(\psi^{k+1,\nu+1} - \psi^{k+1,\nu}) - \Theta^k)/\Delta t = \\ &\operatorname{div}(K(., \psi^{k+1,\nu})\operatorname{grad}(\psi^{k+1,\nu+1} - \gamma z)) + S(., t_{k+1}, \psi^{k+1,\nu}) \quad \text{in } \Omega, \end{aligned} \tag{2.4}$$

$$q_n(., t_{k+1}) + \alpha\psi^{k+1,\nu+1} = \overline{q}(., t_{k+1}, \psi^{k+1,\nu}) \text{ on } \Gamma.$$

The rationale for this formulation is a linearisation of $\Theta(., \psi^{k+1})/\Delta t$, leading to a Newton type iteration in $\Theta(., \psi^{k+1})$, but a fixed point type iteration otherwise.

We are now left with the linear self-adjoint elliptic boundary value problem (2.4). An approximate solution of this is achieved by the *Galerkin approach in S_h*, i. e. $\psi^{k+1,\nu+1} \in S_h$ is defined by

$$
\begin{aligned}
&((\Theta^{k+1,\nu} + C^{k+1,\nu}(\psi^{k+1,\nu+1} - \psi^{k+1,\nu}) - \Theta^k)/\Delta t, \varphi) = \\
&- (K(.,\psi^{k+1,\nu}), \text{grad}\psi^{k+1,\nu+1} \cdot \text{grad}\varphi) + (K(.,\psi^{k+1,\nu}), \gamma\frac{\partial\varphi}{\partial z}) \\
&- < \overline{q}(.,t_{k+1},\psi^{k+1,\nu}), \varphi > + < \alpha\psi^{k+1,\nu+1}, \varphi > \\
&+ (S(.,t_{k+1},\psi^{k+1,\nu}), \varphi) \quad \text{for all } \varphi \in S_h,
\end{aligned}
\tag{2.5}
$$

where $(f,g) := \int_\Omega fgd\underline{x}, \quad < f,g > := \int_\Gamma fgds$.

To evaluate the integrals we have to choose quadrature rules. For the right-hand side we take the approach to substitute e. g. $K(.,\psi^{k+1,\nu})$ by its interpolant $\hat{K}(.,\psi^{k+1,\nu}) \in S_h$ and compute exactly the resulting integrals. Concerning the left-hand side, which corresponds to the time evolution of the process, one can evaluate the mass matrix as it stands or apply a mass lumping procedure in order to maintain a better local mass balance. We use the variants:

(1) Consistent (exact evaluation):

$$\text{mass matrix} \left(\int_\Omega C^{k+1,\nu}\varphi_i\varphi_j d\underline{x}\right)_{ij}, \tag{2.6}$$

(2) mass lumping (according to Neuman [7]):

$$\text{mass matrix diag}\left(\int_\Omega C^{k+1,\nu}\varphi_i d\underline{x}\right), \tag{2.7}$$

(3) mass lumping (by quadrature):

$$\text{mass matrix diag}\left(\int_\Omega C^{k+1,\nu}(P_i)\varphi_i d\underline{x}\right). \tag{2.8}$$

The systems of linear equation are solved directly by Cholesky's method. $\Theta^{k+1,\nu+1} \in S_h$ is now defined by

$$\Theta^{k+1,\nu+1}(P_i) := \Theta(P_i, \psi^{k+1,\nu+1}(P_i)), \tag{2.9}$$

whereas for $C^{k+1,\nu+1} \in S_h$ the following variants are used:

(1) Exact evaluation of C:

$$C^{k+1,\nu+1}(P_i) := C(P_i, \psi^{k+1,\nu+1}(P_i)), \tag{2.10}$$

(2) secant approximation:

$$C^{k+1,\nu+1}(P_i) := \frac{\Theta^{k+1,\nu+1}(P_i) - \Theta^{k+1,\nu}(P_i)}{\psi^{k+1,\nu+1}(P_i) - \psi^{k+1,\nu}(P_i)}, \tag{2.11}$$

(2) is corrected, if the denominator becomes too small.

We have not yet specified the *start of the iteration*.
We consider a cheap and an expensive variant:

$$(1) \qquad \psi^{k+1,0} := \psi^k,\ \Theta^{k+1,0} := \Theta^k,\ C^{k+1,0} := C^k. \tag{2.12}$$

(2) Predictor-corrector approach to (1.5).
Predictor:

$$
\begin{aligned}
&\text{Let } \psi^{k+1/2} \in S_h \text{ be the Galerkin solution of}\\
&C^k(\psi^{k+1/2} - \psi^k)/(\Delta t/2) = \\
&\operatorname{div}(K(.,\psi^k)\operatorname{grad}(\psi^{k+1/2} - \gamma z)) + S(.,t_k,\psi^k) \text{ in } \Omega,\\
&q_n(.,t_{k+1/2}) + \alpha\psi^{k+1/2} = \overline{q}(.,t_k,\psi^k) \text{ on } \Gamma,
\end{aligned}
\tag{2.13}
$$

where $t_{k+1/2} := t_k + \Delta t/2$, and the same quadrature rules as above are used.

$\Theta^{k+1/2}$, $C^{k+1/2}$ are then defined analogously to (2.9) resp. (2.10) or (2.11).

Corrector:

Let $\psi^{k+1,0} \in S_h$ be the Galerkin solution of

$$
\begin{aligned}
&C^{k+1/2}(\psi^{k+1,0} - \psi^k)/\Delta t = \\
&\operatorname{div}(K(.,\psi^{k+1/2})\operatorname{grad}(\psi^{k+1,0} - \gamma z)) + S(.,t_{k+1/2},\psi^{k+1/2}) \text{ in } \Omega,\\
&q_n(.,t_{k+1}) + \alpha\psi^{k+1,0} = \overline{q}(.,t_{k+1/2},\psi^{k+1/2}) \text{ on } \Gamma.
\end{aligned}
\tag{2.14}
$$

Again $\Theta^{k+1,0}$, $C^{k+1,0}$ are defined analogously to (2.9) resp. (2.10) or (2.11).

The *iteration is stopped* in two situations:

i) $\|\psi^{k+1,\nu+1} - \psi^{k+1,\nu}\|_\infty$ is small:
The step size control now decides, whether this step has to be repeated or a new step with Δt^{k+2} can be done. In the latter case

$$\psi^{k+1} := \psi^{k+1,\nu+1},\ \Theta^{k+1} := \Theta^{k+1,\nu+1},\ C^{k+1} := C^{k+1,\nu+1}. \tag{2.15}$$

ii) The number of admissible iterations is exceeded without fulfilling i): This is considered as divergence and the time step is repeated with the new step size $\Delta t^{k+1}/100$.

The reason for this dramatic decrease is to ensure the return to the domain of convergence of the iteration.

The most decisive part of the algorithm is the *step size control.* We choose one, which controls the change in Θ. This may be interpreted as to adjust the time stepping (only) to that (a priori unknown) subregion, where there is a big change in time caused by the time evolution. More specifically, we take a step size control due to Edwards (compare Hauhs [2], Hornung/Messing [6]):

In the $(k+1)$th time step after convergence of the iteration $\Delta\Theta := \|\Theta^k - \Theta^{k+1,\nu+1}\|_\infty$ is compared with a desired change $\Delta\Theta_{\text{opt}}$. If it is twice as big, then the $(k+1)$th time step is repeated with step size $\Delta t^{k+1}/100$. Otherwise a new time step is performed with a step size adapted according to $\Delta\Theta_{\text{opt}}/\Delta\Theta$.

The *flux* $\underline{q} := -\hat{K}(.,\psi^{k+1})\operatorname{grad}(\psi^{k+1} - \gamma z)$ is not continuous. Therefore for a better representation of the flow field, $\underline{q}$ is substituted by a continuous approximation $(q_1^{k+1}, q_2^{k+1}), q_i^{k+1}$ being the best L^2-approximation in S_h to q_i, i. e.

$$
\begin{aligned}
(q_1^{k+1},\varphi) &= (-\hat{K}(.,\psi^{k+1})\frac{\partial}{\partial x}\psi^{k+1},\varphi),\\
(q_2^{k+1},\varphi) &= (-\hat{K}(.,\psi^{k+1})(\frac{\partial}{\partial z}\psi^{k+1}-\gamma),\varphi), \quad \text{for all } \varphi \in S_h.
\end{aligned} \tag{2.16}
$$

Finally, the *mass balance* is monitored. Let

$$
\begin{aligned}
A_{l,k} &:= \int_\Omega \Theta^k - \Theta^l d\underline{x}\\
B_{l,k} &:= \sum_{i=l}^{k-1} \Delta t^{i+1}/2(-\int_\Gamma \underline{q}^i \cdot \underline{n} ds - \int_\Gamma \underline{q}^{i+1} \cdot \underline{n} ds +\\
&\quad + \int_\Omega \hat{S}(.,\psi^i,t_i)d\underline{x} + \int_\Omega \hat{S}(.,\psi^{i+1},t_{i+1})d\underline{x}),
\end{aligned} \tag{2.17}
$$

then we call $A_{l,k} - B_{l,k}$ the *(discrete) mass balance error* in $[t_l, t_k]$ and $(A_{l,k} - B_{l,k})/B_{l,k}$ the *relative* one, respectively. A small mass balance error is a basic consistency check for the approximate solution.

3. Numerical Experiments

We report numerical experiences with two examples, taken from the literature. The first one (from Hornung/Messing [5]) is a rather well-behaved example for horizontal flow in a homogenous soil (Θ, K only depending on ψ), where the exact solution is known. The second one (from Herrling/Leismann [4]) deals with vertical flow, also in a homogenous soil, but the soil characteristics correspond to an actual infiltration experiment by Haverkamp et. a. [3]. The soil is a sand and therefore K varies very strongly over several magnitudes in a small range.

The computations were done on the NORD 540 of the Universität Augsburg in FORTRAN 77-double precision.

3.1 Example 1

$$
\Theta(\psi) = \begin{cases} 2[(\frac{1}{2}\pi)^2 - (\arctan\psi)^2] & \text{for } \psi < 0,\\ \frac{1}{2}\pi^2 & \text{for } \psi \geq 0, \end{cases} \tag{3.1}
$$

$$
K(\psi) = \begin{cases} 2/(1+\psi^2) & \text{for } \psi < 0,\\ 2 & \text{for } \psi \geq 0. \end{cases} \tag{3.2}
$$

The exact solution for horizontal flow ($\gamma = 0$) is

$$
\psi(y,z,t) = \begin{cases} -\frac{1}{2}s & \text{for } s \leq 0,\\ -\tan[(e^s-1)/(e^s+1)] & \text{for } s > 0, \end{cases} \tag{3.3}
$$

where $s := z - y - t$.

$\Omega := (0,1) \times (0,1)$, i. e. at $t = 0$ half of the region and at $t = 1$ the whole region is saturated.

For the following results a uniform triangulation with 441 knots and a constant time step size $\Delta t = 0.01$ are used. The termination tolerance for the iteration is $1\ ' - 6$. The mass balance errors are to be understood cumulative, i. e. in $[0, t]$.

In table 3.1 the different mass matrices are compared. It shows the strong superiority of the lumping procedures.

TABLE 3.1

Example 1, Dirichlet data, exact evalution of C: (2.10), start of iteration by predictor-corrector: (2.13), (2.14).

Comparison	1:	consistent mass matrix	: (2.6),
	2:	mass lumping Neuman	: (2.7),
	3:	mass lumping quadrature	: (2.8).

t	variant	$\|\|\psi - \psi app\|\|_\infty$	mass balance error	relative mass balance error (in %)
0.035	1	$3.02' - 3$	-3.16'-3	-2.22'-1
	2	$7.26' - 5$	1.86'-5	2.14'-3
	3	$7.26' - 5$	1.86'-5	2.14'-3
0.135	1	$2.28' - 3$	-7.53'-3	-2.48'-1
	2	$6.55' - 5$	1.82'-5	8.95'-4
	3	$6.55' - 5$	1.82'-5	8.95'-4
0.255	1	$1.74' - 3$	-1.67'-2	-2.64'-1
	2	$4.19' - 5$	2.36'-5	7.85'-4
	3	$4.19' - 5$	2.36'-5	7.86'-4

Number of iterations: 3 in every case.

From the next table it can be seen, that in this well-behaved situation the different treatments of C perform identical, that the predictor-corrector start in not necessary and that also one iteration step suffices, but only with the predictor-corrector start.

TABLE 3.2

Example 1, Dirichlet data, mass lumping Neuman: (2.7).

Comparison	Predictor-corrector/Newton:	(2.13), (2.14), (2.10),
	Predictor-corrector/secant:	(2.13), (2.14), (2.11),
	— / Newton:	(2.12), (2.10),
	Predictor-corrector/1 Newton step:	(2.13), (2.14), (2.10),
	— / 1 Newton step:	(2.12), (2.10).

t	variant	$\|\psi - \psi_{app}\|_\infty$	mass balance error	relative mass balance error (in %)
0.035	PC/N	$7.26' - 5$	1.86'-5	2.14'-3
	PC/S	$7.26' - 5$	1.86'-5	2.14'-3
	-/N	$7.26' - 5$	1.86'-5	2.14'-3
	PC/1N	$3.91' - 5$	-5.09'-6	-5.82'-4
	-/1N	$3.82' - 4$	-2.05'-4	-2.29'-2
0.135	PC/N	$6.55' - 5$	1.82'-5	8.95'-4
	PC/S	$6.55' - 5$	1.82'-5	8.95'-4
	-/N	$6.55' - 5$	1.82'-5	8.96'-4
	PC/1N	$2.87' - 5$	-8.30'-5	-4.04'-3
	-/1N	$2.20' - 3$	-3.14'-3	-1.33'-1
0.255	PC/N	$4.19' - 5$	2.36'-5	7.85'-4
	PC/S	$4.19' - 5$	2.36'-5	7.85'-4
	-/N	$4.19' - 5$	2.36'-5	7.86'-4
	PC/1N	$2.10' - 5$	-1.61'-4	-5.31'-3
	-/1N	$4.00' - 3$	-1.06'-2	-2.60'-1

Number of iterations: PC/N: 3, PC/S: 3, -/N: 4.
Finally we compare different boundary conditions. It turns out that the accuracy and also the asymptotic behaviour of the error depends heavily on it. This can be expected because of the different stability of the discrete problems. In the extreme for flux conditions the system matrix becomes singular at $t = 1$.

TABLE 3.3

Example 1, mass lumping Neuman: (2.7), C by exact evaluation: (2.10), predictor-corrector start: (2.13), (2.14).

Comparison of different boundary conditions:
1: Dirichlet, 2: mixed with $\alpha = -100$, 3: mixed with $\alpha = -10$, 4: mixed with $\alpha = -1$, 5: Dirichlet for $y = 1$ (always saturated), pure flux elsewhere, 6: Dirichlet for $y = 0$, pure flux elsewhere, 7: pure flux.

t	variant	$\|\psi - \psi\text{app}\|_\infty$	mass balance error	relative mass balance error (in %)
0.035	1	$7.26' - 5$	1.86'-5	2.14'-3
	2	$1.23' - 4$	1.65'-5	1.90'-3
	3	$1.23' - 4$	1.65'-5	1.90'-3
	4	$3.36' - 4$	4.21'-6	4.88'-4
	5	$2.85' - 4$	7.40'-6	8.57'-4
	6	$2.66' - 4$	1.47'-5	1.69'-3
	7	$3.28' - 4$	6.10'-6	7.07'-4
0.135	1	$6.55' - 5$	1.82'-5	8.95'-4
	2	$1.66' - 4$	1.78'-5	8.74'-4
	3	$1.66' - 4$	1.78'-5	8.74'-4
	4	$1.50' - 3$	-2.11'-6	- 1.05'-3
	5	$9.44' - 4$	- 4.68'-6	- 2.31'-4
	6	$4.13' - 4$	1.56'-5	7.67'-4
	7	$1.47' - 3$	-1.26'-5	-6.24'-4
0.255	1	$4.19' - 5$	2.36'-5	7.85'-4
	2	$1.39' - 4$	2.30'-5	7.67'-4
	3	$1.39' - 4$	2.30'-5	7.67'-4
	4	$3.91' - 3$	-5.54'-5	-1.87'-3
	5	$1.23' - 3$	-9.09'-6	-3.03'-4
	6	$2.91' - 4$	1.96'-5	6.51'-4
	7	$3.80' - 3$	-3.63'-5	-1.23'-3

Error monotone increasing in time for 4, 5, 7, globally bounded for 1, 2, 3, 6.

3.2 Example 2

We simulate the artifical groundwater recharge from a ditch in an aquifer, a typical example from engineering practice. $\Omega = (0,2) \times (0,3)$ is the right half of the region, the *boundary conditions* are for $t > 0$

$$\begin{aligned}
&\psi(y,z,t) = 0.2 \quad \text{for } y \in (0,1),\ z = 0 \\
&\text{right half of the ditch ,} \\
&\psi(y,z,t) = z - 2 \text{ for } y = 2,\ z \in (2,3) \\
&\text{(constant ground water table in depth 2 m) ,} \qquad (3.4) \\
&q_n = 0 \quad \text{otherwise} \\
&\text{(symmetry line for } y = 0,\ z \in (0,3)\text{, insultated boundary} \\
&\text{otherwise).}
\end{aligned}$$

The *initial condition* is the hydrostatic equilibrium ($\psi_0(x,z) = z - 2$).

$$\Theta(\psi) = \begin{cases} 1/(4.44444 + 229.46686(-\psi)^{3.96}) + 0.075 & \text{for } \psi < 0 \\ 0.3 & \text{for } \psi \geq 0 \end{cases} \tag{3.5}$$

$$K(\psi) = \begin{cases} 1/(1'4 + 2.5701'7(-\psi)^{4.74}) & \text{for } \psi < 0 \\ 1' - 4 & \text{for } \psi \geq 0 \end{cases} \tag{3.6}$$

The following table shows a simulation, in which the control parameter $\Delta\Theta_{\text{opt}}$ of the step size control and the mass lumping procedure ((2): (2.7). (3): (2.8)) are changed for early times in order to bound the mass balance error. In fact it stabilizes on a rather low level.

TABLE 3.4

Example 2, uniform triangulation with 651 knots, termination tolerance for iteration: 5'-3, maximum of iterations: 20, predictor-corrector start (2.13)/(2.14), secant approximation for C (2.11).

time step number	time t	relative mass balance error (in %)	$\Delta\Theta_{\text{opt}}$	mass lumping
0	000.001	514.53	0.22	(2)
18	20.771	7.98	0.01	(2)
23	24.898	2.91	0.002	(3)
121	82.772	-9.82	0.005	no
133	103.756	-8.17	0.01	further
146	158.266	-5.94	0.005	change
168	199.435	-5.98	0.0075	
183	263.294	-6.29	0.005	
256	500.397	-5.55	no	
379	1002.224	-4.96	further	
485	1501.772	-4.86	change	
583	2002.803	-4.58		
604	2112.000	-4.58		
645	2332.000	-4.67		
663	2400.000	-4.72		
698	2501.672	-4.74		
789	2800.000	-4.69		
830	3000.000	-4.59		

In figure 3.1 the flow field and the lines $\psi \equiv 0$ and $\psi \equiv -0.18968$ (95% saturation) for different times are presented. The differences to the results of [4] (Fig. 3)

FIGURE 3.1

Example 2, flow fields and equipotentials for different times (run of table 3.4).

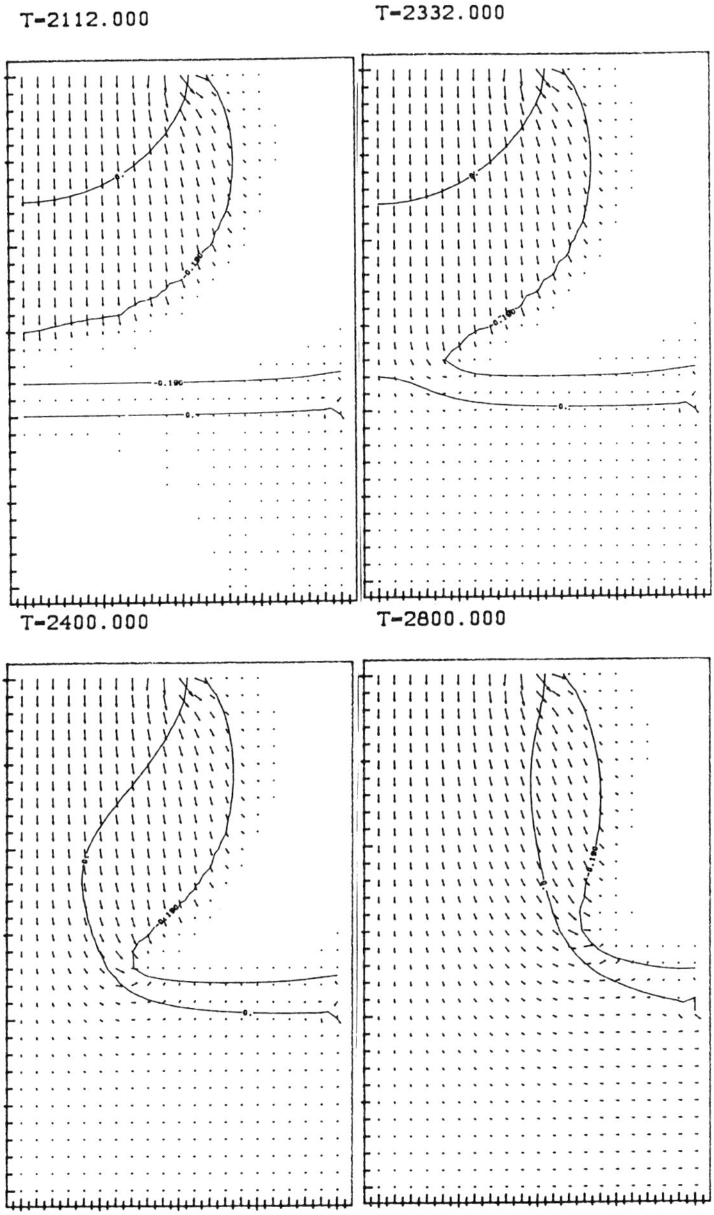

PART II

BOUNDARY VALUE PROBLEMS
FOR
ODE'S AND ELLIPTIC PDE'S

NUMERICAL PATHFOLLOWING BEYOND CRITICAL POINTS IN ODE MODELS

P. Deuflhard, B. Fiedler, P. Kunkel

Abstract.
The paper deals with extensions of a continuation technique, that has recently been suggested and worked out by the authors. For turning point problems, an efficient technique is designed, which may be combined with any boundary value problem solver. For (simple) bifurcation points, several augmented systems are discussed in detail on the basis of stability considerations for ordinary differential equations. For difference or collocation methods, two efficient augmented systems are given. For multiple shooting techniques, an efficient and theoretically satisfactory treatment could only be found for "not too stiff" ODEs.

0. Introduction

Numerical pathfollowing beyond critical points (turning or bifurcation points) is an important task in real life applications. In ODE models, the presently most popular algorithmic approaches seem to be the ones of Seydel [17,18] (code BIFPAC) and the one of Doedel [8] (code AUTO). The latter approach is based on pseudo-arclength continuation as suggested by Keller [11]. The present paper aims at the extension of techniques from [7], that have been developed recently for algebraic systems. In this approach, tangent (or Euler) continuation is combined with a Gauss-Newton method including steplength control.

Even though the pathfollowing techniques to be given below can, in principle, be combined with *any* ODE boundary value problem (BVP) solver, the actual implementation has been done in the context of multiple shooting (MS) techniques [20,6,13]. In section 1, pathfollowing beyond turning points is treated. Algorithmic details are worked out and numerical comparisons are given. The computation of (simple) bifurcation points as solutions of different augmented systems is discussed in section 2 in terms of stability of the ODE's.

1. Pathfollowing Beyond Turning Points

Consider a τ-dependent family of boundary value problems (BVP's) for n ordinary differential equations (ODE's):

$$\begin{aligned} &a)\quad x' - f(x,\tau) = 0 \\ &b)\quad r(x(a),x(b),\tau) = 0\ , \end{aligned} \tag{1.1}$$

where τ varies in a finite interval $[\tau_{min},\tau_{max}]$. Following closely the notation in [6], let $W(t_2,t_1)$ denote the Wronskian matrix of the associated variational equation

$$\begin{aligned} &a)\quad \frac{d}{dt} W(t,t_0) = f_x(x(t),\tau)W(t,t_0) \\ &b)\quad W(t_0,t_0) = I\ , \end{aligned} \tag{1.2}$$

where the τ-dependence is dropped for convenience. Moreover, let

$$\begin{aligned} &A := \frac{\partial r}{\partial x(a)}\ ,\quad B := \frac{\partial r}{\partial x(b)}\ , \\ &E(t) := AW(a,t) + BW(b,t) = E(a)W(a,t)\ . \end{aligned} \tag{1.3}$$

If E is nonsingular (at any point $t\in[a,b]$), then the BVP (1.1) has a locally unique solution x^*. At a turning point (x^*,τ^*), det(E) vanishes for all $t\in[a,b]$. Note that, since $\det(W)>0$, a turning point (x^*,τ^*) is a simultaneous turning point for all $(x^*(t),\tau^*)$, $t\in[a,b]$. As a consequence, it is sufficient to study branches $(x(a;\tau),\tau)$.

1.1. Algorithmic Details

Determinant computation.

In any BVP solver, certain projections and a linearization (or vice versa) are performed, which lead to the solution of a sequence of large sparse linear systems. Let J be the Jacobian (N,N)-matrix of such a system. Then, under mild assumptions on the projection, any zero of det(J) is also a zero of $\det(E_h)$, where E_h is an approximation of E

- see the rather general treatment in section 1 of [3]. Hence, in order to detect a turning point with respect to τ and use the superlinearly convergent procedure for turning points from [7], one is merely required to monitor det(J) and to make sure that the projection varies sufficiently smoothly with τ. In multiple shooting techniques with *condensing* in the linear equation solver (to be called MSC hereafter), det (E_h) is directly available.

Tangent computation.

This process requires the computation of the kernel vector $\bar{t}$ of the extended Jacobian (N,N+1)-matrix (with a τ-column included). For MS this matrix reads (notation similar to [6]):

$$J_{ext}\bar{t} := \begin{bmatrix} G_1 & -I & & & g_1 \\ & \cdot & \cdot & & \cdot \\ & & \cdot & \cdot & \cdot \\ & & G_{m-1} & -I & g_{m-1} \\ A & & & B & g_m \end{bmatrix} \begin{pmatrix} \bar{t}_1 \\ \bar{t}_2 \\ \\ \bar{t}_m \\ \\ \bar{t}_\tau \end{pmatrix} = 0 \qquad (1.4)$$

$$j=1,\ldots,m-1: \; G_j := \frac{\partial x(t_{j+1}|x_j,\tau)}{\partial x_j}, \quad g_j := \frac{\partial x(t_{j+1}|x_j,\tau)}{\partial \tau}$$

$$A,B \text{ from } (1.3), \; g_m := \frac{\partial r(x(a),x(b),\tau)}{\partial \tau} .$$

Remark. For pure BV embedding $f_\tau \equiv 0$ implies $g_j=0$ for $j=1,\ldots m-1$.

In multiple shooting techniques with *global* linear solver (to be called MSG hereafter) the kernel vector $\bar{t}$ is easily computed following the suggestions of [9], p.6. In MSC one computes the (n,n+1)-matrix [E,g], where

$$E := A+BG_{m-1}\cdot\ldots\cdot G_1 \equiv E_h(a)$$
$$(1.5.a)$$
$$g := g_m+B(g_{m-1}+\ldots+G_{m-1}\cdot\ldots\cdot G_2 g_1) .$$

Then QR-decomposition on [E,g] leads to the algorithm:

$$Q[E,g]\Pi = [R,S]$$
$$v := R^{-1}S \qquad (1.5.b)$$
$$\begin{pmatrix} \bar{t}_1 \\ \bar{t}_\tau \end{pmatrix} := \Pi \begin{pmatrix} -v \\ 1 \end{pmatrix}$$

$$\bar{t}_{j+1} := G_j \cdot \bar{t}_j + g_j \cdot \bar{t}_\tau \qquad j=1,\ldots,m-1 \tag{1.5.c}$$

$$\text{normalization of } \bar{t} \quad . \tag{1.5.d}$$

In order to guarantee sufficient accuracy of the computed values $\tilde{t}$, the *iterative refinement sweep* (IRS) technique from [6] is adapted. The only new item compared with [6] is the refinement on $\tilde{t}_1, \tilde{t}_\tau$: this is done by refining $\tilde{v}$.

Gauss-Newton (GN) method.

Let $x=(x_1,\ldots,x_m)$, $y=(x,\tau)$ denote the unknown node values to be computed by a given BVP solver. Then any Newton-type algorithm will be able to compute any corrections $\Delta\hat{y}$ satisfying

$$J_{ext}\, \Delta\hat{y} + F = 0 \tag{1.6}$$

Note that this system is underdetermined. Following [7], one is required to compute GN corrections

$$\Delta y := -J_{ext}^{+} F \; . \tag{1.7}$$

For small systems, J^+ is conveniently realized by QR-decomposition. For *large, sparse* systems, the following technique is generally applicable, once the kernel vector $\bar{t}$ from (1.4) is available:

$$\Delta y = \Delta\hat{y} - \frac{(\bar{t},\Delta\hat{y})}{(\bar{t},\bar{t})}\,\bar{t} \tag{1.8}$$

Here $(\cdot,\cdot)$ denotes a given inner product that, in turn, specifies the Moore-Penrose pseudoinverse J_{ext}^{+}.

Proof.

For convenience, let $(u,v)=u^T v$ and $(\bar{t},\bar{t})=1$. Then

$$-J^+F = J^+J\Delta\hat{y} = (I_{N+1}-\bar{t}\bar{t}^T)\Delta\hat{y} = \Delta\hat{y}-(\bar{t},\Delta\hat{y})\bar{t} \; .$$

The modification to other inner products and to unnormalized $\bar{t}$ in (1.8) is immediate.

Note that (1.8) is also applicable in a Hilbert space setting - such as *quasilinearization.*

An immediate consequence of (1.8) is the *orthogonality property*

$$(\Delta y,\bar{t}) = 0 \tag{1.9}$$

In MSC one typically obtains

$$\Delta\hat{y} := -J^{-}_{ext}F \quad , \tag{1.10}$$

where J^{-}_{ext} denotes a generalized inverse based on $[E,g]^{+}$. The associated generalized GN method has been suggested in [4] for the computation of periodic orbits. Both GN iterations based on J^{-} or J^{+}, respectively, converge locally and quadratically (under mild assumptions). However, J^{-} replaces the orthogonality property (1.9) by

$$\Delta x_1^T \bar{t}_1 + \Delta\tau \cdot \bar{t}_\tau = 0 \ . \tag{1.11}$$

Numerical experience shows that the angle between $\Delta\hat{y}$ from (1.10) and $\bar{t}$ may be undesirably small. Moreover, the J^{-}-version would also imply inconvenient changes in the steplength prediction strategy. Therefore the J^{+}-version seems to be preferable - compare the examples of section 1.2.1.

Choice of inner product.

In the implementation to be presented, the inner product $(\cdot,\cdot)$ in $\mathbb{R}^{N+1}$ was chosen as the trapezoidal sum approximation of the L^2-product. In MS this reads

$$(u,v) := \frac{0.5}{b-a}\,[u_1 v_1 \cdot (t_2-t_1) + u_m v_m (t_m - t_{m-1}) + \\ + \sum_{j=2}^{m-1} u_j v_j (t_{j+1} - t_{j-1})] + u_\tau v_\tau \ , \tag{1.12}$$

where $u_j, v_j \in \mathbb{R}^n$, $u_\tau, v_\tau \in \mathbb{R}^1$. As usual , the quantities u,v are understood to be scaled. The idea behind this choice is to achieve at least some independence of the coarse mesh selection. Note that this choice induces special norms, which, in turn, affect the convergence criteria and the steplength control for the tangentcontinuation from [7].

Jacobian approximations.

The quantities (G_j, g_j) in J_{ext} are either computed by finite difference approximations or, if (f_x, f_τ) is available, by numerical integration of the associated variational equations. In order to save computing time, one may apply *rank-1 updates* of the form

$$[G_j, g_j]^{k+1} := [G_j, g_j]^k - F_j^{k+1} \cdot \frac{\Delta y_j^T}{\|\Delta y_j\|_2^2} \tag{1.13}$$

where $\Delta y_j := (\Delta x_j, \Delta\tau)$. With these rank-1 updates, the total Jacobian correction is generally no longer of rank one. As a consequence, the GN iterates $\{y^k\}$ will no longer just vary within an n-dimensional hyperplane H orthogonal to $\bar{t}(y^0)$. In principle it is easy to confine the iterates to H by just fixing $\bar{t} := \bar{t}(y^0)$ throughout the iteration. A short series of numerical experiments, however, revealed that such a confinement does not pay off: rather, the varying kernel vectors $\bar{t}(y^k)$ seem to be clearly preferable. Also, a careful examination shows that this slight modification (compared with [7]) has only a negligible effect on the steplength prediction strategy, since $\bar{y} - \hat{y} \doteq y^0$ (notation of [7]).

Convergence criterion.

The pointwise rank-1 updates (1.13) seem to lead to a rather good approximation of the total Jacobian J_{ext}. As a consequence, the monotonicity restriction could be relaxed (parameters $\Theta_{max}=1$, $\rho=0.25$ of [7] throughout the examples of section 1.2).

1.2. Numerical Comparisons

All subsequent computations have been made in FORTRAN double precision on the IBM 3081D of the Computing Center of the University of Heidelberg. The common required relative accuracy throughout the examples was EPS=1.D-5.

1.2.1. GN Techniques in MSC

In this kind of approach, a generalized GN method based on the special pseudoinverse J^- in (1.10) seems to be natural. Such an approach has been favored in [4] for periodic orbit computation in autonomous ODE's. By means of formula (1.8) with (1.5) and (1.10) the GN method based on the Moore-Penrose pseudoinverse can easily be realized. In Table 1, comparison runs of J^-- and J^+-versions with the code PERIOD due to [4,23] are given. Obviously, the J^+-version reduces the computational amount significantly. Just this version is anyway the natural one for MSG.

Table 1. Comparison of GN methods for periodic orbit computation.

example	GN iterations (damped) J^-	J^+	number of trajectories J^-	J^+
nerve membrane [4], stable orbit	14(5)	7(0)	37	19
Lorenz problem [4], unstable orbit	17(12)	11(6)	61	35

1.2.2. Continuation Techniques

The above considerations led to the implementation of a FORTRAN code MULCON (mnemotechnically for MULtiple shooting with CONtinuation), which, at present, realizes MSC. The alternative MSG implementation, which has also been discussed in section 1.1, will be presented elsewhere. The new code MULCON is compared with the following two codes

AUTO: due to Doedel [8,9], pseudo-arclength continuation method combined with collocation,

BIFPAC: due to Seydel [19], explicit change of parametrization combined with MS techniques.

In AUTO and MULCON the Jacobian matrix was computed via the variational equations, whereas BIFPAC only realizes the finite difference option. In MULCON, a special adaptation of the extrapolation integrator DIFEX1 [5] was used, whereas BIFPAC applies the Runge-Kutta code RKF7 in the standard case. In this situation, the computing times of different codes are not meaningful for the relative evaluation of the competing continuation techniques. More insight is gained by comparing robustness, reliability, and distribution of continuation points.

Example. Bratu problem [8]
In [8], the following one-dimensional analogue of the originally two-dimensional Bratu problem is given as a driver example:

$$y'' + \tau e^y = 0, \quad y(0) = y(1) = 0 . \tag{1.14}$$

The code AUTO was run with the specifications (m=5, ncol=4), as given

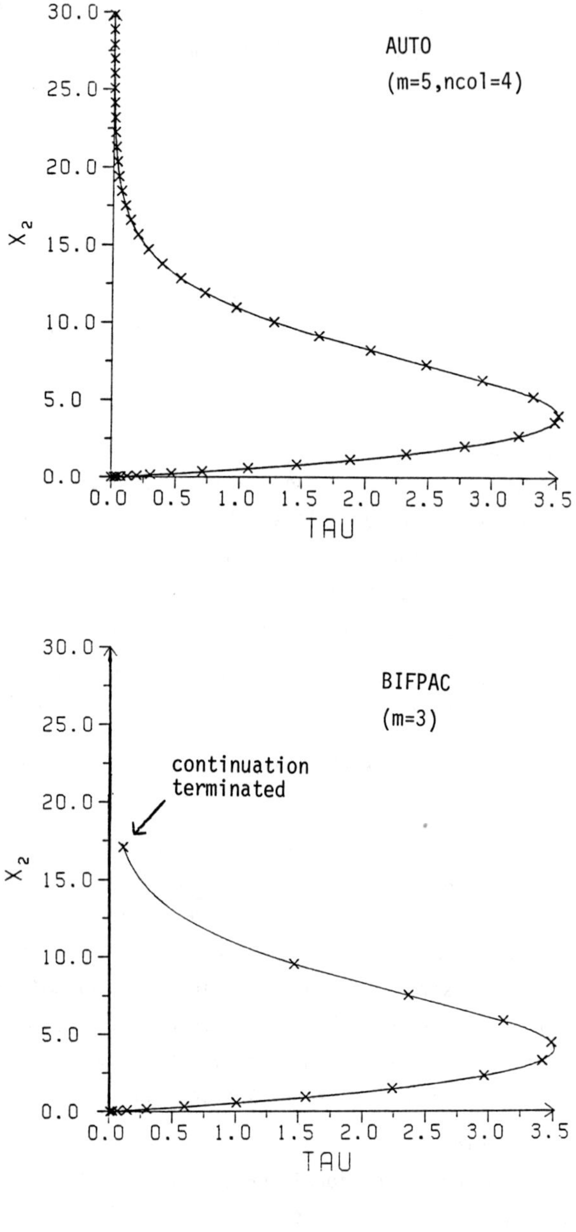

Fig.1. Selected continuation points from different codes in the Bratu problem.

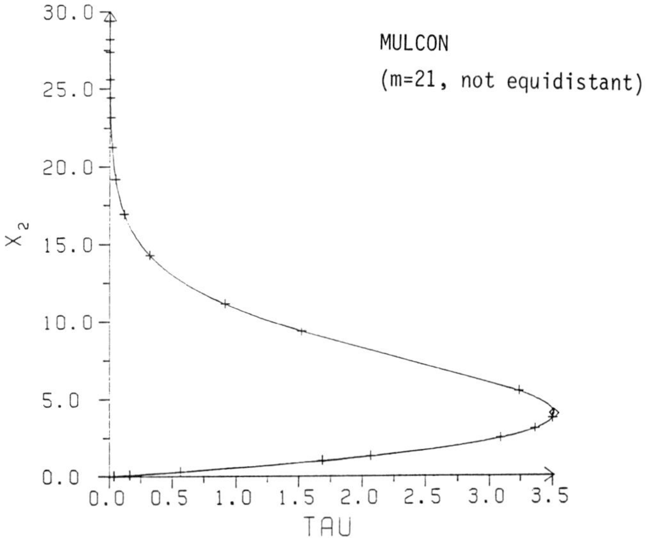

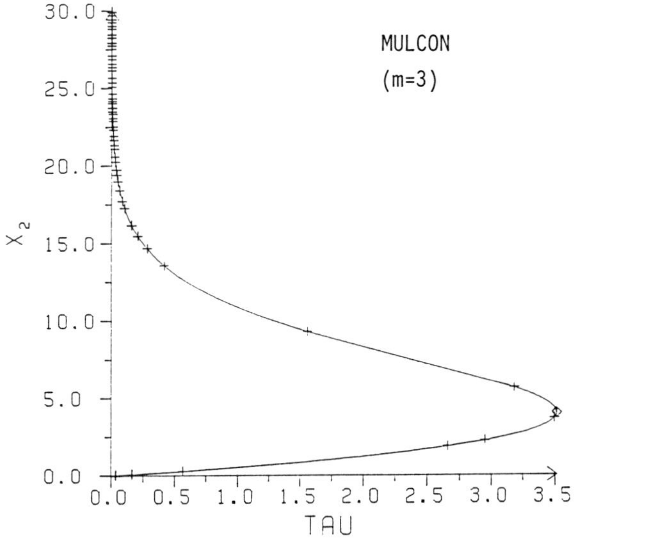

<u>Fig.1.</u> continued

in [8]. The MS codes BIFPAC and MULCON were run with several numbers m of nodes. A comparison of the distribution of selected continuation points is presented in Fig. 1. (For convenience, the plots of the path for the AUTO and BIFPAC parts of Fig. 1 were also made by cubic Hermite interpolation based on the MULCON plot output.) From Fig. 1, the following observations can be made:

(a) The pseudo-arclength continuation in AUTO is robust and reliable, but too cautious - and therefore slow.

(b) The continuation procedure in BIFPAC is not robust: it just terminated where the problem turned to be sensitive.

(c) The tangent continuation in MULCON is robust, reliable, and flexible - and therefore fast. In the sensitive part of the path an increase of multiple shooting nodes speeds up the continuation process.

The computing times of the multiple shooting codes were comparable along the part, which BIFPAC solved. The computing time of AUTO was significantly less - a fact, which demonstrates that collocation is preferable in this special problem. The computation of the turning point was easily realized in both AUTO and MULCON (standard option). In BIFPAC, this task requires additional programming by the user and was therefore not performed.

Summary of section 1.

Among the competing BVP approaches, MSC is known to be most economic in terms of storage - a feature, which is important in large scale computing. The present MSC version of MULCON seems to be promising for pathfollowing beyond turning points in those BVPs, that are suited for multiple shooting - with non-stiff or stiff integration involved.

2. Discussion of Augmented Systems for Simple Bifurcation Points

In this section the numerical determination of both perfect and imperfect simple bifurcation points in ODE models is discussed. The τ-dependent problem is written as

$$F : X \times \mathbb{R} \to Z$$

$$F(x,\tau) = 0$$

At a *simple* bifurcation point $y^*=(x^*,\tau^*)$ the Jacobian DF has 2-dimensional kernel, co-rank 1, and D^2F non-degenerate in the kernel.

On the basis of theoretical consideration and numerical evidence presented in [7], the augmented system of Moore [15] seems to be the first choice. However, the subsequent discussion will show that for multiple shooting with *stiff* ODE's the alternative systems of Seydel [17] and Weber [21] deserve careful examination.

Augmented system of Moore [15].

The abstract system reads

$$\begin{aligned} &a)\quad F(y) - \alpha z = 0 \\ &b)\quad (z,DF(y)) = 0 \\ &c)\quad (z,z) = 1 \end{aligned} \tag{2.1}$$

where $(\cdot,\cdot)$ denotes the inner product in the Hilbert space Z. The unknowns are (y,z,α). For sufficiently small imperfection parameter $\alpha^*\in\mathbb{R}$, (2.1.a) is equivalent to

$$(F,F) = \min = \alpha^{*2} \tag{2.1'.a}$$

subject to the constraints (2.1.b,c). Upon specifying F according to (1.1) and $X=H^1$, $Z=L^2\times\mathbb{R}^n$ with the canonical inner product, standard calculation leads to the following explicit form of (2.1):

$$\begin{aligned} a)\quad & x' - f(x,\tau) - \alpha\lambda = 0 \\ & \lambda' + f_x(x,\tau)^T\lambda = 0 \\ & \xi_1' - f_\tau^T(x,\tau)\lambda = 0 \\ & \xi_2' - \lambda^T\lambda = 0 \end{aligned} \tag{2.2}$$

$$\begin{aligned} b)\quad & r(x(a),x(b),\tau) - \alpha\zeta = 0 \\ & A^T\zeta - \lambda(a) = 0 \\ & B^T\zeta + \lambda(b) = 0 \\ & \xi_1(a) = \xi_2(a) = 0 \\ & r_\tau^T\zeta - \xi_1(b) = 0 \\ & \zeta^T\zeta + \xi_2(b) = 1 \end{aligned}$$

Note that

$$\begin{aligned} \xi_1(b) &= \int_a^b f_\tau(x(t),\tau)^T \lambda(t)dt \ , \\ \xi_2(b) &= \int_a^b \lambda^T(t)\lambda(t)dt \quad . \end{aligned} \tag{2.2}$$

The above system involves a topological perturbation in the space Z as a whole. As an alternative, one might as well keep the ODE unperturbed and just perturb the boundary conditions. This means to specify

$$F(x(a),\tau) = r(x(a),x(b;x(a),\tau,a),\tau)$$
$$X = Z = \mathbb{R}^n \quad ,$$

where $x(t;x_0,\tau,a)$ is the solution of (1.1.a) with $x(a;x_0,\tau,a)=x_0$ fixed. Instead of (2.2), one now obtains:

$$\begin{aligned} &\text{a)} \quad x' - f(x,\tau) = 0 \\ &\qquad s' - f_x(x,\tau)s - f_\tau(x,\tau) = 0 \\ &\text{b)} \quad r(x(a),x(b),\tau) - \alpha\zeta = 0 \\ &\qquad s(a) = 0 \\ &\qquad [E(a),r_\tau + Bs(b)]^T\zeta = 0 \\ &\qquad \zeta^T\zeta = 1 \end{aligned} \tag{2.3}$$

Herein

$$s(t) = x_\tau(t;x(a),\tau,a)$$

denotesthe sensitivity.

Remark.

There is some arbitrariness in the selection of x(a) as independent variable in r, which results in some asymmetry in the above augmented boundary conditions. Upon selecting any $t_0 \in [a,b]$ and $x_0 := x(t_0;x_0,\tau,t_0)$ as variable, one obtains

$$s(t_0) = 0$$
$$[E(t_0),r_\tau + As(a)+Bs(b)]^T\zeta = 0 \quad .$$

For special applications, a deliberate choice of t_0 might be advantageous.

An equivalent form of (2.3), which better shows the similarity with (2.2), is:

$$\begin{aligned}
\text{a)}\quad & x' - f(x,\tau) = 0 \\
& \lambda' + f_x(x,\tau)^T\lambda = 0 \\
& \xi' - f_\tau(x,\tau)^T\lambda = 0 \\
\text{b)}\quad & r(x(a),x(b),\tau) - \alpha\zeta = 0 \\
& A^T\zeta - \lambda(a) = 0 \\
& B^T\zeta + \lambda(b) = 0 \\
& \xi(a) = 0 \\
& r_\tau^T\zeta - \xi(b) = 0 \\
& \zeta^T\zeta = 1
\end{aligned} \tag{2.4}$$

Among the BVP solvers, difference [12] or collocation [1] methods may well be applied to both (2.2) and (2.4). Multiple shooting techniques for these two systems, however, will only be efficient, if the ODE system (1.1.a) is "non-stiff", since the adjoint variational equation for λ has the reversed stability behavior. At first glance, system (2.3) seems to circumvent this restriction: the evaluation of $E(a)=A+BW(b,a)$ just requires the forward variational equation for W, which is anyway necessary in combination with Newton's method (recall from (1.4) that $G_j=W(t_{j+1},t_j)$). However, the difficulty arises in the evaluation of the remaining part of the augmented Jacobian. If one computes the tensor

$$\frac{\partial E}{\partial x_a}$$

via forward variational equations, then n^2 linear inhomogeneous ODE systems must be solved - a prohibitive amount of work. On the other hand, one might hope to evaluate

$$\frac{\partial}{\partial x_a}\,(E^T\zeta)$$

by integrating just n ODE systems. In fact, this is possible, but only by solving n adjoint variational equations. To see this, just verify that $E^T(t)\zeta$ satisfies the adjoint equation. Therefore, the multiple shooting treatment of bifurcations in the presence of stiff ODE systems remains open - at least in terms of Moore's augmented system.

Alternative augmented systems.

At this point, interest focusses on those augmented systems that require variational equations only in "forward direction". The two principal possibilities are due to Seydel [17] and Weber [21]. In explicit form, Seydel's system reads:

$$
\begin{aligned}
&\text{a)}\quad x'-f(x,\tau) = 0\\
&\qquad h'-f_x(x,\tau)\,h = 0\\
&\text{b)}\quad r(x(a),x(b),\tau) = 0\\
&\qquad Ah(a)+Bh(b) = 0\\
&\qquad \|h\| = 1
\end{aligned}
\tag{2.5}
$$

At simple bifurcation points, the augmented Jacobian of this system is necessarily singular. As a consequence, Newton's method will converge only linearly and only for starting points within some convergence cone - for details see e.g. [2]. For the same reason, imperfect bifurcation points cannot be treated in this manner. In his more recent work [18], Seydel actually recommends to avoid the explicit accurate computation of bifurcation points.

The other alternative is the augmented system of Weber [21], which in abstract form reads:

$$
\begin{aligned}
&F(x,\tau) = 0\\
&D_xF(x,\tau)h_1 = 0\\
&D_xF(x,\tau)h_2+D_\tau F(x,\tau) = 0\\
&\|h_1\| = 1\\
&(h_1,h_2) = 0
\end{aligned}
\tag{2.6}
$$

Unlike (2.5), this system uses the f_τ-information, which is necessary to determine the directions of the bifurcating branches. This system still becomes singular at a bifurcation point. For the treatment of imperfect bifurcations, Weber [22] suggests to apply a physically meaningful 2-parameter embedding in the spirit of Keener [10]. In contrast to Moore, Weber does not give a universal unfolding. Such an unfolding can be constructed in any direction z_0 not contained in the range of degenerate DF. This leads to

$$
\begin{aligned}
&F(x,\tau)-\alpha z_0 = 0 \;, \quad z_0 \notin \mathrm{im}(DF(y^*)) \\
&DF(x,\tau)k_1 = 0 \\
&DF(x,\tau)k_2 = 0 \\
&||k_1|| = 1 \\
&||k_2|| = 1 \\
&(k_1,k_2) = 0 \\
&(k_0,k_2) = 0 \quad , \quad k_0 \notin (\ker DF(y^*))^{\perp}
\end{aligned}
\tag{2.7}
$$

where the notation $k_i=(h_i, \delta_i)$ is used. For a perfect bifurcation ($\alpha^*=0$), the associated augmented Jacobian is nonsingular. The natural choice for k_0 is the last tangent vector $\bar{t}$ computed in the pathfollowing process. The somewhat critical choice of z_0 has to be combined with the pivot strategies for J.

The above system is larger than the ones of Moore or Seydel. The additional expense, however, is inevitable, if only forward variational equations are to be integrated. Indeed, unfolding the problem in this way requires the computation of the full *2-dimensional* kernel of $DF(y^*)$.

Summary of section 2.

The computation of simple bifurcation points by means of the augmented systems of Moore is recommended for

(a) global BVP solvers such as difference and collocation methods,

(b) multiple shooting techniques, if the original ODE system to be integrated is "not too stiff" .

For stiff multiple shooting problems, an alternative augmented system needs to be considered, which, however, seems to be too costly. Therefore this realization might be efficient only for those problems, where compensating advantages of multiple shooting predominate.

References

[1] U. Ascher, I. Christiansen, R.D. Russell: A Collocation Solver for Mixed Order Systems of Boundary Value Problems. Math.Comp. 33, 659-679 (1979).

[2] D.W. Decker, H.B. Keller, C.T. Kelley: Convergence Rates for Newton's Method at Singular Points. SIAM J. Numer.Anal. 20, 296-314 (1983).

[3] P. Deuflhard: Nonlinear Equation Solvers in Boundary Value Problem Codes. In: Childs et al. (ed.): Codes for Boundary-Value Problems in Ordinary Differential Equations (1979). Springer Verlag.

[4] P. Deuflhard: Computation of Periodic Solutions of Nonlinear ODE's. BIT 24, 456-466 (1984).

[5] P. Deuflhard: Recent Progress in Extrapolation Methods for Ordinary Differential Equations. SIAM Rev., 27, 505-535 (1985)

[6] P. Deuflhard, G. Bader: Multiple Shooting Techniques Revisited. Univ. Heidelberg, SFB 123: Tech.Rep. 163 (1982).

[7] P. Deuflhard, B. Fiedler, P. Kunkel: Efficient Numerical Pathfollowing Beyond Critical Points. Univ. Heidelberg, SFB 123: Tech.Rep. 273 (1984).

[8] E.J. Doedel: AUTO-version Sept. 84 (VAX/VMS) Concordia Univ., Montreal: Comp.Sci. Dep.: Tech.Rep. (1984).

[9] E.J. Doedel, J.P. Kernevez: Software for Continuation Problems in Ordinary Differential Equations with Applications (Sept. 84).

[10] J.P. Keener, H.B. Keller: Perturbed Bifurcation Theory. Arch.Rat. Mech.Anal. 50, 159-175 (1973).

[11] H.B. Keller: Numerical Solution of Bifurcation and Nonlinear Eigenvalue Problems. In [16], 359-384 (1977).

[12] M.Lentini, V. Pereyra: An adaptive finite difference solver for nonlinear two-point boundary value problems with mild boundary layers. SIAM J.Numer.Anal. 14, 91-111 (1977).

[13] M. Lentini, M. Osborne, R. Russell: The Close Relationships between Methods for Solving Two-Point Boundary Value Problems.SIAM J. Numer.Anal. 22, 280-309 (1985).

[14] H.D. Mittelmann, H.Weber (ed.): Bifurcation Problems and their Numerical Solution. Birkhäuser/Boston: ISNM 54 (1980).

[15] G. Moore: The Application of Newton's Method to Simple Bifurcation and Turning Point Problems. Univ. Bath: Ph.D. Thesis (1979).

[16] P.H. Rabinowitz (ed.): Applications of Bifurcation Theory. Academic Press, New York, San Francisco, London (1977).

[17] R. Seydel: Numerical computation of branch points in ordinary differential equations. Numer.Math. 32, 51-68 (1979).

[18] R. Seydel: Branch Switching in Bifurcation Problems for Ordinary Differential Equations. Numer.Math. 41, 93-116 (1983).

[19] R. Seydel: BIFPACK- a Program Package for Calculating Bifurcations SUNY, Buffalo (1983).

[20] J. Stoer, R. Bulirsch: Introduction to Numerical Analysis. Springer (1980).

[21] H. Weber: Numerische Behandlung von Verzweigungsproblemen bei gewöhnlichen Differentialgleichungen. Numer.Math. 32, 17-29 (1979).

[22] H. Weber: Shooting Methods for Bifurcation Problems in Ordinary Differential Equations. In [14], 185-210 (1980).

[23] R. Winzen: Numerische Behandlung periodischer Lösungen dynamischer Systeme mit Paramterabhängigkeit. University of Heidelberg: Diploma thesis (1984).

COMPUTING BIFURCATION DIAGRAMS FOR LARGE NONLINEAR VARIATIONAL PROBLEMS

Helmut Jarausch and Wolfgang Mackens

We report on a method to reduce the computational effort of computing bifurcation diagrams for large nonlinear parameterdependent systems of variational type. Included are techniques to find a point on a branch, to trace solution branches, to detect singularities, to compute turning points, simple and double nondegenerate bifurcation points and to calculate emanating directions from bifurcation points. The performance of the method is demonstrated at two examples.

1. The Condensed System

We want to trace solution branches $(u,\lambda) \in \mathbb{R}^n \times \mathbb{R}$ of

(1) $A\,u = F(u,\lambda)$ where $F \in C^2(\mathbb{R}^n \times \mathbb{R},\ \mathbb{R}^n)$,
F_u is symmetric and
A is symmetric and positive definite

Equation (1) is the Euler-equation of the variational problem

(2) $\frac{1}{2}\, u^T A\, u \;-\; V(u,\lambda) = \text{stationary}$
where $V(u,\lambda)$ is a scalar potential
with $V_u = F$.

Typically eq. (1) arises from a Finite Element or a Finite Difference discretization of a given continuous problem. Since standard solution-branch analyzing techniques are quite expensive when applied to big systems we want to condense (1) into a much smaller system. This small system is intended to approximate the solution-set of the full system closely, including singular points.
Note that the linear eigenvalue problem

$$(3) \quad \mu A v = F_u v \quad (\mu \in \mathbb{R})$$

has an eigenvalue $\mu = 1$ at a singular point of (1). It is therefore necessary to 'extract' those parts of the solution which correspond to eigenvectors with eigenvalues around 1 and put them into the small system. For reasons to be explained later on we take all those parts corresponding to eigenvalues around and bigger than 1 in modulus. In many applications there are only a few such eigenvalues, see the remarks in section 4 .

Note that our assumptions guarantee that there is a complete set of A-orthonormal eigenvectors $z_1,\dots,z_n$ of (3) with corresponding eigenvalues μ_i such that

$$|\mu_1| \geq |\mu_2| \geq \dots |\mu_n|$$

Definition 1

Fix a constant $0 < \gamma < 1$ and choose m to be the smallest integer such that $|\mu_i| \leq \gamma$ for all $i > m$. The corresponding system of column vectors $Z = (z_1,\dots,z_m)$ will be called a support system. Since the vectors z_i are A-orthonormal we have the following A-orthogonal complementary projectors

$$(4) \quad P \equiv Z Z^T A \quad \text{and} \quad Q \equiv Id - P$$
$$\text{such that} \quad PQ = QP = 0 .$$

Note that the range of P has dimension m which is much smaller the the dimension n of the full system (1). Typically we found $m \leq 8$. Using these projectors we split the solution space accordingly

$$u = p + q \quad \text{where} \quad p = Pu \quad \text{and} \quad q = Qu \ .$$

We endow $\mathbb{R}^n$ with the geometry induced by A and define

$$\langle u,v\rangle \equiv u^T A\, v \quad \text{and} \quad \|u\|_A \equiv \sqrt{\langle u,u\rangle}$$

We rewrite equation (1) as

$$(5) \quad R(u,\lambda) \equiv u - A^{-1} F(u,\lambda) = 0$$

and introduce the residuals

$$r_P \equiv \langle PR,PR\rangle \quad (\text{ P-residuum })$$
$$r_Q \equiv \langle QR,QR\rangle \quad (\text{ Q-residuum })$$

Note the relation $\langle R,R\rangle = r_P + r_Q$.

We can now split (5) equivalently into a condensed system of dimension m only, the

P-SYSTEM

$$(6a) \quad P\, R(p+q,\lambda) = 0 \quad \text{or with} \quad c \equiv Z^T A u \quad \text{and thus} \quad p = Zc$$

$$(6b) \quad f_P^*(c,\lambda\ ;q) \equiv c - Z^T F(\ Zc + q,\ \lambda\ ) = 0$$

$$\text{where} \quad f_P^*(.,.;q) : \mathbb{R}^m \times \mathbb{R} \dashrightarrow \mathbb{R}^m,$$

and a big system of dimension n-m, the

Q-SYSTEM or 'supported Picard system'

$$(7a) \quad Q\, R(p+q,\lambda) = 0 \quad \text{or written out}$$

$$(7b) \quad q = f_q(q\ ;p,\lambda) \equiv Q\, A^{-1} F(p+q,\lambda) \quad \text{for} \quad q \in Q\mathbb{R}^n \ .$$

This system is nearly as large as the full system but it turns out to be <u>contractive</u> with rate γ, which can be chosen quite small independent of the dimension n of the discrete system. Furthermore iterating the Q-system involves the approximate solution of positive definite linear systems with the constant matrix A. Any preconditioning work like sophisticated (block-) preconditioners or even a (direct) decomposition of A may pay off.

We summarize some properties of these systems in the following lemma, see [10] for more details.

<u>Lemma 1</u>

(i) Equations (6) and (7) are decoupled up to first order, i.e. $df_p^*/dq = 0$ and $df_q/dp = 0$.

Therefore iterating one equation will not invalidate the other one up to first order.

(ii) $(df_p^*/dc\ ,\ df_p^*/d\lambda\ )$ is singular (rank deficient) iff $R'(u)$ is singular (rank deficient).

(iii) For fixed p and λ the fixed-point mapping $f_q(\ .\ ;p,\lambda)$ is contractive with Lipschitz-constant γ (see Definition 1). Therefore eq. (7) is solvable by fixed-point iteration. The contraction rate γ can be chosen as small as one likes (at the expense of increasing the size of the support system). Normally less than 8 support vectors were sufficient to get γ below 0.5 within our applications.

2. <u>The λ-dependency of the Q-system</u>

While Lemma 1 states that equation (7) is invariant (up to first order) against changes of p we still have to take the λ-dependency into account. Differentiating (7) gives

(8) $dq/d\lambda = Q\,A^{-1}\,F_u\ dq/d\lambda + Q\,A^{-1}\,F_\lambda$, i.e.

$q_\lambda \equiv dq/d\lambda$ can be computed from a fixed point equation with

the same contraction rate γ as for eq. (7) .
With the aid of q_λ we can give the final definition of the small (condensed) system:

Definition 2

Assume for a given tuple (c',q',λ') we have computed q_λ from eq. (8) up to a prescribed precision. We now define the condensed system to be

$$f_p(c,\lambda\ ;q') \equiv c - Z^T F(Zc + q' + (\lambda-\lambda') q_\lambda , \lambda)$$

3. A Prototype Algorithm

In order to find a solution branch, then to follow it and to compute turning points or bifurcation points and finally to escape from a bifurcation point one proceeds in principle as follows:
Given an approximate partial eigensystem $Z = (z_1,\ldots,z_m)$ and q_λ from eq. (8)

Determine $(\Delta c , \Delta\lambda)$ by applying algorithms for the above tasks to the condensed system $f_p(c,\lambda\ ;q)$ and update

$$\begin{aligned} c &:= c + \Delta c \\ \lambda &:= \lambda + \Delta\lambda \qquad (9) \\ q &:= q + \Delta\lambda\, q_\lambda \end{aligned}$$

Only if necessary update q further by iterating the Q-system (7).

4. Quality of the Support-System

For the efficiency of an algorithm based on this prototype algorithm two crucial questions have to be answered.

First
How large is the dimension m of the small system.
We have found $m < 8$ in most of our applications. We have

no general answer to this question but the following remarks apply.

- The dimension of the small system is nearly independent of the dimension n of the discrete system (1) but depends on the underlying continuous problem only.

- The number of eigenvalues of eq. (3) which are larger then +1 is usually small in the region of interest, because for (v,μ) an eigen-vector/value of (3) with $\mu > +1$ we have

$$v^T (A - F_u) v = (1-\mu) v^T A v < 0 \quad \text{, i.e.}$$

v is a direction of negative curvature of the given variational problem (2). In applications one is often interested in those parts of the solution branch where there are not too many of these directions of negative curvature.

- In the other case $\mu < -1$ the matrix A is not a good enough 'preconditioner' since we could add the positive semi-definite rank 1 matrix $\mu v v^T$ to A and thus make the new matrix A even more positive definite. Since $\mu v v^T$ is a full matrix one would not actually add it to A but keep it separate in case of a c-g-type solver for the linear positive definite systems to solve. For other solvers one would construct a suitable sparse update of A. In some cases even a diagonal matrix will do.

<u>Second</u>

Even more important is the question of how much accuracy do we need for the approximate eigenvectors z_i and for q_λ. Using results of [12] we have a precise instrument to control the precision of the support vectors that we need:

<u>Lemma 2</u>

Let $Z = (z_1,\ldots,z_m)$ be a set of A-orthonormal vectors approximating the dominant eigenvectors of $A^{-1}F_u$ such that the generalized Rayleigh-quotient $M' \equiv Z^T F_u Z$ is diagonal and $\| A^{-1}F_u Z - Z M' \|_A \le \epsilon$, then there exists a symmetric perturbation matrix E such that

$$\|E\| = \epsilon\ , \qquad [\ A^{-1}F_u + E\ ]\ Z = Z\ M'$$

$\|\ M - M'\ \| \leq \epsilon$ where M is the diagonal matrix of the exact eigenvalues of $A^{-1}F_u$

$\|df_p/dq\| \leq \epsilon \qquad \|df_q/dc\|_A \leq \epsilon$ (decoupling)

$df_p/dc = Id - M'$ and

$\|df_q/dq\|_A \leq |\mu_{m+1}| + \epsilon \leq |\mu'_{m+1}| + 2\ \epsilon$ where

μ'_{m+1} can be computed by using a larger Ritz projection which is used anyway when the support vectors are reiterated - see below.

The quality of q_λ can be checked by substituting it into equation (8) and taking the A-norm of the residuum.

In practical computations the support system Z is initialized to random vectors that have been A-orthonormalized. This initial system is then improved by a <u>refreshment</u> of the support system which consists of a single simultaneous power iteration wich reads as follows:

- power step $\qquad Z := A^{-1}\ F_u()\ Z$

(10) - Ritz projection $\qquad B \equiv Z^T\ F_u()\ Z \quad (\ = Z^T A\ (A^{-1}F_u)\ Z\ )$

- diagonalization of the small (mxm) matrix B giving $B = V\ M\ V^T$

- lifting $\qquad Z := Z\ V$

Note that one can do the a-posteriori estimate of the last lemma simultaneously without extra work.

Now the decoupling properties (see Lemma 2) and the contraction rates of equations (7) and (8) are constantly monitored to signal if the support system is no longer good enough. If necessary we include a new refreshment step of the support system or an iteration of eq. (8). Thus we adapt the extended support system (Z,q_λ) during the solution process.

5. Going Back to the Solution Branch

To get onto a solution branch we perform a sequence of dogleg-steps consisting of optimal linear combination of the Gauss-Newton step for f_p and the gradient step for $\frac{1}{2}f_p^T f_p$ to get the optimal descent in a given trust region. This is done similarly as in the λ-fixed case which has been treated in [7,10]. Global convergence has been shown for that case.

6. Following the Solution Branch

Starting at a given 'check-point' (c,λ,q) on the full arc we follow the solution arc of the small system $f_p(c,\lambda;q) = 0$ and adapt (c,λ,q) as given by (9). The quantities Δc and $\Delta\lambda$ are computed by applying the branch-following scheme of Deuflhard et al [5] to the small system. Additionally we set up a trust-region like arc-length for which the small arc does not deviate too much from the 'full' arc. This deviation can be checked from Q-residuum r_Q with the aid of the contraction rate γ of the Q-system. If the arc-following subroutine for the small system is going to exceed this arc-length (since the last check-point) it is interrupted. We then go back to the full arc (see the last section) which usually implies some iterations of the Q-system only. Note that if (t_c,t_λ) is the tangent of the small system - normalized according to

$$\|t_c\|_2^2 + (1 + \|q_\lambda\|_A^2)\, t_\lambda^2 = 1 \quad \text{then}$$

$$(Z\, t_c + t_\lambda\, q_\lambda\ ,\ t_\lambda)$$

is the normalized tangent of the full system. Thus the projection of the full tangent to the range of P coincides with the small tangent. Furthermore the update (9) performs a true tangent step for the full system - at least if q_λ were exact. We even have second derivative information available in the small system - see section 8.

7. Detection of Singularities

Following the solution arc we do a Ritz-projection of F_u onto the fixed support system Z according to eq. (10). The computation of this projection matrix B and its eigenvalues (B was smaller than 10 x 10 in our examples) is very cheap. The arc-following subroutine signals the possibility of a singular point whenever (an) eigenvalue(s) of B come(s) close to +1 or crosses +1. Thus a singular point is easily detected and its multiplicity can be predicted. To analyze these points further we switch to specialized routines.

8. Computation of singular points, general remarks

We are able to compute turning points, simple bifurcation points and double bifurcation points, where for each of these adequate nondegeneracy conditions are assumed.

One observes that predictor-corrector path-following routines very often become trapped at bifurcation points producing a spider web between the different branches. To avoid such an undesirable behaviour we compute bifurcation points accurately by direct methods and determine directions of emanating branches, afterwards.

As discussed in Section 7 the detection of singular points is no problem, since the relevant part of the spectrum of $A^{-1}F_u$ is at our disposal.

To actually compute a singular point one has to use a specific routine for each of the different kinds of singularity. The decision process on the kind of singularity is supported by our method:

The Jacobian of the small f_p-system

$$(df_p/dc \ , \ df_p/d\lambda)$$

is known during branch following and the part df_p/dc is nearly diagonal . Thus it is easy to decide whether

$$df_p/d\lambda \quad \epsilon \quad \text{Range}(df_p/dc)$$

for example.

Having thus made up our minds concerning the nature of the singularity we use direct iteration methods within the condensed system to determine it. These direct methods are integrated as follows into a full space iteration scheme:

ALGORITHM S

WHILE NOT CONVERGED DO

1. Calculate a singular point of the desired type for the condensed system $f_p(c,\lambda;q)=0$. (Notice that bifurcation points of the small system are likely to be perturbed (unfolded) during the iteration even if the original bifurcation point of the full system is not. Hence the applied method to compute bifurcation points must be able to compute unfolded ones.)

2. Reduce r_Q by iterating the Q-system (7) .

3. Refresh the supports Z and q_λ .

OD .

Remarks:

a) If $r_Q=0$ and the reduction data Z and q_λ are exact at (c',λ',q') then all the derivatives of the condensation $f_p(c,\lambda;\ q')$ from Definition 2 up to the second order coincide at (c',λ') with the corresponding ones of the Lyapunov-Schmidt reduction

$$f_{LS}(c,\lambda) := c - Z^T F(Zc+q(c,\lambda),\ \lambda\ ) = 0\ ,$$
$$\text{with } q(c,\lambda) \text{ solving } \ q = QA^{-1}F(Zc + q,\ \lambda)$$

of the original system.

b) If, as is usually the case when candidates for singular points are identified during branch following, r_Q is small and Z and q_λ are sufficiently precise (these qualities are monitored during Algorithm S by the two last steps of each cycle), f_p should deliver good approximations to these derivatives of f_{LS}. Thus one can hope (and it turns out numerically that this is true) that those singularities of f_{Ls} and hence of the original system (c.f. [2,11] e.g.) are computable by Algorithm S which are determined by second derivative information.

c) Work on concise proofs of convergence for iterations of type S is in progress.

The specific algorithms that we use for the various subtasks are outlined in the following four sections.

9. Turning-point approximation

We want to compute a turning-point y_t of the solution set f_p^{-1} of

$$g(y) \equiv f_p(c,\lambda;q') = 0 \in \mathbb{R}^m, \quad y \equiv (c,\lambda) \in \mathbb{R}^{m+1} .$$

A (λ-)turning point is an element of f_p^{-1} where the tangent of f_p^{-1} has a vanishing λ-component.

We apply Newton's iteration to the system

$$(11) \qquad \begin{matrix} g(y) = 0 \\ e_{m+1}^T \; T(y) = 0 \end{matrix} ,$$

where $T(y)$ is the unit tangential vector field defined by

$$(12) \qquad \begin{matrix} Dg(y)T(y) = 0 \\ T(y)^T T(y) = 1 \end{matrix} , \qquad \det \begin{pmatrix} Dg(y) \\ T(y)^T \end{pmatrix} > 0 .$$

As usual we assume that the turning point is a regular point of g, i.e. rank $Dg(y_t) = m$. Then (12) makes sense near

y_t (cf. [17],e.g.)

If y_t is a simple turning point of f_p^{-1} (i.e. if $e_{m+1}^T DT(y)T(y) \neq 0$), then y_t is an isolated solution of system (11).

The first m rows of the Jacobian of (11) consist of $Dg(y)$ and hence are known. For the last row the components of the gradient G of $e_{m+1}^T T(y)$ are computed from

$$G_j \equiv d\{e_{m+1}^T T(y)\}/dy_j = - Dg(y)^+ \, (d\{Dg(y)\}/dy_j) \, T(y)$$

by use of finite differencing

$$G_j \approx -h^{-1} \, e^T_{m+1} \, Dg(y)^+ Dg(y_1,..,y_{j-1},y_j+h,y_{j+1},..,y_m)T(y) \; .$$

The vector $v \in \mathbb{R}^{m+1}$ from

$$\begin{pmatrix} Dg(y_k) \\ G^T(y_k) \end{pmatrix} v = \begin{pmatrix} g(y_k) \\ e_{m+1}^T T(y_k) \end{pmatrix}$$

which is needed for the Newton step $y_{k+1} \equiv y_k - v$ is computed via

$$w \equiv Dg(y_k)^+ g(y_k) \; ; \; v \equiv w + \frac{(e_{m+1}^T T(y_k) - G(y_k)^T w)}{G(y_k)^T T(y_k)} * T(y_k).$$

Notice that the tangent T and the generalized inverse Dg^+ have to be computed only once per step.

10. Computation of simple bifurcation points

A simple nondegenerate bifurcation point y_b of f_p^{-1} is characterized by the following three conditions (cf.[1,5]), where we use again the abbreviation $g(y) \equiv f_p(c,\lambda;q')$.

i) $g(y_b) = 0$

ii) rank $Dg(y_b) = m-1$

iii) the quadratic form $S \, D^2 g(y_b)_{|N \times N}$ is nondegenerate and indefinite.

Here S denotes the orthogonal projector of $\mathbb{R}^m$ onto the Corange of $Dg(y_b)$ and N= Ker $Dg(y_b)$.

Let $h_1,\ldots,h_m \in \mathbb{R}^m$ and $k_1,\ldots,k_{m+1} \in \mathbb{R}^{m+1}$ be sets of orthogonal vectors such that

$$
(13) \quad
\begin{array}{lcl}
\text{span } \{ h_1,\ldots,h_{m-1} \} & \approx & \text{Range } Dg(y_b) , \\
\text{span } \{ h_m \} & \approx & \text{Corange } Dg(y_b) , \\
\text{span } \{ k_1,\ldots,k_{m-1} \} & \approx & \text{Coker } Dg(y_b) , \\
\text{span } \{ k_m,k_{m+1} \} & \approx & \text{Ker } Dg(y_b) .
\end{array}
$$

Such approximations may be calculated by means of QR- decomposition or SVD of the (small!) matrix $Dg(y)$ for y close to y_b.

We put $H_1 \equiv (h_1,\ldots,h_{m-1})$, $H_2 \equiv (h_m)$,
$K_1 \equiv (k_1,\ldots,k_{m-1})$ and $K_2 \equiv (k_m, k_{m+1})$.

If the approximation (13) is sufficiently good then one can show that $y_b= K_1\alpha_b + K_2\beta_b$, where $(\alpha_b,\beta_b) \in \mathbb{R}^{m-1}x\mathbb{R}^2$ is an isolated solution of the system

$$(14a) \quad H_1^T\, g(K_1\alpha + K_2\beta) = 0 \quad \in \mathbb{R}^{m-1}$$

$$(14b) \quad (r_{22}(\alpha,\beta)-r_{21}(\alpha,\beta)r_{11}(\alpha,\beta)^{-1}r_{12}(\alpha,\beta))^T= 0 \in \mathbb{R}^2 ,$$

$$\text{with} \quad r_{ij}(\alpha,\beta) \equiv H_i^T\, Dg(K_1\alpha + K_2\beta)\, K_j \ , \ i,j= 1,2 .$$

It is this system that we solve.

Remarks:

a) What is interesting about system (14) is the fact that it can be solved by one step block Gauss-Seidel-Newton iteration doing a Newton step with (14a) to improve and, after having updated α, a Newton step with (14b) to improve β within each iteration sweep. If second derivatives have to be approximated by finite differencing (which is usually assumed, c.f. [5].e.g.), only two extra evaluations of $Dg(y)$

(differencing in β-direction) are required per step.

b) The convergence of the iteration is linear. However, it has been proved in [15] that the rate is proportional to $\| H_1^T Dg(y_b)K_2 \|$ which is equal to zero if the approximations (13) are exact. Hence updating of (13) during the iteration will lead to superlinear convergence.

c) The iteration may be understood as a numerical version of the Lyapunov-Schmidt reduction at simple bifurcation point as described in [1] . If the system has a slightly perturbed bifurcation point this point is still a solution of (14). Notice that (14) does not require $y_b \in f_p^{-1}$ since $H_2^T g(y_b)$ (the unfolding "parameter" of our iteration) may be different from zero.

11. Emanating branches at simple bifurcation points

Having found a bifurcation point we compute the condensation data Z and q_λ and the vectors k_m, k_{m+1}, h_m to high precision. Then tangents to the emanating branches are given by

$$\begin{pmatrix} Z & q_\lambda \\ 0 & 1 \end{pmatrix} (k_m, k_{m+1}) \beta \quad \in \quad \mathbb{R}^{n+1}$$

where the unit vector $\beta \in \mathbb{R}^2$ solves the quadratic equation

$$(15) \quad \beta^T \begin{pmatrix} h_m^T D^2 g(y_b)(k_m, k_m) & h_m^T D^2 g(y_b)(k_m, k_{m+1}) \\ h_m^T D^2 g(y_b)(k_m+1, k_m) & h_m^T D^2 g(y_b)(k_m+1, k_{m+1}) \end{pmatrix} \beta = 0$$

(cf. [13,6,5] e.g.) .

12. Computing double bifurcation points

For double bifurcation points (where rank $Dg = m-2$) we proceed similarly to the case of simple bifurcation points. Again we use system (14) with the difference that instead of

(13) now

$$\begin{aligned}
\text{span } H_1 &= \text{span } \{ h_1, \dots, h_{m-2} \} &\approx \quad & \text{Range } Dg(y_b) \quad , \\
\text{span } H_2 &= \text{span } \{ h_{m-1}, h_m \} &\approx \quad & \text{Corange } Dg(y_b) \quad , \\
\text{span } K_1 &= \text{span } \{ k_1, \dots, k_{m-2} \} &\approx \quad & \text{Coker } Dg(y_b) \quad , \\
\text{span } K_2 &= \text{span } \{ k_{m-1}, k_m, k_{m+1} \} &\approx \quad & \text{Ker } Dg(y_b) \quad .
\end{aligned}$$

Since (14b) incorporates now a total of six components, the system is overdetermined. Instead of a Newton step for this part of the iteration we use a Gauss-Newton step to reduce the sum of squares of the elements of (14b) to zero. Convergence properties of this iteration will be dealt with elsewhere.

13. Branch switching at double bifurcation points

The procedure is again very similar to the simple bifurcation point case. Instead of (15) we have to solve the two quadratic forms

(16.1) $\quad \beta^T C_l \beta = 0 \ , \ l=1,2$

for a unit vector $\beta \in \mathbb{R}^3$, where the elements c^l_{ij} of $C_l \in \mathbb{R}^{3\times3}$ are given by

$$c^l_{ij} = h^T_{m-2+l} \ D^2g(y_b) \ (k_{m-2+i}, k_{m-2+j}) \ , \ i,j = 1,2,3 \ .$$

It is known that the 'characteristic rays'

$$\begin{pmatrix} Z & q_\lambda \\ 0 & 1 \end{pmatrix} (k_{m-1}, k_m, k_{m+1}) \ \beta \qquad \text{for such } \beta\text{-vectors}$$

determine directions of emanating branches if certain nondegeneracy conditions (c.f. [16,19,3,4]) are fulfilled. These conditions demand in principle that all solutions of (16.1 and 16.2) on the unit sphere be simple and isolated.
For this nondegenerate situation case we use predictor-corrector pathfollowing to trace the one dimensional solution set of (16.1) on the unit sphere. During this trace we

monitor the sign of the quadratic form from (16.2). In this way very reliable starting values for a Newton iteration are produced.

14. Numerical Results

We have applied our method to several equations. We report on two of them here. The first equation has been suggested by Mittelmann [18]. It is

$$- \Delta u + 10 u = \lambda e^{u} \quad \text{on} \quad [0,1] \times [0,1]$$

with Neumann boundary conditions. This example has the advantage that it has a ´constant branch´ on which the solution is a constant function. Thus there is no discretization error on that branch. Furthermore this branch is given by $\lambda = 10 u e^{-u}$. The only turning point on that branch is at $u = 1$ and $\lambda = 10/e$.

We have used a Finite Element discretization on a 17×17 grid with linear elements which reduces to the usual 5-point difference operator in the interior of the domain.

Since the eigenvalues of this discrete operator are known to be

$$\rho(m,n) = 2 \left((1-\cos(n\pi h))/h^2 + (1-\cos(m\pi h))/h^2 \right)$$

the bifurcation points on that branch are given by

$$u = 1 + \rho(m,n)/10 \quad \text{and}$$
$$\lambda = 10 u e^{-u} .$$

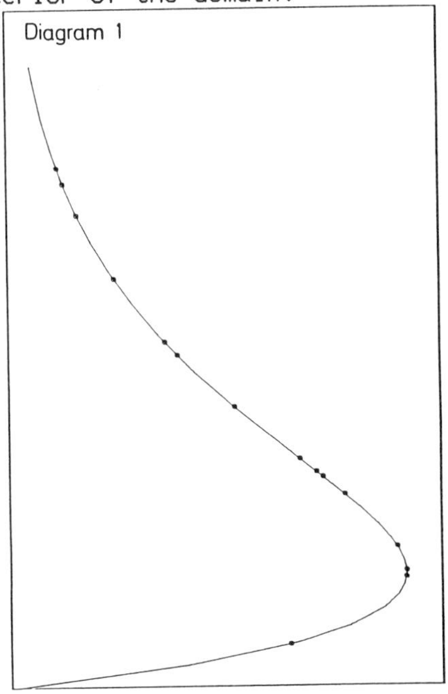

Diagram 1 shows this ´constant´ arc where those points are marked where the arc-following of the small system has been interrupted because a singularity has been detected or because the arc-length-trust-region has been hit. In that case the algorithm goes back to the full arc - see sections 5 and 6. The turning point has been computed by few iterations of the routine described in section 9 and is correct up to 8 decimal places. The bifurcation points labeled 'a' and 'b' in diagram 2 correspond to (m=n=1) and (m=0,n=1 or m=1,n=0), respectively. Both bifurcation points have been computed by the scheme given in sections 10 and 12 up to 5 decimal places (and could be computed with even higher precision if one likes). The bifurcation point 'a' is a simple one and posed no problems. The bifurcation point 'b' was detected to be double which conforms to the above analysis. It turned out that the usual nondegeneracy conditions for double bifurcation points are not satisfied here. The two quadratic forms C_1 and C_2 given in section 12 are of the form

$$\begin{matrix} 0 & 0 & a \\ 0 & 0 & 0 \\ a & 0 & 0 \end{matrix} \qquad \text{and} \qquad \begin{matrix} 0 & 0 & 0 \\ 0 & 0 & b \\ 0 & b & 0 \end{matrix}$$

and thus the escape directions cannot be computed using second derivative information only. We have found the 4 arcs in addition to the 'constant' arc by using a bilinear initial guess of similar symmetry type as the eigenfunctions of the Laplacian at that point and used our very robust 'back to the arc routine' mentioned in section 5. We then verified (numerically) that these arcs run into this bifurcation point, indeed. But we cannot claim that there are no more branches.

All other bifurcation points 'c','d','e' and 'f' have been detected by the arc-following routine - refined by the bifurcation routines and escaped from by the routines given in sections 8 to 12. Several turning points were detected but not marked in order not to overload the diagram 2. We now give a short description of the different branches.

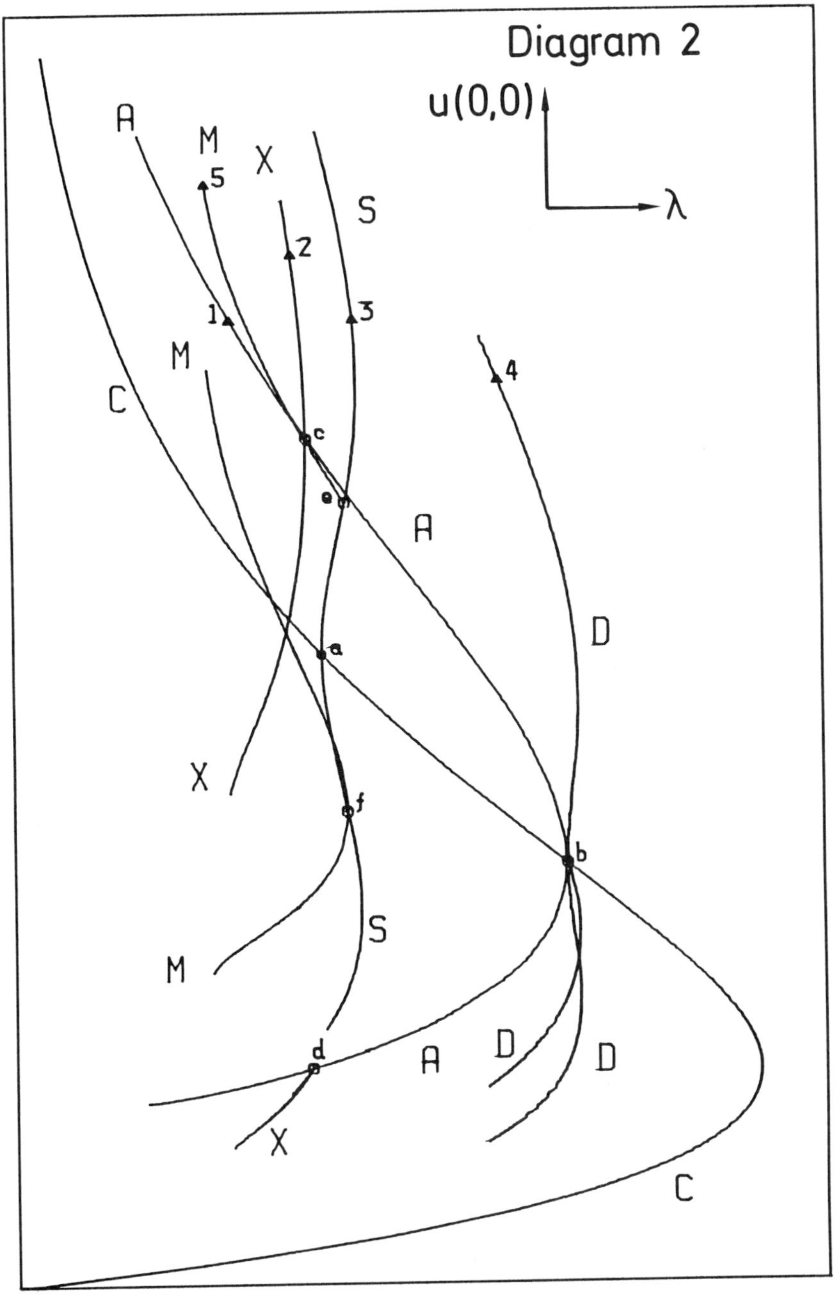

Diagram 2
u(0,0)
λ
A
M
X
S
C
M
X
A
D
S
M
A
D
D
X
C
1
2
3
4
5
a
b
c
d
e
f

Legend to diagram 2

The horizontal axis corresponds to λ with a maximum value of 4.0, the vertical axis to the value of the solution at the origin (0,0) with a maximum value of 6.0 . Above the constant branch u(0,0) coincides with the maximum-norm of u.

We have labeled the branches of different symmetry type by 'A','C','D','M','S' and 'X'. Some intersections are caused by the projection; the true bifurcation points are labeled by 'a' to 'f'. At the bifurcation point 'c' only the branches 'A' and 'X' meet but not branch 'M'. On each branch we show a typical solution - labeled '1' to '5' . These are shown in separate figures. The maximum height of these solutions is about 5 in all cases - the z-axis has been plotted with a different scale.

Branch

- C is the above mentioned 'constant' branch - we have omitted a plot for obvious reasons
- D has a symmetry accross one diagonal of the square. There are two such branches - see solution 4 and that solution rotated by 90°.
- A is symmetric with respect to a parallel line to an axis through the center of the square. Again there are two such branches, a solution on one of these is labeled solution '1'.
- S is doubly symmetric with respect to both diagonals of the square - see solution '3'. The front corner and the back corner have the same z-coordinate. This branch emenates from branch 'C' at 'a'.
- M has 'monkey-saddle-type' solutions as is shown by solution '5' . These branches emenate from branch 'S' at the points 'e' or 'f' .
- X has solutions of quite surprising type shown as solution '2'. These branches emenate from points 'c' or 'd'.

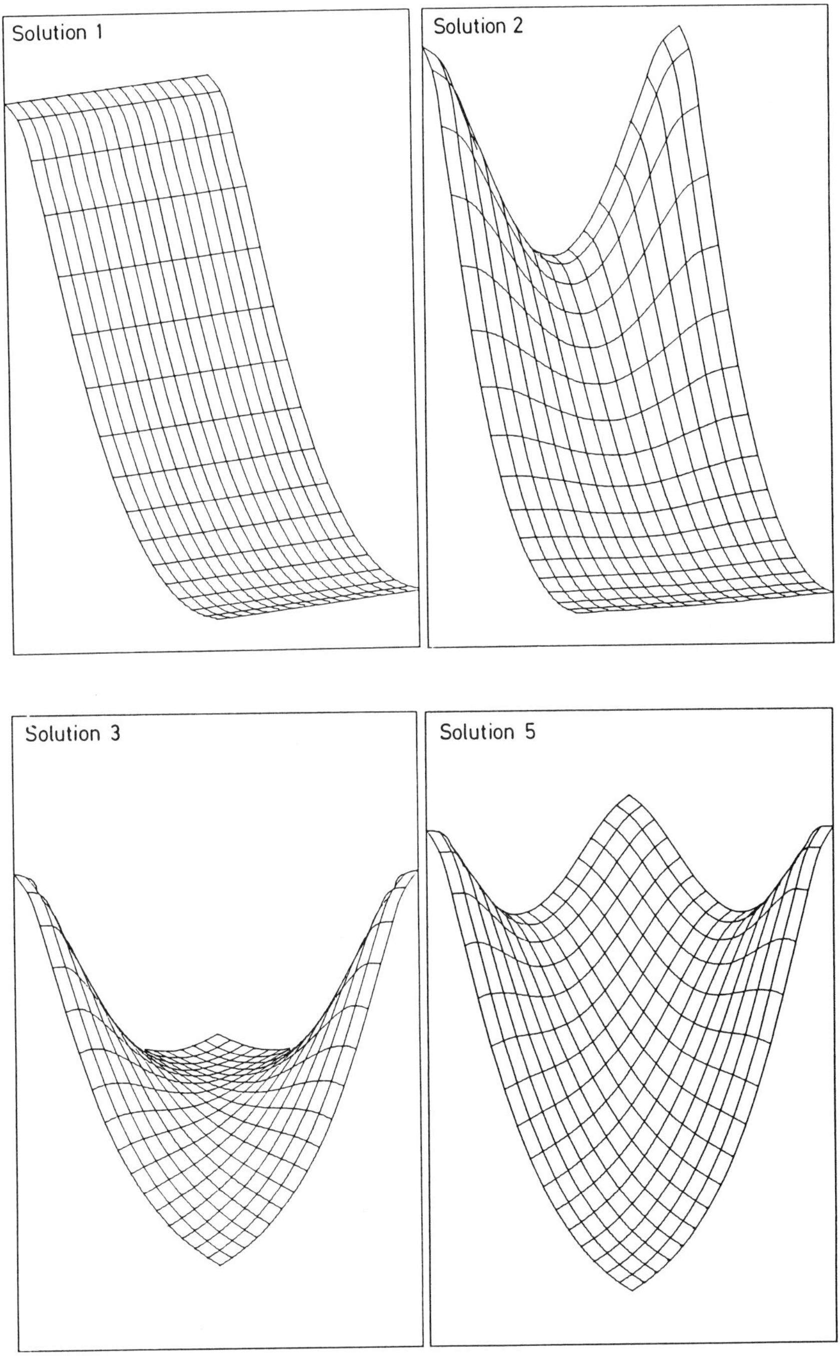
Solution 1
Solution 2
Solution 3
Solution 5

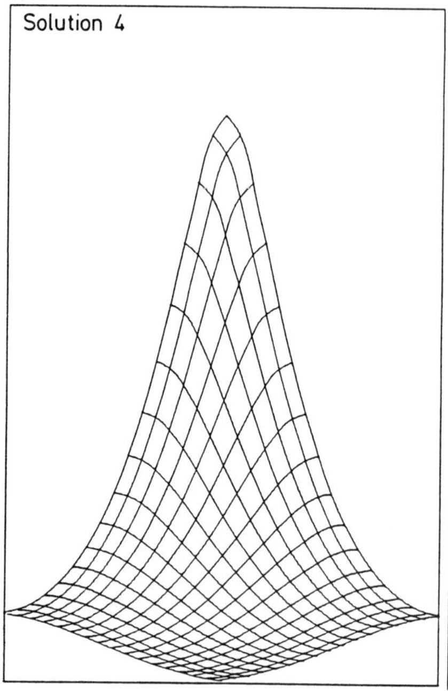

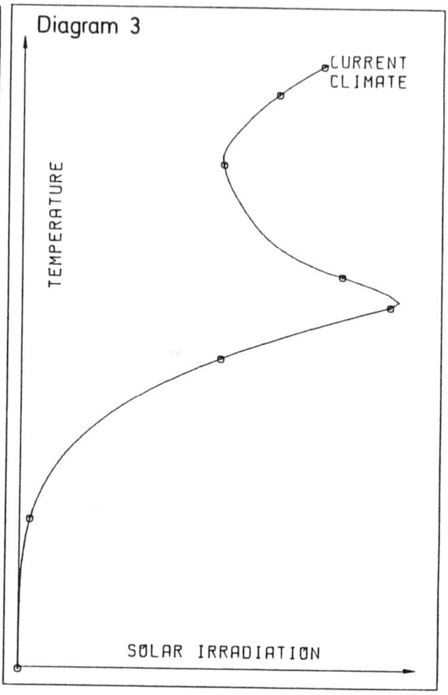

In the second numerical example we compute a 2-D climate model of Budyko-Sellers type. The task is to compute the steady state solution of the averaged temperature distribution on the globe described by a diffusive energy balance model. This is given by the following equation

$$-\mathrm{div}\ (\ k(x)\ \nabla u(x)\ ) = \lambda\ Q(x)\ [1-\alpha(x,u)] - \sigma(u)\ u^4$$

where

$k(x)$	is the heat conductivity
$\lambda Q(x)$	is the incoming solar radiation (λ=1 today)
$\alpha(x,u)$	is a simple albedo-model given by
$=$	$\{\ b-c_1\ \min(\ u-0.0065h(x),\ u_M\ )\ \}_c$

where $h(x)$ is the height over sea-level

$$\text{and}\quad \{y\}_c \equiv \begin{cases} \eta & \text{for } y < 0.25 \\ 0.85 & \text{for } y > 0.85 \\ y & \text{otherwise} \end{cases}$$

with $\eta = 0.25$ over the continents and $\eta = 0.1$ over the sea.

$\sigma(u)\ u^4$ is the Boltzmann - radiation where $\sigma(u)$ models a 'greenhouse'-effect:

$$\sigma(u) \equiv 1 - 0.5 \tanh(c\ u^6)$$

The very simple albedo model is just prelimary and will be replaced by more eloborated ones.

The resulting bifurcation diagram is shown by diagram 3 where the current climate is marked on the uppermost branch.

Acknowlegdement

The linear solver on the surface of a sphere for the climate model has been worked out by Prof. Dr. H. Yserentant - see his contribution to this volume.

We would like to thank Dr. Hossfeld of the nuclear research center at Juelich for granting us computing time on the CRAY XMP for the climate model.

15. References

[1] Brezzi, F., Rappaz, T., Raviart, P.: Finite dimensional approximation of nonlinear problems. Part III. Simple bifurcation points, Numer. Math. 38, 1-30, 1981.

[2] Beyn, W.-J.: A note on the Lyapunov-Schmidt reduction. Manuscript, Univ. of Konstanz, 1982.

[3] Beyn, W.-J.: Defining equations for singular solutions and numerical applications. pp. 42-56 in [13].

[4] Descloux, J., Rappaz, J.: Approximation of solution branches of nonlinear equations. R.A.I.R.O. Anal. Num. 16, 319-349, 1982.

[5] Deuflhard, P., Fiedler, B., Kunkel, P.: Efficient numerical pathfollowing beyond critical points. Preprint no 278, Univ. of Heidelberg, SFB 123, 1984.

[6] Iooss, G., Joseph, D.D.: Elementary stability and bifurcation theory, Springer, New York, 1980.

[7] Jarausch, H., Mackens, W.: A fast globally convergent scheme to compute stationary points of elliptic variational problems. Report 15 of the Institut fuer Geometrie und Praktische Mathematik, R.W.T.H., Aachen, 1982.

[8] Jarausch, H., Mackens, W.: Computing solution branches by use of a Condensed Newton - Supported Picard iteration scheme. ZAMM 64, T282-T284, 1984.

[9] Jarausch, H., Mackens, W.: Numerical treatment of bifurcation branches by adaptive condensation. pp. 296-309 in [13].

[10] Jarausch, H., Mackens, W.: Solving large nonlinear equations by an adaptive condensation process. Report 29 of the Institut fuer Geometrie und Praktische Mathematik, R.W.T.H., Aachen, to appear.

[11] Jepson, A.D., Spence, A.: Singular points and their computation. pp. 195-209 in [13].

[12] Kahan, W., Parlett, B.N., Jiang, E.: Residual bounds on approximate eingensystems of nonnormal matrices. SIAM J. Numer. Anal. 19, 407-484, 1982.

[13] Keller, H.B.: Numerical solution of bifurcation and nonlinear eigenvalue problems. pp. 359-384 in P.H. Rabinowitz[ed.]: Applications of bifurcation theory. Academic Press, New York, 1977.

[14] Kuepper, T., Mittelmann, H.D., Weber, H.[eds.]: Numerical methods for bifurcation problems, ISNM 70, Birkhaeuser Verlag, Basel 1984.

[15] Mackens, W.: A note on an adaptive Lyapunov-Schmidt reduction at secondary bifurcation points. Preprint of the Institut fuer Geometrie und Praktische Mathematik, R.W.T.H., Aachen, 1985.

[16] McLeod, J.B., Sattinger, D.: Loss of stability and bifurcation at a double eigenvalue. J. Funct. Anal. 14, 62-84, 1973.

[17] Melhem, R.G., Rheinboldt, W.C.: Comparison of methods for determining turning points of nonlinear equations. Computing 29, 201-226, 1982.

[18] Mittelmann, H.D.: Multi-level continuation techniques for nonlinear boundary value problems with parameter dependence. Report of Arizona State University, 1985.

[19] Rappaz, J., Raugel, G.: Finite dimensional approximation of bifurcation problems at a multiple eigenvalue. Report no 71, Centre de Mathematiques appliquees, Ecole Polytechnique, Palaiseau, 1981.

EXTINCTION LIMITS FOR PREMIXED LAMINAR FLAMES IN A STAGNATION POINT FLOW

V. Giovangigli[1]
M. D. Smooke[2]

1. Introduction.

Conclusions derived from the solution of premixed laminar flames in a stagnation point flow are important in the determination of chemically controlled extinction limits, in the ability to characterize the combustion processes occurring in turbulent flames and in the study of pollutant formation. Experimentally these flames can be produced by a single reactant stream impinging on an adiabatic wall or by two counterflowing reactant streams emerging from two coaxial jets. In the neighborhood of the stagnation point produced by these flows, a chemically reacting boundary layer is established. Along the stagnation point streamline the governing equations can be reduced to a system of coupled nonlinear two-point boundary value problems. In the single reactant stream configuration only one reaction zone is produced. If the exit velocity and the equivalence ratio of the fuel-air mixture of each jet are equal, then in the two reactant stream problem a double flame is produced with a plane of symmetry through the stagnation point and parallel to the two jets (see Figure 1).

A number of experimental and analytic studies of premixed flames in a stagnation point flow have appeared recently in the literature [1-9]. In several of these investigations a single reactant jet was utilized [1,2,4-7] while in others two counterflowing reactant streams were employed [1,3,8,9]. In most of the cases it was found that the Lewis number played an important role in the behavior of these flames near extinction. Computationally, for a given equivalence ratio, the input flow velocity can be varied and one can obtain a relationship between the strain rate (the velocity gradient) and the peak temperature. In general, as the input flow velocity is increased and the flame nears extinction, the peak temperature decreases. It is in the neighborhood of the extinction point, however, that computation of the solution often becomes difficult. In particular, at the extinction point one can show that the Jacobian of the system is singular. If solutions are

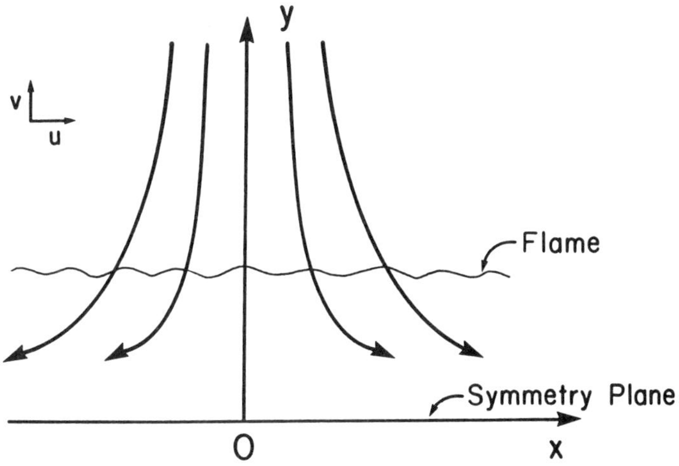

Figure 1. Schematic of the stagnation point flow configuration.

desired in this region, the computational procedure must be modified to account for the singular behavior of the system.

Procedures enabling the calculation of bifurcation and limit points for systems of nonlinear equations have been discussed, for example, by Keller [10], Jepson and Spence [11], Chan [12], Seydel [13] and Heinemann, Overholser and Reddien [14-15]. In particular, in the work of Heineman et al. a version of Keller's arclength continuation method was used to calculate the multiple steady-states of a model one-step, nonadiabatic, premixed laminar flame [14] and a premixed, nonadiabatic, hydrogen-air system [15]. In the hydrogen-air problem the cold boundary temperature was taken as the bifurcation parameter. In the premixed flames we consider, the strain rate (or more precisely the inverse of the strain rate) is the natural bifurcation parameter.

The goal in this paper is to generalize the ideas used in [14-15] so that extinction limits of premixed laminar flames in a stagnation point flow can be calculated. We focus our attention on doubly premixed laminar flames produced by two counterflowing coaxial jets. By applying appropriate boundary conditions at the plane of symmetry, the model we consider is, in principle, equivalent to that of a single reactant stream impinging on an adiabatic wall with slip. We investigate the extinction properties of premixed methane-air flames. However, to simplify the computations, we apply a global one-step kinetics model in which the Lewis numbers are assumed equal to one. We realize, of course, that such a one-step procedure cannot predict adequately the effects of nonunit Lewis numbers of the deficient reactants on extinction. Nevertheless, the approach we take illustrates the effectiveness of the numerical bifurcation procedure in calculating the multiple steady-states of these particular flames.

2. Problem Formulation.

Our model for counterflowing premixed flames assumes the flow to be laminar, stagnation point flow. Hence, the governing boundary layer equations for mass, momentum, chemical species and energy can be written in the form

$$\frac{\partial(\rho u x^\alpha)}{\partial x} + \frac{\partial(\rho v x^\alpha)}{\partial y} = 0, \tag{2.1}$$

$$\rho u \frac{\partial u}{\partial x} + \rho v \frac{\partial u}{\partial y} + \frac{\partial p}{\partial x} = \frac{\partial}{\partial y}\left(\mu \frac{\partial u}{\partial y}\right), \tag{2.2}$$

$$\rho u \frac{\partial Y_k}{\partial x} + \rho v \frac{\partial Y_k}{\partial y} + \frac{\partial}{\partial y}(\rho Y_k V_{ky}) - \dot{w}_k W_k = 0, \quad k = 1, 2, \ldots, K, \tag{2.3}$$

$$\rho u c_p \frac{\partial T}{\partial x} + \rho v c_p \frac{\partial T}{\partial y} - \frac{\partial}{\partial y}\left(\lambda \frac{\partial T}{\partial y}\right) + \sum_{k=1}^{K} \rho Y_k V_{ky} c_{pk} \frac{\partial T}{\partial y} + \sum_{k=1}^{K} \dot{w}_k W_k h_k = 0. \tag{2.4}$$

where α represents a geometric factor ($\alpha = 0$ for cartesian coordinates and $\alpha = 1$ for cylindrical coordinates). For the remainder of this paper we set $\alpha = 0$. The system is closed with the ideal gas law,

$$\rho = \frac{p\overline{W}}{RT}. \tag{2.5}$$

In these equations x and y denote independent spatial coordinates; T, the temperature; Y_k, the mass fraction of the kth species; p, the pressure; u and v the tangential and the transverse components of the velocity, respectively; ρ, the mass density; W_k, the molecular weight of the kth species; $\overline{W}$, the mean molecular weight of the mixture; R, the universal gas constant; λ, the thermal conductivity of the mixture; c_p, the constant pressure heat capacity of the mixture; c_{p_k}, the constant pressure heat capacity of the kth species; $\dot{w}_k$, the molar rate of production of the kth species per unit volume; h_k, the specific enthalpy of the kth species; μ the viscosity of the mixture and V_{ky} is the diffusion velocity of the kth species in the y direction. The free stream (tangential) velocity at the edge of the boundary layer is given by $u_e = ax$ where a is the strain rate.

Upon introducing the notation

$$f' = \frac{u}{u_e}, \tag{2.6}$$

$$V = \rho v, \tag{2.7}$$

the boundary layer equations can be transformed into a system of ordinary differential equations valid along the stagnation-point streamline $x = 0$. For a system in rectangular coordinates, we have

$$\frac{dV}{dy} + a\rho f' = 0, \tag{2.8}$$

$$\frac{d}{dy}\left(\mu\frac{df'}{dy}\right) - V\frac{df'}{dy} + a(\rho_e - \rho(f')^2) = 0, \tag{2.9}$$

$$-\frac{d}{dy}(\rho Y_k V_k) - V\frac{dY_k}{dy} + \dot{w}_k W_k = 0, \quad k = 1, 2, \ldots, K, \tag{2.10}$$

$$\frac{d}{dy}\left(\lambda\frac{dT}{dy}\right) - c_p V\frac{dT}{dy} - \sum_{k=1}^{K} \rho Y_k V_k c_{pk}\frac{dT}{dy} - \sum_{k=1}^{K} \dot{w}_k W_k h_k = 0. \tag{2.11}$$

At the plane of symmetry ($y = 0$) the boundary conditions are given by

$$V = 0, \tag{2.12}$$

$$\frac{df'}{dy} = 0, \tag{2.13}$$

$$\frac{dY_k}{dy} = 0, \quad k = 1, 2, \ldots, K, \tag{2.14}$$

$$\frac{dT}{dy} = 0, \tag{2.15}$$

and as $y \to \infty$ by

$$f' = 1, \tag{2.16}$$

$$Y_k = Y_{k_e}, \quad k = 1, 2, \ldots, K, \tag{2.17}$$

$$T = T_e. \tag{2.18}$$

The mass fractions $Y_{k_e}, k = 1, 2, \ldots, K$ and the temperature T_e at the edge of the boundary layer are specified quantities.

3. One-Step Model.

Most complex combustion systems ordinarily involve large numbers of chemical species. These species are related through a detailed kinetics mechanism involving many elementary chemical reactions. Solution of the governing equations in such systems reduces to the solution of an ordinary or partial differential equation for each species mass fraction. In some applications the determination of the appropriate chemical reactions and their respective rate constants can be a difficult task. It can be made simpler, however, by postulating a single global reaction for the system. This can also be useful when the size of the system to be solved (for the computer being used) results in a computationally infeasible problem. We realize, of course, that a global reaction mechanism does not provide detailed information on the system's minor species–information that is often needed in assessing the detailed structure of a reacting system (see e.g., [16] and [17]).

Determination of overall, global, reaction rates for flames has been investigated, for example, by Levy and Weinberg [18,19], Westbrook and Dryer [20] and Coffee, Kotlar and Miller [21]. In the paper by Coffee et al., the equations governing freely propagating, premixed, laminar flames with detailed kinetics and complex transport were solved for the temperature and the species mass fractions. From the calculated temperature, a heat release profile was obtained as a function of the independent spatial coordinate. The reaction rate parameters could then be obtained by a two parameter least squares fit to this data. In this paper we utilize these calculated reaction rate parameters in the numerical solution of one-step, premixed, laminar methane-air flames in a counterflow geometry.

We assume the fuel and the oxidizer obey a single overall irreversible reaction of the type

$$\text{Fuel } (F) + \text{Oxidizer } (O) \rightarrow \text{Products } (P), \tag{3.1}$$

in the presence of an inert gas (N). We have

$$\nu_F F + \nu_O O \rightarrow \nu_P P, \tag{3.2}$$

where ν_F, ν_O and ν_P are the stoichiometric coefficients of the fuel, the oxidizer and the product, respectively. In addition, in the one-step model we consider, we neglect thermal diffusion and we assume that the ordinary mass diffusion velocities can be written in terms of Fick's law, i.e.,

$$V_k = -\frac{D_k}{Y_k}\frac{dY_k}{dy}, \quad ,k = 1, 2, \ldots, K, \tag{3.3}$$

where D_k is the diffusion coefficient of the kth species into the mixture. We also take the quantities $c_p = c_{p_k}, \rho\lambda, \rho^2 D_k$ and $\rho\mu$ to be constant.

If, for purposes of the discussion that follows, we introduce the Lewis number of each species

$$\text{Le}_F = \frac{\lambda}{\rho D_F c_p}, \quad \text{Le}_O = \frac{\lambda}{\rho D_O c_p}, \tag{3.4a}$$

$$\text{Le}_P = \frac{\lambda}{\rho D_P c_p}, \quad \text{Le}_N = \frac{\lambda}{\rho D_N c_p}, \tag{3.4b}$$

and the Prandtl number

$$\mathrm{Pr} = \frac{\mu c_p}{\lambda}, \tag{3.5}$$

then the governing equations in (2.8-2.11) become

$$\frac{dV}{dy} + a\rho f' = 0, \tag{3.6}$$

$$\mathrm{Pr}\frac{d}{dy}\left(\frac{\lambda}{c_p}\frac{df'}{dy}\right) - V\frac{df'}{dy} + a(\rho_e - \rho(f')^2) = 0, \tag{3.7}$$

$$\frac{1}{\mathrm{Le}_F}\frac{d}{dy}\left(\frac{\lambda}{c_p}\frac{dY_F}{dy}\right) - V\frac{dY_F}{dy} - W_F\nu_F\dot{w} = 0, \tag{3.8}$$

$$\frac{1}{\mathrm{Le}_O}\frac{d}{dy}\left(\frac{\lambda}{c_p}\frac{dY_O}{dy}\right) - V\frac{dY_O}{dy} - W_O\nu_O\dot{w} = 0, \tag{3.9}$$

$$\frac{1}{\mathrm{Le}_P}\frac{d}{dy}\left(\frac{\lambda}{c_p}\frac{dY_P}{dy}\right) - V\frac{dY_P}{dy} + W_P\nu_P\dot{w} = 0, \tag{3.10}$$

$$\frac{1}{\mathrm{Le}_N}\frac{d}{dy}\left(\frac{\lambda}{c_p}\frac{dY_N}{dy}\right) - V\frac{dY_N}{dy} = 0, \tag{3.11}$$

$$\frac{d}{dy}\left(\frac{\lambda}{c_p}\frac{dT}{dy}\right) - V\frac{dT}{dy} + \frac{(W_F\nu_F h_F + W_O\nu_O h_O - W_P\nu_P h_P)}{c_p}\dot{w} = 0, \tag{3.12}$$

where

$$\dot{w} = -\frac{\dot{w}_F}{\nu_F} = -\frac{\dot{w}_O}{\nu_O} = \frac{\dot{w}_P}{\nu_P}, \tag{3.13}$$

is the rate of progress of the reaction and where we have made use of the fact that $\sum_{k=1}^{K} Y_k V_k = 0$. From equation (3.11) we see that the inert gas profile $Y_N = Y_{N_e} =$ constant.

If we now assume that the Lewis numbers and the Prandtl number are equal to one and if we introduce the heat release per unit mass of the fuel Q where

$$Q = h_F + \frac{W_O\nu_O}{W_F\nu_F}h_O - \frac{W_P\nu_P}{W_F\nu_F}h_P, \tag{3.14}$$

we can derive the following Shvab-Zeldovich relations

$$Y_F = Y_{F_e} - \frac{c_p}{Q}(T - T_e), \tag{3.15}$$

$$Y_O = Y_{O_e} - \frac{c_p}{Q}\frac{W_O\nu_O}{W_F\nu_F}(T - T_e), \tag{3.16}$$

$$Y_P = Y_{P_e} + \frac{c_p}{Q}\frac{W_P\nu_P}{W_F\nu_F}(T - T_e). \tag{3.17}$$

With the rate of progress given by an Arrhenius type relation, the heat release per unit volume can be written in the form

$$q = QW_F\nu_F\dot{w} = Q(\rho Y_F)^{\nu_F}(\rho Y_O)^{\nu_O}A\exp(-E/RT), \tag{3.18}$$

and the one-step model reduces to the solution of

$$\frac{dV}{dy} + a\rho f' = 0, \tag{3.19}$$

$$\frac{d}{dy}\left(\frac{\lambda}{c_p}\frac{df'}{dy}\right) - V\frac{df'}{dy} + a(\rho_e - \rho(f')^2) = 0, \tag{3.20}$$

$$\frac{d}{dy}\left(\frac{\lambda}{c_p}\frac{dT}{dy}\right) - V\frac{dT}{dy} + \frac{q}{c_p} = 0, \tag{3.21}$$

with the boundary conditions at $y = 0$ given by

$$V = 0, \tag{3.22}$$

$$\frac{df'}{dy} = 0, \tag{3.23}$$

$$\frac{dT}{dy} = 0, \tag{3.24}$$

and as $y \to \infty$ by

$$f' = 1, \tag{3.25}$$

$$T = T_e. \tag{3.26}$$

We observe that, as a result of (3.15-3.18) and the ideal gas law, the expression for q in (3.21) is strictly a function of the temperature and, in addition to the stoichiometry of the global reaction, it depends upon the heat release per unit mass of the fuel, the edge temperature and mass fractions and the two Arrhenius parameters. We also point out that, as the strain rate approaches zero, the flame will shift further and further from the plane of symmetry so that (up to an arbitrary translation) it eventually approaches a freely propagating premixed laminar flame. As a result, in the calculations reported in Section 6, we use the values of the heat release Q and the two Arrhenius constants A and E that were obtained by Coffee et al. [21] from a set of detailed chemistry calculations of freely propagating premixed laminar flames. In their investigation the adiabatic flame temperature of a detailed chemistry model was used to determine the value of the heat release per unit mass. The parameters A and E were then determined by a least squares fit such that

$$A \exp(-E/RT^*) \approx q^*/Q(\rho(T^*)Y_F(T^*))^{\nu_F}(\rho(T^*)Y_O(T^*))^{\nu_O}, \tag{3.27}$$

where (T^*, q^*) are the temperatures and heat releases from the detailed chemistry model.

4. Method of Solution.

Solution of the equations in (3.19-3.26) proceeds by an adaptive nonlinear boundary value method. The solution procedure has been discussed in detail elsewhere [22] and we outline only the essential features here. Our goal is to obtain a discrete solution of the governing equations on the mesh $\mathcal{M}$

$$\mathcal{M} = \{0 = y_0 < y_1 < \ldots < y_m = L\}, \tag{4.1}$$

where $h_j = y_j - y_{j-1}, j = 1, 2, \ldots, m,$ and where the value of L is taken large enough so that the zero flux boundary conditions are satisfied to an acceptable level of accuracy. We approximate spatial derivatives with finite difference expressions. Specifically, we write

$$\frac{d}{dy}\left(a(y)\frac{dg}{dy}\right)_{y_j} \approx \left(\frac{2}{y_{j+1} - y_{j-1}}\right)\left(a_{j+1/2}\partial g_{j+1} - a_{j-1/2}\partial g_j\right), \tag{4.2}$$

where we define $g_j = g(y_j), j = 0, 1, \ldots, m$ and

$$g_{j+1/2} = \frac{(g_{j+1} + g_j)}{2}, \quad j = 0, 1, \ldots, m-1, \tag{4.3}$$

$$\partial g_{j+1} = \frac{(g_{j+1} - g_j)}{h_{j+1}}, \quad j = 0, 1, \ldots, m-1. \tag{4.4}$$

First derivatives are differenced using upwind difference expressions.

Newton's Method

With the continuous differential operators replaced by expressions similar to those in (4.2-4.4), we convert the problem of finding an analytic solution of the governing equations to one of finding an approximation to this solution at each point of the mesh $\mathcal{M}$. We seek the solution U_h^* of the nonlinear system of difference equations

$$F(U_h^*) = 0. \tag{4.5}$$

For an initial solution estimate U^0 that is sufficiently "close" to U_h^*, the system of equations in (4.5) can be solved by Newton's method. We write

$$J(U^k)\left(U^{k+1} - U^k\right) = -\lambda_k F(U^k), \quad k = 0, 1, \ldots, \tag{4.6}$$

where U^k denotes the kth solution iterate, λ_k the kth damping parameter ($0 < \lambda \leq 1$) [23] and $J(U^k) = \partial F(U^k)/\partial U$ the Jacobian matrix. A system of linear block tridiagonal equations must be solved at each iteration for corrections to the previous solution vector. For many problems the cost of forming (either analytically or numerically) and factoring the Jacobian matrix can be a significant part of the cost of the total calculation. In such problems we apply a modified Newton method in which the Jacobian is re-evaluated only periodically [24].

Adaptive Gridding

Solution of boundary value problems by finite difference methods requires that a mesh be determined a priori. Many of the methods that have been used to determine adaptive grids for two-point boundary value problems can be interpreted in terms of equidistributing a positive weight function over a given interval [25,26]. We say that a mesh $\mathcal{M}$ is equidistributed on the interval $[0, L]$ with respect to the non-negative function W and the constant C if

$$\int_{y_j}^{y_{j+1}} W \ dy = C, \quad j = 0, 1, \ldots, m-1. \tag{4.7}$$

The major differences in the various approaches center around the choice of the weight function and whether or not the mesh is coupled with the calculation of the dependent solution components (see, e.g., [27,28]). We determine the mesh (see [22]) by employing a weight function that equidistributes the difference in the components of the discrete solution and its gradient between adjacent mesh points. Upon denoting the vector of N dependent solution components by $\tilde{U} = [\tilde{U}_1, \tilde{U}_2, \ldots, \tilde{U}_N]^T$, we seek a mesh $\mathcal{M}$ such that

$$\int_{y_j}^{y_{j+1}} \left|\frac{d\tilde{U}_i}{dy}\right| dy \ \leq \ \delta \left| \max_{0 \leq y \leq L} \tilde{U}_i - \min_{0 \leq y \leq L} \tilde{U}_i \right| \quad \begin{array}{c} j = 0, 1, \ldots, m-1 \\ i = 1, 2, \ldots, N \end{array}, \tag{4.8}$$

and

$$\int_{y_j}^{y_{j+1}} \left|\frac{d^2\tilde{U}_i}{dy^2}\right| dy \ \leq \ \gamma \left| \max_{0 \leq y \leq L} \frac{d\tilde{U}_i}{dy} - \min_{0 \leq y \leq L} \frac{d\tilde{U}_i}{dy} \right| \quad \begin{array}{c} j = 1, 2, \ldots, m-1 \\ i = 1, 2, \ldots, N \end{array}, \tag{4.9}$$

where δ and γ are small numbers less than one and the maximum and minimum values of $\tilde{U}_i$ and $d\tilde{U}_i/dy$ are obtained from a converged numercial solution on a previously determined mesh.

A potential problem of such an equidistribution procedure is the formation of a mesh that may not be smoothly varying. For example, the ratio of consecutive mesh intervals may differ by several orders of magnitude. This can affect the accuracy of the method as well as the convergence properties of the Newton iteration. As a result, we impose the added constraint that the mesh produced by employing (4.8) and (4.9) be locally bounded, i.e., the ratio of adjacent mesh intervals must be bounded above and below by constants. We require

$$\frac{1}{A} \leq \frac{h_j}{h_{j-1}} \leq A, \quad j = 2, 3, \ldots, m, \tag{4.10}$$

where A is a constant ≥ 1. This smooths out rapid changes in the size of the mesh intervals.

In employing the adaptive mesh algorithm, we first solve the boundary value problem on a coarse mesh and obtain the maximum and minimum values of $\tilde{U}_i$ and $d\tilde{U}_i/dy$. The inequalities in (4.8-4.10) are then tested and if any of them is not satisfied, a grid point is inserted at the midpoint of the interval in question. Once a new mesh has been obtained, the previously converged numerical solution is interpolated onto the new mesh. The problem is solved on the new mesh and the process continues until (4.8-4.10) are satisfied.

5. Arclength Continuation.

We observe that the system of equations in (4.5) can be written in the form

$$F(U, 1/a) = 0, \tag{5.1}$$

where specific reference to the parametric dependence on the inverse of the strain rate has been made. (The inverse of the strain rate is used, as opposed to the strain rate itself, due to its relationship to the Damköhler number [9]). As the value of a increases and the flame nears extinction, the maximum value of the temperature decreases. It is near the extinction limit, however, that the numerical calculations become increasingly difficult. In particular, at the extinction point the Jacobian of the system is singular. To alleviate the computational difficulties, a modified form of the governing equations is solved [10]. We introduce the reciprocal of the strain rate as a new dependent variable. The vector of dependent variables $(U, 1/a)^T$ can now be considered functions of a new independent parameter s. If we define

$$Z(s) = (U(s), 1/a(s))^T, \tag{5.2}$$

then the new problem we want to solve is given by

$$G(Z, s) = 0, \tag{5.3}$$

where

$$G(Z, s) = \begin{bmatrix} F(U(s), 1/a(s)) \\ N(U(s), 1/a(s), s) \end{bmatrix} = \begin{bmatrix} F(Z(s)) \\ N(Z(s), s) \end{bmatrix}, \tag{5.4}$$

and where N is an arbitrary normalization. As is often the case, (see, e.g., [10]) the normalization is chosen such that s approximates the arclength of the solution branch in the space $(||U||, 1/a)$ (see Figure 2). As an example, if $(U(s_0), 1/a(s_0))^T$ is a known solution, then two commonly used normalizations are

$$N_1 = \theta \left(\frac{dU(s_0)}{ds} \right)^T (U(s) - U(s_0)) +$$

$$(1-\theta) \frac{d}{ds} \left(\frac{1}{a(s_0)} \right) \left(\frac{1}{a(s)} - \frac{1}{a(s_0)} \right) - (s - s_0) = 0, \tag{5.5}$$

and

$$N_2 = \theta ||U(s) - U(s_0)||^2 + (1-\theta) \left(\frac{1}{a(s)} - \frac{1}{a(s_0)} \right)^2 - (s - s_0)^2 = 0, \tag{5.6}$$

where $\theta \in (0, 1)$.

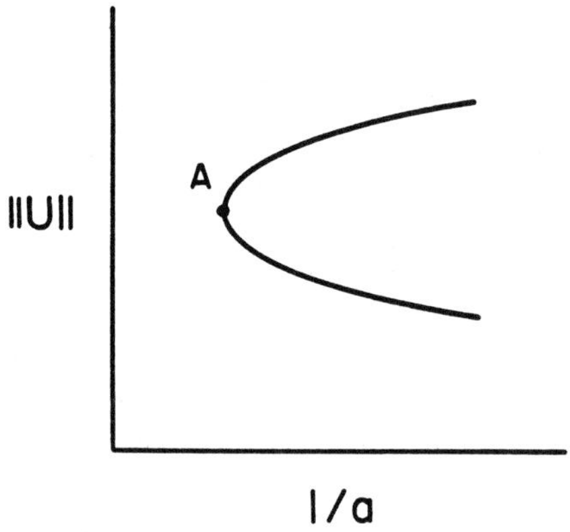

Figure 2. Bifurcation diagram illustrating a simple turning point (A).

The Jacobian of the new system can be written in the form

$$J(Z,s) = G_Z(Z,s) = \begin{bmatrix} F_U(U(s), 1/a(s)) & F_{1/a}(U(s), 1/a(s)) \\ N_U(U(s), 1/a(s), s) & N_{1/a}(U(s), 1/a(s), s) \end{bmatrix}, \tag{5.7}$$

where $F_U, F_{1/a}, N_U$ and $N_{1/a}$ denote the appropriate partial derivatives. It can be shown that at a simple turning point, even though F_U is singular J is not. In addition, given a solution $Z(s)$ we can determine a new predicted value for $Z(s + \delta s)$ by forming

$$Z(s + \delta s) = Z(s) + \frac{dZ}{ds}\delta s, \tag{5.8}$$

where dZ/ds is determined from

$$J\frac{dZ}{ds} = \begin{bmatrix} 0 \\ N_s(Z(s), s) \end{bmatrix}. \tag{5.9}$$

Several points are worth discussing. As a result of the differencing used in (4.2-4.4), the Jacobian of the system in (4.5) is block tridiagonal. However, if after introduction of the reciprocal of the strain rate as a dependent variable along with the extra normalization condition, the quantities $F_{1/a}$ and N_U are not of the proper form, the block tridiagonal structure of the Jacobian can be destroyed. For the normalizations considered in (5.5) and (5.6) this is ordinarily the case. Although solution of the system of linear equations corresponding to (4.6) can proceed by methods discussed in [12], we would like to keep the basic block tridiagonal structure of the Jacobian. In this way we can utilize the solution method used in solving adiabatic, premixed, laminar flames [29], burner-stabilized, premixed, laminar flames [22], counterflow, laminar, diffusion flames [30] and the extinction problems we consider here.

The procedure we follow is similar to that used in the solution of adiabatic, premixed, laminar flames [29]. For each value of the parameter s, we want to obtain the corresponding value of the strain rate and the remaining dependent solution components. We point out that for each value of the pseudo-arclength the strain rate is constant. Hence, it satisfies the trivial differential equation

$$\frac{d}{dy}\left(\frac{1}{a}\right) = 0. \tag{5.10}$$

We can maintain the block tridiagonal structure of the Jacobian in (5.7) if we introduce the reciprocal of the strain rate as a dependent variable at m of the $m+1$ grid points and if we specify a normalization condition at the remaining grid point that does not introduce nonzero Jacobian entries outside of the three block diagonals. The success of this procedure depends upon the choice of the normalization condition.

In premixed flame extinction studies (see, e.g. [3,7]) the maximum temperature is used often as the ordinate in bifurcation curves. The maximum temperature is both a measurable quantity and one of practical interest. In the counterflowing premixed flames we consider here, the maximum temperature is attained at the symmetry plane $y = 0$. Hence, it is natural to introduce the temperature at the first grid point along with the reciprocal of the strain rate as the dependent variables in the normalization condition. We do not include the remaining dependent variables. In this way the block tridiagonal structure of the Jacobian can be maintained.

The final form of the governing equations we solve is given by

$$\frac{dV}{dy} + a\rho f' = 0, \tag{5.11}$$

$$\frac{d}{dy}\left(\frac{\lambda}{c_p}\frac{df'}{dy}\right) - V\frac{df'}{dy} + a(\rho_e - \rho(f')^2) = 0, \tag{5.12}$$

$$\frac{d}{dy}\left(\frac{\lambda}{c_p}\frac{dT}{dy}\right) - V\frac{dT}{dy} + \frac{q}{c_p} = 0. \tag{5.13}$$

$$\frac{d}{dy}\left(\frac{1}{a}\right) = 0, \tag{5.14}$$

with the boundary conditions at $y = 0$ given by

$$V = 0, \tag{5.15}$$

$$\frac{df'}{dy} = 0, \tag{5.16}$$

$$\frac{dT}{dy} = 0, \tag{5.17}$$

$$\frac{dT(0,s_0)}{ds}(T(0,s) - T(0,s_0)) + \frac{d}{ds}\left(\frac{1}{a}(0,s_0)\right)\left(\frac{1}{a}(0,s) - \frac{1}{a}(0,s_0)\right) - (s - s_0) = 0, \tag{5.18}$$

and as $y \to \infty$ by

$$f' = 1, \tag{5.19}$$

$$T = T_e. \tag{5.20}$$

6. Numerical Results.

In this section we apply the one-step kinetics model and the modified arclength continuation procedure to determine the extinction limits as a function of the strain rate for counterflowing premixed methane-air flames. In all cases values for the specific heats, the heat releases, the thermal conductivities, the pre-exponential constants and the activation energies used in (5.11-5.20) are taken from Coffee et al. [21]. For problems in which intermediate values are required, we apply a piecewise linear interpolation procedure to obtain the appropriate parameter values.

We performed a sequence of calculations for both lean and rich methane-air flames. For each methane-air mixture the modified arclength continuation procedure was implemented to obtain profiles of the maximum temperature versus the inverse of the strain rate. The adaptive boundary value solution method (see Section 4) was used first to obtain a solution for a methane-air mixture consisting of 6.5% methane (mole fraction) with an initial strain rate of $a = 500\ \text{sec}^{-1}$. The Euler continuation procedure discussed in (5.8-5.9) was used then to help obtain solutions (both physical and unphysical) as the parameter s (and hence the strain rate) changed. The computations were performed on a domain of one cm..

For each strain rate calculation 60-80 adaptively chosen grid points were used. However, as the strain rate adjusted, the location of the flame front shifted as well. To prevent the adaptive gridding procedure from using all of the grid points of a previous strain rate calculation, we implemented a "skeleton" grid procedure similar to the one used in [31] to restrict the growth in the number of mesh points. Typically, one half of the mesh points from a previous stain rate calculation was used initially in each new strain rate calculation.

Once the 6.5% calculation was complete, we repeated the procedure for mixtures containing 6.85, 7.5, 8.5, 9.5 (stoichiometric), 10.5, 11.5, 12.5, 12.85 and 13.5% methane (mole fraction). In each case we obtained "C-shaped" extinction curves. The corresponding curves are plotted in Figures 3 and 4. As the figures illustrate, as we move along the upper branch in the direction of increasing strain rate, the peak temperature decreases. Ultimately, as the value of $dT_{max}/d(1/a) \rightarrow \infty$, the flame extinguishes. We can, however, continue past the extinction point with the arclength procedure. We find that, as the strain rate begins to decrease, the peak temperature continues to fall. Temperature profiles (both physical and unphysical) for an 11.5% methane (mole fraction) flame are illustrated in Figures 5 and 6.

If we collect the results of the ten methane-air flame calculations, we can plot the value of the strain rate at extinction (at $x = 1$ cm.) versus the equivalence ratio. These results are contained in Figure 7. We observe that the peak value of the strain rate occurred for slightly rich conditions–the flame with the highest calculated temperature. In addition, if we define the extinction distance as the distance from the plane of symmetry to the point at which the temperature is equal to 50% of its maximum temperature, we can obtain a plot of extinction distance versus equivalence ratio similar to the plot in Figure 7. These results are shown in Figure 8. It is worthwhile to point out that the results shown in Figure 8 differ qualitatively from those reported by Sato [3] for lean methane-air mixtures. This is due primarily to the fact that we have assumed the Lewis numbers equal one.

As a final point, we illustrate in Figure 9 two C-shaped curves for a rich 11.5% methane (mole fraction) flame. The lower curve corresponds to one of the physical and unphysical branches illustrated in Figure 4. The upper C-shaped curve, however, represents a set of solutions that are thermodynamically impossible. The 11.5% methane (mole fraction) flame has an equivalence ratio greater than one ($\phi \approx 1.24$). As a result, some of the fuel remains after combustion takes place. We point out, however, that there is an adiabatic flame temperature that is consistent with this system under zero strain rate conditions. The peak temperatures of the upper C-shaped curve are all above this adiabatic value. This higher temperature results from the fact that there is no constraint on the system in (5.1) or (5.3) that restricts the mass fractions from becoming negative. As a result, numerical errors could cause the mass fraction of the oxidizer to become negative with the result that more fuel could be consumed (and more heat was released) than is thermodynamically possible. In such a case, the solution could change from the physical branch of the lower C-shaped curve to the unphysical upper curve. Temperature profiles for both the upper and lower C-shaped curves with the strain rate equal to 1500 sec^{-1} are illustrated in Figure 10.

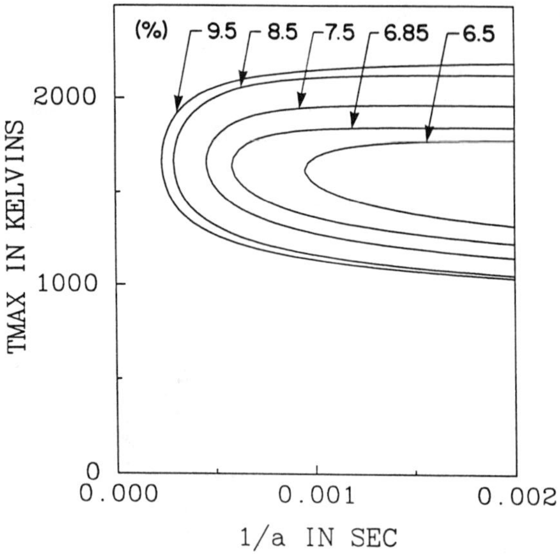

Figure 3. C-shaped extinction curves for lean (6.5, 6.85, 7.5, 8.5%) and stoichiometric (9.5%) methane-air flames.

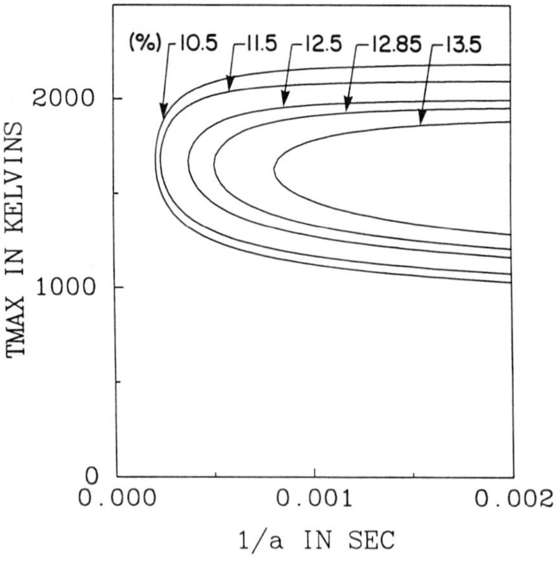

Figure 4. C-shaped extinction curves for rich (10.5, 11.5, 12.5, 12.85, 13.5%) methane-air flames.

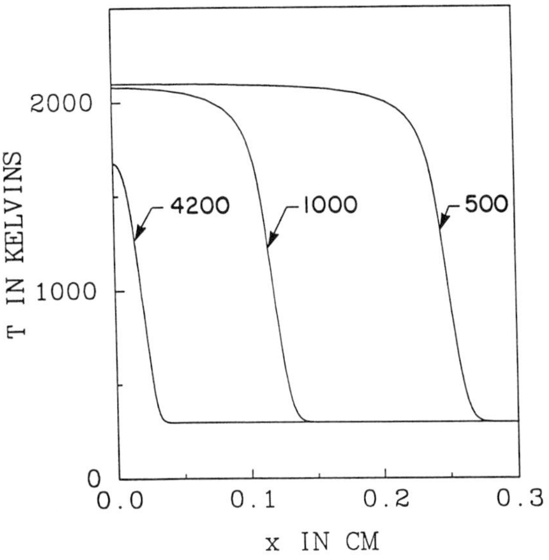

Figure 5. Temperature profiles (physical) for an 11.5% methane-air flame. Profiles for strain rates of 500, 1000 and 4200 sec^{-1} are illustrated.

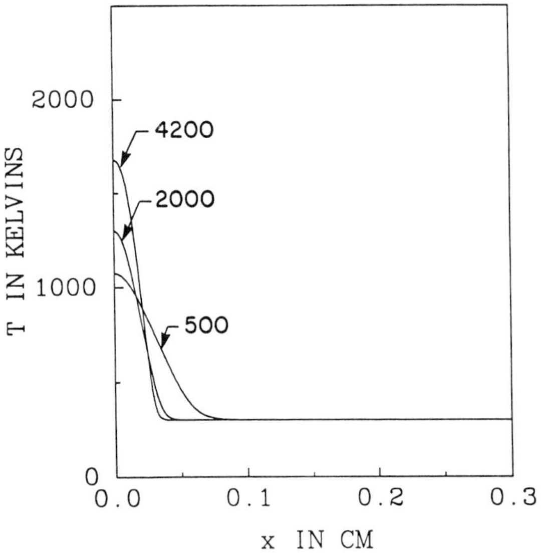

Figure 6. Temperature profiles (unphysical) for an 11.5% methane-air flame. Profiles for strain rates of 500, 2000 and 4200 sec^{-1} are illustrated.

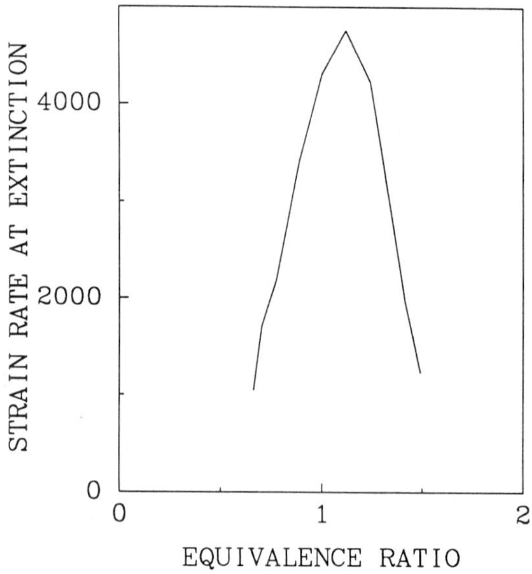

Figure 7. Illustration of the strain rate (in sec^{-1}) at extinction versus the methane-air equivalence ratio.

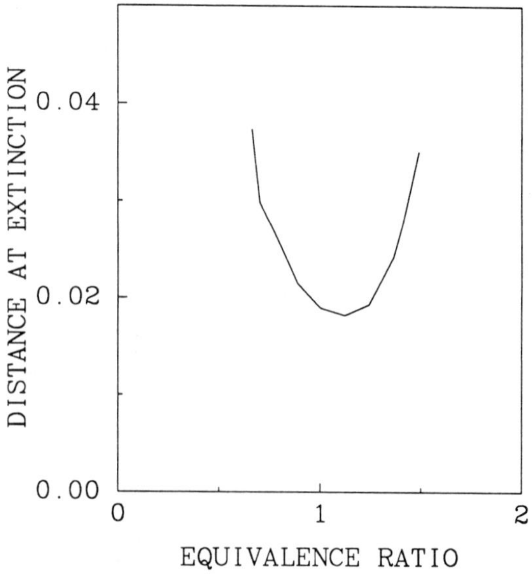

Figure 8. Illustration of the distance (in cm) at extinction versus the methane-air equivalence ratio.

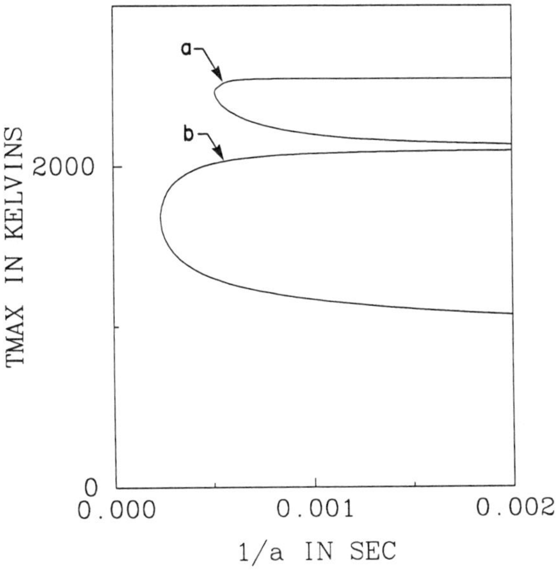

Figure 9. C-shaped curves for an 11.5% methane-air flame. The upper C-shaped curve (a) corresponds to unphysical (thermodynamic) solutions. The lower C-shaped curve (b) is identical to the corresponding curve in Figure 4.

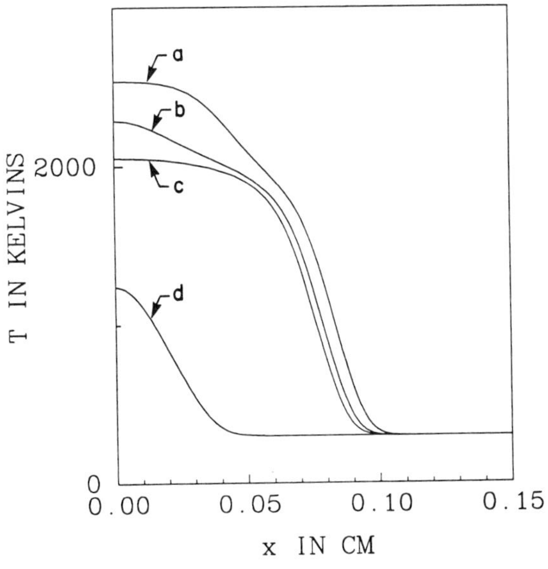

Figure 10. Temperature profiles for an 11.5% methane-air flame with $a = 1500$ sec^{-1}. The two uppermost curves (a,b) correspond to unphysical (thermodynamic) solutions. The remaining curves correspond to the physical (c) and unphysical (d) solutions in Figure 4.

7. Conclusion.

We have applied an arclength continuation procedure to calculate extinction limits for a set of premixed hydrogen-air and methane-air flames in a stagnation point flow. Profiles for the peak temperature versus the strain rate were obtained for a variety of incoming fuel-air mixtures. To simplify the computations, we considered a one-step global kinetics model in which all the Lewis numbers were assumed equal to one. We realize of course that such a procedure cannot predict adequately all of the detailed behavior resulting from a complex kinetics calculation. We have, however, illustrated the applicability of the numerical bifurcation procedure in calculating extinction limits for premixed flames in a stagnation point flow. In a subsequent paper we will apply the methods discussed in this study to obtain extinction limits of hydrogen-air and methane-air flames in which the chemistry is governed by a detailed multistep mechanism.

This work was supported by the Office of Naval Research

[1] Laboratoire de Mécanique Théorique, Université Paris
and Laboratoire d'Energetique, École Centrale, Chatenay-Malabry, France.

[2] Department of Mechanical Engineering, Yale University, New Haven, CT.

REFERENCES

[1]. S. Ishizuka and C. K. Law, *An Experimental Study on Extinction and Stability of Stretched Premixed Flames*, Nineteenth Symposium (International) on Combustion, The Combustion Institute, Pittsburgh, (1982), p. 327.

[2]. C. K. Law, S. Ishizuka and M. Mizomoto, *Lean-Limit Extinction of Propane-Air Mixtures in the Stagnation-Point Flow*, Eighteenth Symposium (International) on Combustion, The Combustion Institute, Pittsburgh, (1981), p. 1791.

[3]. J. Sato, *Effects of Lewis Number on Extinction Behavior of Premixed Flames in a Stagnation Flow*, Nineteenth Symposium (International) on Combustion, The Combustion Institute, Pittsburgh, (1982), p. 1541.

[4]. G. I. Sivashinsky, *On a Distorted Flame Front as a Hydrodynamic Discontinuity*, ACTA Astronautica, **3** (1976), p. 889.

[5]. J. Buckmaster, *The Quenching of a Deflagration Wave Held in Front of a Bluff Body*, Seventeenth Symposium (International) on Combustion, The Combustion Institute, Pittsburgh, (1979), p. 835.

[6]. J. Sato and H. Tsuji, *Behavior and Extinction of a Premixed Flame in a Stagnation Flow for General Lewis Numbers (in Japanese)*, Sixteenth Symposium on Combustion, Japan, (1978), p. 13.

[7]. J. Sato and H. Tsuji, *Extinction of Premixed Flames in a Stagnation Flow Considering General Lewis Number*, Comb. Sci. and Tech., **33** (1983), p. 193.

[8]. H. Tsuji and I. Yamaoka, *Structure and Extinction of Near-Limit Flames in a Stagnation Flow*, Nineteenth Symposium (International) on Combustion, The Combustion Institute, Pittsburgh, (1982), p. 1533.

[9]. P. A. Libby and F. A. Williams, *Strained Premixed Laminar Flames with Two Reaction Zones*, Comb. Sci. and Tech., **37** (1984), p. 221.

[10]. H. B. Keller, *Numerical Solution of Bifurcation and Nonlinear Eigenvalue Problems*, in Applications of Bifurcation Theory, P. Rabinowitz, Ed., Academic Press, New York, (1977), p. 359.

[11]. A. Jepson and A. Spence, *Singular Points and Their Computations*, in Numerical Methods for Bifurcation Problems, T. Kupper, H. Mittelmann and H. Weber, Eds., Birkhauser Verlag, Basel, (1984).

[12]. T. Chan, *Techniques for Large Sparse Systems Arising from Continuation Methods*, in Numerical Methods for Bifurcation Problems, T. Kupper, H. Mittelmann and H. Weber, Eds., Birkhauser Verlag, Basel, (1984).

[13]. R. Seydel, *Numerical Computation of Branch Points in Nonlinear Equations*, Numer. Math., **3** (1979), p. 339.

[14]. R. F. Heinemann, K. A. Overholser and G. W. Reddien, *Multiplicity and Stability of Premixed Laminar Flames: An Application of Bifurcation Theory*, Chem Eng. Sci., **34** (1979), p. 833.

[15]. R. F. Heinemann, K. A. Overholser and G. W. Reddien, *Multiplicity and Stability of the Hydrogen-Oxygen-Nitrogen Flame: The Influence of Chemical Pathways and Kinetics on Transitions Between Steady States*, AIChE Journal, **26** (1980), p. 725.

[16]. J. A. Miller, R. E. Mitchell, M. D. Smooke and R. J. Kee, *Toward a Comprehensive Chemical Kinetic Mechanism for the Oxidation of Acetylene: Comparison of Model Predictions with Results from Flame and Shock Tube Experiments*, Nineteenth Symposium (International) on Combustion, Pittsburgh, (1982), p. 181.

[17]. J. A. Miller, M. D. Smooke, R. M. Green and R. J. Kee, *Kinetic Modeling of the Oxidation of Ammonia in Flames*, Comb. Sci. and Tech., **34** (1983), p. 149.

[18]. A. Levy and F. J. Weinberg, *Optical Flame Structure Studies: Some Conclusions Concerning the Propagation of Flat Flames*, Seventh Symposium (International) on Combustion, Pittsburgh, (1959), p. 296.

[19]. A. Levy and F. J. Weinberg, *Optical Flame Structure Studies: Examination of Reaction Rate Laws in Lean Ethylene-Air Flames*, Comb. and Flame, **3** (1959), p.229.

[20]. C. K. Westbrook and F. L. Dryer, *Simplified Reaction Mechanisms for the Oxidation of Hydrocarbon Fuels in Flames*, Comb. Sci. and Tech., **27** (1981), p. 31.

[21]. T. P. Coffee, A. J. Kotlar and M. S. Miller, *The Overall Reaction Concept in Premixed, Laminar Steady-State Flames. I. Stoichiometries*, Comb. and Flame., **54** (1983), p. 155.

[22]. M. D. Smooke, *Solution of Burner-Stabilized Pre-Mixed Laminar Flames by Boundary Value Methods*, J. Comp. Phys., **48** (1982), p. 72.

[23]. P. Deuflhard, *A Modified Newton Method for the Solution of Ill-Conditioned Systems of Nonlinear Equations with Application to Multiple Shooting*, Numer. Math., **22** (1974), pp. 289.

[24]. M. D. Smooke, *An Error Estimate for the Modified Newton Method with Applications to the Solution of Nonlinear Two-Point Boundary Value Problems*, J. Opt. Theory and Appl., **39** (1983), p. 489.

[25]. J. Kautsky and N. K. Nichols, *Equidistributing Meshes with Constraints*, SIAM J. Sci. Stat. Comput., **1** (1980), pp. 499.

[26]. R. D. Russell, *Mesh Selection Methods*, Proceedings of the Conference for Working Codes for Boundary Value Problems in ODE's, B. Childs et al., Eds., Springer-Verlag, New York, (1979).

[27]. V. Pereyra and E. G. Sewell, *Mesh Selection for Discrete Solution of Boundary Value Problems in Ordinary Differential Equations*, Numer. Math., **23** (1975), pp. 261.

[28]. A. B. White, *On Selection of Equidistributing Meshes for Two-Point Boundary Value Problems*, SIAM J. Numer. Anal., **16** (1979), pp. 472.

[29]. M. D. Smooke, J. A. Miller and R. J. Kee, *Determination of Adiabatic Flame Speeds by Boundary Value Methods*, Comb. Sci. and Tech., **34** (1983), p. 79.

[30]. M. D. Smooke, J. A. Miller and R. J. Kee, *Solution of Premixed and Counterflow Diffusion Flame Problems by Adaptive Boundary Value Methods*, in Numerical Boundary Value ODEs, U. M. Ascher and R. D. Russell, Eds., Birkhauser, Basel, (1985).

[31]. M. D. Smooke and M. L. Koszykowski, *Fully Adaptive Solutions of One-Dimensional Mixed Initial-Boundary Value Problems with Applications to Unstable Problems in Combustion*, to be published in SIAM J. Sci. and Stat. Comp., (1985).

A NUMERICAL METHOD FOR CALCULATING COMPLETE THEORETICAL SEISMOGRAMS IN VERTICALLY VARYING MEDIA

Uri Ascher and Paul Spudich

Abstract

A numerical method is presented for calculating complete theoretical seismograms. The earth models are assumed to have velocity, density and attenuation profiles which are arbitrary piecewise-continuous functions of depth only. Solutions for the stress-displacement vectors in the medium are expanded in terms of orthogonal cylindrical functions. The resulting two-point boundary value problems are integrated by a variety of schemes, automatically chosen to suit the type of wave solution. These schemes include collocation for low-frequency and for highly evanescent waves, various special methods for highly oscillatory waves and for turning waves, and combinations of those. A Fourier-Bessel transform is then applied to get back into the space-time domain.

1. Introduction

One basic problem in computational seismology is to calculate displacements at the earth surface due to an earthquake or a man-made explosion, for a given earth model. The governing system of PDEs (in time and 3 space variables) is rather complex and often some restrictions are placed on the model. One such restriction, which has been often considered, is to allow seismic velocities, density and attenuative factors to be functions of depth only, see for instance Aki & Richards [1] or Kennett [13]. Then the displacement and the traction across a horizontal plane (i.e. the z-component of the stress tensor) can be expanded in a cylindrical coordinate system, see Takeuchi & Saito [23]. Due to the assumed horizontal material homogeneity, this results in a double integral (a Fourier-Bessel transform), with the azimuthal dependence essentially disappearing. Still, each integrand evaluation in the frequency-wavenumber plane corresponds to two linear systems of coupled boundary value ordinary differential equations (BVPs) describing SH and P-SV waves as a function of depth z.

In a typical calculation, there may sometimes be 20000 such BVPs to solve. Care must therefore be taken to solve them very efficiently. A saving grace is that their solution is independent and can be done entirely *in parallel*.

In this paper we concentrate on the numerical solution of these BVPs. Most of the material presented here is contained either in Spudich & Ascher [22] or in Ascher

& Spudich [6]. Our purpose here is to expose the problem and our numerical procedures for its solution to an audience not necessarily acquainted with the seismological literature. We also concentrate on the numerical aspects and leave the seismological discussion as much as possible to [22] and [6].

Consider then a half-space in which P wave velocity α, S wave velocity β and density ρ are piecewise-continuous functions of depth into the half-space, z, only. For $0 \leq z < z_>$, α,β and ρ are arbitrary, but for $z \geq z_>$ they are constant, $\alpha(z) = \alpha_>$, $\beta(z) = \beta_>$, $\rho(z) = \rho_>$, as is depicted in Fig. 1. We allow α and β to be complex functions of z, thereby modelling the effects of anelastic attenuation, see Schwab and Knopoff [21] and [13]. We also allow ω to be complex, following Phinney [17].

Applying a Fourier-Bessel transform, let ω be a given frequency and k a horizontal wavenumber and define slowness

$$p = k/\omega. \tag{1.1}$$

Then we get differential equations of the form

$$\partial_z \mathbf{b} = \omega A(z)\mathbf{b} \qquad 0 < z < z_> \tag{1.2}$$

where A depends on p but is independent of ω. For SH waves we have a problem of order $n = 2$

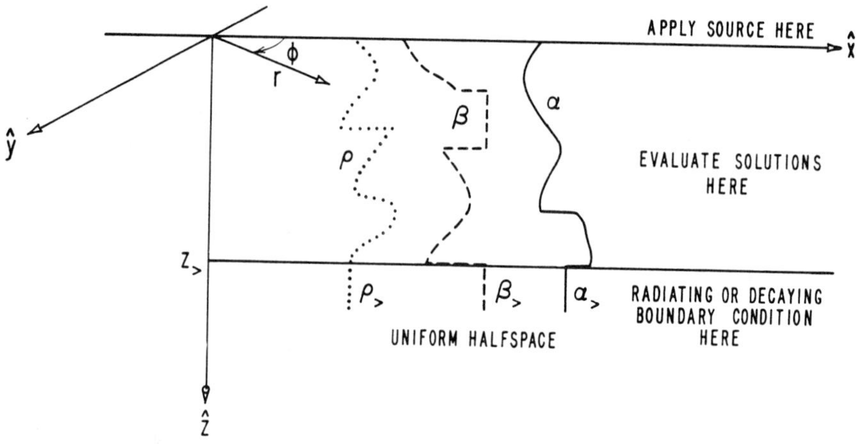

Figure 1. General problem geometry.

$$\mathbf{b} \equiv \mathbf{b}_H \qquad A \equiv A_H = \begin{pmatrix} 0 & \mu^{-1} \\ \mu p^2 - \rho & 0 \end{pmatrix} \qquad \mu = \rho\beta^2 \tag{1.3}$$

whereas for P-SV waves we get $n = 4$ and

$$\mathbf{b} \equiv \mathbf{b}_p \qquad A \equiv A_p = \begin{pmatrix} 0 & (\rho\alpha^2)^{-1} & q_1 & 0 \\ -\rho & 0 & 0 & p \\ -p & 0 & 0 & \mu^{-1} \\ 0 & -q_1 & q_2 & 0 \end{pmatrix} \tag{1.4}$$

$$q_1 = p\,(1 - 2\beta^2/\alpha^2)$$

$$q_2 = 4\mu p^2(1 - \beta^2/\alpha^2) - \rho.$$

The system (1.2) is supplemented by $n/2$ boundary conditions at each of the two ends, $z = 0$ and $z = z_>$, as described in [22]. Basically, at $z_>$ we impose radiation conditions, requiring only downgoing waves for $z \geq z_>$. At $z = 0$ we apply three orthogonal forcing term conditions following Olson et al [16], which allow us to construct the necessary Green's functions for the medium, using their symmetry to exchange the locations of the observer and the source. This trades an unpleasant (explosion) source term in the middle of our domain for a harmless inhomogeneity in the boundary values.

Next we briefly discuss the integration in ω and k (or p) to yield back solutions in time and space. The efficient integration of this Fourier-Bessel transform is obviously quite crucial, and an attempt to minimize the number of evaluation points (each of which corresponds to solving BVPs for (1.3) and (1.4)) should be made. This will be dealt with in more detail elsewhere. For the frequency integration we have used a fast Fourier transform (FFT). Handling the (Hankel) transform for k is more debatable and we have used a discrete wavenumber summation as in Alekseev & Mikhailenko [2] and in [16]. Other approaches as in Apsel and Luco [3] or Frazer and Gettrust [11] seem promising.

We now focus on the BVPs. Due to the complexity of the problem, even with the simplifying assumptions above, most efforts have concentrated on some particular aspects of it, e.g. near field or far field approximations, low frequency or high frequency (ray theory) approximations, body waves or surface waves, etc. Most of the traditional work for solving the BVPs has been made along two approaches. In one approach, initial value numerical techniques are used. But these techniques may become unstable for high frequencies. In the other approach, transformations of the boundary value problem are employed. This approach has been quite successful for certain models. For example, at high frequencies and in regions having small gradients in material properties, asymptotic expansions have been successfully used to describe P and S wave propagation (Richards [18], Chapman [10], Richards & Frasier [19]). However, in some cases the accuracy of the approximation deteriorates (or, alternatively, good approximations are harder to obtain) when the medium is not sufficiently close to being a stack of uniform layers (i.e. when there are regions of large velocity gradients). Also, the uniform quality of the approximation is hard to maintain and quantitative error control has rarely been attempted.

In [22] a collocation method was used for this problem. The distinguishing features of the method are (i) that it remains stable even for high frequencies (and is rather effective for highly evanescent waves) and (ii) that it is general, so it can handle regions of large gradients, cases where the frequency is not very large (where ray approximations, for instance, may be invalid), and cases with multiple turning points. The resulting code therefore had the advantage of being general, flexible and robust: The velocity- and density- depth profiles may be arbitrary piecewise continuous functions of depth; the method can be used in any distance range, either near field or far field, to calculate all possible body waves, surface waves and leaky modes; and in addition, phase velocity and group velocity windowing can be applied.

The main flaw of our method was its long computation time: The collocation schemes do not perform well for highly oscillatory solutions (high frequency body waves), because polynomial approximation, on which they are based, is not very suitable in such circumstances. Here we describe improvements we have made in this regard, see also [6]. We have used the expansions in [18], [8] and [13] to construct a specialized multiple shooting scheme, bypassing oscillatory solution details. Schemes are also constructed for P wave turning points and for leaky modes. All these schemes are then used as building blocks for an adaptive method. The domain in z is adaptively subdivided into subintervals and in each subinterval an appropriate scheme is chosen.

In the next two sections we describe the resulting numerical method and give an example.

2. Numerical solution of the boundary value problem

The solution behaviour of the BVPs described above varies a great deal for different values of p and ω. Hence the numerical scheme employed has to be versatile enough to handle the variety of wave types. At the same time, the numerical scheme has to be reasonably efficient, because it is invoked many times. We therefore begin with some analysis of the BVP.

2.1. Solution behaviour for various values of frequency and slowness

In (1.2) we have a homogeneous system of ordinary differential equations. To see its behaviour, we attempt to decouple the fundamental solutions. Thus, consider a transformation

$$\mathbf{v}(z) := D^{-1}(z)\mathbf{b}(z) \tag{2.1}$$

where $D(z)$ is a piecewise smooth, nonsingular matrix function satisfying on some interval $[\varsigma_0, \varsigma_1]$ in some reasonable norm,

$$\|D(z)\| \, \|D^{-1}(z)\| \leq const. \qquad \varsigma_0 \leq z \leq \varsigma_1 \tag{2.2}$$

and the constant in (2.2) is of moderate size. This transforms (1.2) into

$$\partial_z \mathbf{v} = [\omega D^{-1} A D - D^{-1} \partial_z D] \mathbf{v} \equiv C(z) \mathbf{v} \tag{2.3}$$

and we would like to choose $D(z)$ so that $C(z)$ is (essentially) diagonal, or at least upper triangular. In that case, the eigenvalues of C are called *kinematic eigenvalues*. The information regarding stability of the BVP (not of the numerical algorithm) is contained in these kinematic eigenvalues.

Now, note that for the matrices A of (1.3) and (1.4) there exists a piecewise-smooth transformation matrix $E(z)$ such that

$$E^{-1} A E = \Lambda \qquad 0 < z < z_> \tag{2.4}$$

with $\Lambda(z)$ a diagonal eigenvalue matrix, $\Lambda(z) = diag\,\{\lambda_1(z), \ldots, \lambda_n(z)\}$. For SH,

$$\Lambda = diag\,\{\nu_\beta, -\nu_\beta\} \tag{2.5}$$

while for P-SV

$$\Lambda = diag\,\{\nu_\alpha, \nu_\beta, -\nu_\alpha, -\nu_\beta\} \tag{2.6}$$

with

$$\nu_\alpha = (p^2 - \alpha^{-2})^{1/2}, \qquad \nu_\beta = (p^2 - \beta^{-2})^{1/2}, \tag{2.7}$$

and we choose

$$Re\,(\omega\nu) \geq 0; \qquad Im\,(\omega\nu) \leq 0 \quad when \quad Re\,(\omega\nu) = 0. \tag{2.8}$$

Note that $\nu_\alpha(z) \neq \nu_\beta(z)$ for all z, because $\mathrm{Re}(\alpha(z)) > \mathrm{Re}(\beta(z))$ by physical considerations.

It is natural to ask whether these eigenvalues of A (times ω) are close to the kinematic eigenvalues, i.e., whether the eigenvector matrix E can be used for decoupling. Thus, consider choosing $D \equiv E$ in (2.1)-(2.3). If $\nu_\alpha(z)$ and $\nu_\beta(z)$ are bounded away from 0 on $[\varsigma_0, \varsigma_1]$ then (2.2) holds. If, further, α, β and ρ are constant for $\varsigma_0 \leq z \leq \varsigma_1$ (i.e. we have a uniform layer) then the matrix C in (2.3) is diagonal and the differential equations decouple. The components of $\mathbf{v}(z)$ then correspond to up- and downgoing waves. The boundary conditions at $z_>$ are in fact derived in this way by requiring that the upgoing waves vanish for $z \geq z_>$. Moreover, the fundamental matrix $V(z, \varsigma_0)$ satisfying (2.3) and $V(\varsigma_0; \varsigma_0) = I$ is given by

$$V(z; \varsigma_0) = \exp\{\omega \Lambda (z - \varsigma_0)\} \tag{2.9}$$

Once a fundamental matrix $V(z;\varsigma_0)$ is found, the *propagator matrix* $B(z;\varsigma_0)$ for (1.2) is given by

$$B(z;\varsigma_0) = D(z)V(z;\varsigma_0)D^{-1}(\varsigma_0). \tag{2.10}$$

Then we have, in general,

$$\mathbf{b}(z) = B(z;\varsigma_0)\mathbf{b}(\varsigma_0). \tag{2.11}$$

If ν_β, say, has a significant real part, then in (2.9) there are solution modes which grow and decay exponentially, like $\exp\{\pm Re(\nu_\beta)\omega(\varsigma_1-\varsigma_0)\}$. These modes are mixed up in (2.10), hence also in (2.11), and are responsible for the observed instability of shooting techniques, for highly evanescent waves (ω large). When considering a stack of uniform layers, expressions like $B(\varsigma_1;\varsigma_0)$ for each layer can be recursively aggregated for $0 \leq z \leq z_>$; but this again leads to well-known numerical instabilities (Haskell [12]). Kennett & Kerry [15] avoided these instabilities by integrating components of decoupled equations separately, building recursions only in stable directions.

The situation is less rosy when the medium in $\varsigma_0 \leq z \leq \varsigma_1$ is not uniform, for then the matrix $C(z)$ in (2.3) is not quite diagonal or upper triangular. The generalization of (2.9),

$$Q(z) = \exp\{\omega \int_{\varsigma_0}^{z} \Lambda(\varsigma)d\varsigma\} \tag{2.12}$$

does not satisfy (2.3) precisely, and an asymptotic expansion

$$V(z;\varsigma_0) = (\sum_{j=0}^{\infty} P_j(z)\omega^{-j})Q(z) \tag{2.13}$$

is needed. In practice it is too expensive to calculate more than the two leading terms of the sum in (2.13). The omitted remainder introduces an error which depends on the size of ω^{-1} and $E^{-1}\partial_z E$ relative to Λ. This is discussed in §2.3.

An additional difficulty occurs in the vicinity of a turning point, where $\nu_\alpha(z) \approx 0$ or $\nu_\beta(z) \approx 0$. In such regions, condition (2.2) fails to hold if $D(z)$ is the eigenvector matrix $E(z)$, and another transformation is needed. The kinematic eigenvalues are quite different than the eigenvalues of ωA and, since they are not readily available, we seek a transformation which decouples the S- and P- waves. In Kennett & Illingworth [14], following Chapman [9], a transformation satisfying (2.2) is given which brings A into the forms

$$H_\beta = \begin{pmatrix} 0 & p \\ \nu_\beta^2/p & 0 \end{pmatrix} \quad \textit{and} \quad \begin{pmatrix} H_\alpha & 0 \\ 0 & H_\beta \end{pmatrix} \tag{2.14}$$

for SH and P-SV problems, respectively. This allows development of a Langer approximation, which is a uniform asymptotic expansion that has the general form

(2.13), with the exponentials in Q of (2.12) replaced by Airy functions. Again, however, the sum in (2.13) has to be truncated after at most two terms and the error introduced is sometimes too large. Moreover, the Airy functions may not be used in the case of close multiple turning points. Use of such transformations is described in §2.5.

The collocation method introduced for these BVPs in [22] has two desirable properties. The method is accurate for low and intermediate values of ω, where the truncated sum of (2.13) may be a poor approximation. The method also performs well if used properly for highly evanescent waves, see Ascher & Weiss [7], Ascher & Bader [4]. The latter holds even though no explicit separation of up- and downgoing waves is performed. On the other hand, as mentioned earlier, the collocation method does not perform efficiently when the solution is highly oscillatory, i.e. when $\nu_\alpha(z)$ or $\nu_\beta(z)$ are (almost) purely imaginary and ω is large. At the same time, construction of solutions through (2.10)-(2.13) does not suffer from numerical instability precisely under these circumstances! Thus, a hybrid method suggests itself.

In the following sections we discuss a general framework for such a hybrid method (§2.2), followed by descriptions of applicable schemes (§§2.3-2.5) and, finally, we give a general algorithm (§2.6).

2.2. A general framework for the numerical method

For given values of ω and p, the interval in z is subdivided by a mesh

$$0 = z_1 < z_2 < \cdots < z_N < z_{N+1} = z_> \tag{2.15}$$

which is constructed to satisfy the following properties:

(a) All points of material discontinuity (i.e. interfaces between layers, where $\alpha(z)$, $\beta(z)$, $\rho(z)$ or their derivatives may vary discontinuously) are included in the mesh. Thus we may assume, e.g., that

$$A(z) \in C^2[z_i, z_{i+1}), \qquad 1 \le i \le N. \tag{2.16}$$

(b) On each subinterval (or element) (z_i, z_{i+1}), precisely one of the cases of solution behaviour, discussed before (and in the following sections), holds.

(c) Each element size

$$h_i := z_{i+1} - z_i \qquad 1 \le i \le N \tag{2.17}$$

is sufficiently small so that the numerical method used on it produces the solution to a desired accuracy.

An algorithm for constructing such a mesh is described in §2.6. Here, assuming that we have one, we seek a numerical solution $\{\mathbf{b}_j\}_{j=1}^{N+1}$ such that for a given tolerance δ,

$$|\mathbf{b}_j - \mathbf{b}(z_j)| \le \delta, \qquad 1 \le j \le N+1. \tag{2.18}$$

This numerical solution is obtained by requiring that $\mathbf{b}_1$ and $\mathbf{b}_{N+1}$ satisfy the boundary conditions for $\mathbf{b}(0)$ and $\mathbf{b}(z_>)$ respectively, and that on each mesh element a relation of the form

$$R_j \mathbf{b}_{j+1} = S_j \mathbf{b}_j \qquad 1 \leq j \leq N \tag{2.19}$$

hold. Here, R_j and S_j are $n \times n$ matrices and R_j is nonsingular. Recall from [7] that the collocation schemes can be written in the form (2.19) with $R_j = I$ and $S_j = \Gamma_j$ calculated directly. More generally, given (2.19) we write

$$\Gamma_j = R_j^{-1} S_j . \tag{2.20}$$

Comparing (2.19) to (2.11), we expect Γ_j to approximate the propagator matrix $B(z_{j+1}; z_j)$, unless the solution on the element (z_j, z_{j+1}) contains highly evanescent waves. Conversely, any method which generates approximations to $B(z_{j+1}; z_j)$ for some subdivision of the form (2.15), can obviously be written in the form (2.19).

The system of $n(N+1)$ linear equations (2.19) plus the boundary equations has a familiar form and is solved e.g. as described in [22, §2.4].

2.3. A numerical method for rapidly oscillating solutions (body waves)

Consider the problem (1.2) for $z_j \leq z \leq z_{j+1}$, with j fixed, $1 \leq j \leq N$, assuming that ω is (positive and) large, and that $\nu_\alpha(z)$ and $\nu_\beta(z)$ are almost purely imaginary. Thus, there may be many wavelengths between z_j and z_{j+1} and we want a method which would connect the information at z_j to that at z_{j+1}, without closely following all the oscillations in between. General finite difference or collocation methods are inadequate for this purpose when applied directly to (1.2).

On the other hand, the situation here is easier than in the highly evanescent case in that approximating the propagator matrix $B(z_{j+1}; z_j)$ numerically makes sense: The growth factor of the fundamental solution components is at most

$$\exp\{h_j \max_{z_j \leq z \leq z_{j+1}} [Re\,(\omega \nu_\beta)]\}$$

which is bounded in this case by a constant of moderate size. In other words, the condition number of the initial value problem which the propagator matrix solves is not significantly worse than that of the given boundary value problem.

Thus, we seek to construct a matrix Γ_j to approximate $B(z_{j+1}; z_j)$ using (2.13), (2.12). This leads to asymptotic expansions in (large) ω. The idea is not new but its use in our general setting is novel. We now describe the simplest of these constructions, following [18] (see also [9] and Scheid [20]).

With $\varsigma_0 = z_j$ in (2.13) we write

$$V(z; z_j) = (\sum_{l=0}^{\infty} P_l(z) \omega^{-l}) \exp\{\omega Q_0(z) + Q_1(z)\} \tag{2.21}$$

where P_l are bounded matrices, Q_0 and Q_1 are diagonal matrices and $Re\,(Q_0(z)) = O\,(\omega^{-1})$. This is possible to do because $A\,(z)$ can be smoothly diagonalized and its eigenvalues are well-separated with only small real parts. Substituting into (2.3) with $D \equiv E$ and equating coefficients for $\omega, \omega^0, \omega^{-1}, \ldots$ we get

$$P_0 Q_0{}' = \Lambda P_0 \tag{2.22a}$$

$$P_l{}' + P_l Q_1{}' - \tilde{E} P_l = \Lambda P_{l+1} - P_{l+1} Q_0{}' \qquad l = 0,1,\ldots \tag{2.22b}$$

where

$$\tilde{E}(z) = -E^{-1}(z) E'(z), \qquad (\;)' \equiv \partial_z\,. \tag{2.23}$$

To satisfy (2.22a) we take

$$P_0 = I, \qquad Q_0(z) = \int_{z_j}^{z} \Lambda(\varsigma)\, d\varsigma \tag{2.24}$$

(cf. (2.12)). Subsequently, we see from (2.22b) with $Q_0{}' = \Lambda$ that the diagonal elements of $P_l{}' + P_l Q_1{}' - \tilde{E} P_l$ must be zero for each $l \geq 0$. For $l = 0$ this gives

$$Q_1{}' = diag\,\{\tilde{E}_{11}, \ldots, \tilde{E}_{nn}\}, \qquad Q_1(z_j) = 0, \tag{2.25}$$

while the nondiagonal entries for $l = 0$ give

$$(P_1)_{ik} = (\lambda_k - \lambda_i)^{-1} \tilde{E}_{ik} \qquad i \neq k. \tag{2.26}$$

When satisfying the diagonal entries of (2.22b) for $l = 1$ we obtain

$$(P_1)_{ii} = \int_{z_j}^{z} \sum_{k \neq i} \tilde{E}_{ik} (P_1)_{ki} = \int_{z_j}^{z} \sum_{k \neq i} (\lambda_k - \lambda_i)(P_1)_{ik} (P_1)_{ki} \tag{2.27}$$

So, an $O\,(\omega^{-2})$ approximation to $V(z\,;z_j)$ is obtained by

$$\hat{V}(z\,;z_j) = (I + \omega^{-1} P_1) \exp\{\omega Q_0(z) + Q_1(z)\} \tag{2.28}$$

Using (2.10), an $O\,(\omega^{-2})$ approximation to $B(z_{j+1};z_j)$ is obtained by Γ_j of (2.20) with

$$R_j = (I - \omega^{-1} P_1(z_{j+1}^-)) E^{-1}(z_{j+1}^-) \tag{2.29a}$$

$$S_j = \exp\{\omega Q_0(z_{j+1}^-) + Q_1(z_{j+1}^-)\}\{I - \omega^{-1}P_1(z_j^+)\}E^{-1}(z_j^+) \qquad \mathbf{(2.29b)}$$

Moreover, by (2.26) it can be seen that the magnitude of the neglected $O(\omega^{-2})$ term is related to that of $E(z)$. This corresponds to the fact that in zones of large velocity gradients, a significant portion of the waves gets reflected back to the surface $z = 0$, and the above approximation makes sense only for higher frequencies. On the other extreme, if the material properties α, β and ρ are uniform in (z_j, z_{j+1}) then (2.29) give the propagator matrix exactly, for any ω.

Explicit formulae for (2.29) applied to (1.3), (1.4) have been calculated. Note that the matrices E, P_1, Q_0 and Q_1 are independent of ω. Thus, a significant computational efficiency could be gained if solutions for many large values of ω with the same p are desired. For if we use the same mesh element (z_j, z_{j+1}), then most of the assembly of the finite difference matrices S_j and R_j can be done once, independently of ω. And indeed, for sufficiently large ω, say $\omega \geq \omega_0$, the mesh can be taken independently of ω. It can be easily seen that subdividing (z_j, z_{j+1}) into smaller subintervals and using the above approximation on each of the new subintervals produces the same result as applying the approximation to the whole element (z_j, z_{j+1}), if $\alpha(z)$, $\beta(z)$ and $\rho(z)$ are continuously differentiable throughout (z_j, z_{j+1}).

Based on the above considerations we determine for a particular mesh subinterval (z_j, z_{j+1}) the feasibility of using the approximation (2.29) as follows: Given an accuracy tolerance δ, require that

$$Re(\omega\nu_\beta)\, h_j \leq 4, \quad |\omega\nu_\beta| \geq |\beta'|\, \delta^{-1/2}, \quad |\omega\rho^{-1}| \geq |\rho'|\, \delta^{-1/2} \qquad (2.30a)$$

$$|\omega\nu_\alpha| \geq |\alpha'|\, \delta^{-1/2}, \qquad (2.30b)$$

$$Re(\omega\nu_\alpha)\, h_j \leq 6, \quad Re(\nu_\alpha) \leq Im(\nu_\alpha), \qquad (2.30c)$$

hold both at z_j and at z_{j+1}. If these conditions are satisfied, and there is no turning point between z_j and z_{j+1} (which is easy to check: simply verify that $Re(\beta - p^{-1})$ and $Re(\alpha - p^{-1})$ do not change sign), then we may use (2.29). Whether or not we actually use it depends on the element size h_j it would require compared to the element size that a collocation method would need at the same depth z_j, see §2.6. These conditions effectively prevent an unacceptable roundoff error accumulation.

2.4. Collocation and the midpoint rule

Let us recall the collocation method at Gaussian points. Basically, on a mesh element (z_j, z_{j+1}) with l collocation points, the constructed solution is a polynomial of degree l satisfying the differential equations exactly at the collocation points. The values of these polynomials on neighboring mesh elements are matched at the interfacing mesh points to form a continuous piecewise polynomial approximation on $[0, z_>]$, by requiring relations like (2.19) to hold.

The simplest instance of such a collocation scheme is with one collocation point taken at the midpoint of each element. This is known as the midpoint scheme (or the box scheme) and we have for (1.2),

$$R_j = I - \frac{1}{2}\,\omega h_j A\,(z_{j+1/2}) \tag{2.31a}$$

$$z_{j+1/2} = \frac{1}{2}(z_j + z_{j+1}),\ \ 1 \le j \le N$$

$$S_j = I + \frac{1}{2}\,\omega h_j A\,(z_{j+1/2}) \tag{2.31b}$$

Compared to a higher order collocation scheme, the construction of Γ_j (of (2.20)) here is much simpler (and cheaper); however the precision is lower, so smaller step sizes h_j are needed to achieve a given accuracy.

Consider next the selection of a mesh (2.15) in order to approximately satisfy the uniform accuracy requirements (2.18) using collocation. The principle behind a successful mesh selection (which attempts to make the number of mesh elements N as small as possible) is that of error equidistribution, i.e. choosing h_j such that the error contribution from the j-th element is approximately constant for each j, $1 \le j \le N$. Such a selection process requires some knowledge of the solution profile and so in general can only be performed adaptively (see [5] for an example of such an algorithm). However, for the P-SV and SH problems we can obtain some a priori knowledge of the solution, which allows us to select an appropriate mesh in advance and hence to compute the solution more cheaply, as described below.

The properties of the matrix $A\,(z\,)$ of (1.2) plus the boundary conditions at $z_>$ allow us to consider just downgoing waves. Thus consider one differential equation of the form

$$y\,' = -\omega\nu_\beta(z\,)y\,, \qquad y\,(0) = 1; \qquad Re\,(\omega\nu_\beta) \ge 0 \tag{2.32}$$

and we further assume that $\nu_\beta(z\,)$ is constant on each element $[z_j\,, z_{j+1})$ of the mesh to be constructed. These simplifying assumptions are particularly appropriate when $Re\,(\omega\nu_\beta)$ is large, which is also the case when the collocation method should be used and a good mesh selection is important.

The solution of (2.32) is

$$y\,(z\,) = \exp\{-\sum_{i=2}^{j} \omega\,(\nu_\beta(z_{i-1}) - \nu_\beta(z_i\,))\ z_i\ - \omega\nu_\beta(z_j\,)\ z\,\} \qquad z_j \le z \le z_{j+1} \tag{2.33}$$

So we can estimate

$$|\ y\,(z_j\,)\ | \ \approx \exp\{-\sum_{i=1}^{j-1} h_i\, Re\,(\omega\nu_\beta(z_i\,))\} \tag{2.34}$$

$$| y^{(2l)}(z_j) | \equiv | \partial_z^{2l} y(z_j) | = | \omega \nu_\beta(z_j) |^{2l} \, | y(z_j) |$$

Now, the error contribution from the j-th element is (as in Gaussian quadrature) proportional to

$$h_j^{2l} \, | y^{(2l)}(z_j) |$$

So the principle of error equidistribution requires choosing h_j so that

$$h_j^{2l} \, | \nu_\beta(z_j) |^{2l} \exp\{ -\sum_{i=1}^{j-1} h_i \, Re\,(\omega \nu_\beta(z_i)) \} = h_1^{2l} \, | \nu_\beta(0) |^{2l} \tag{2.35}$$

In practice we choose for the SH equations

$$h_j := h_1 \min\{1, | \frac{\nu_\beta(0)}{\nu_\beta(z_j)} | \} \exp\{ \frac{1}{2l} \sum_{i=1}^{j-1} h_i \, Re\,(\omega \nu_\beta(z_i)) \} \tag{2.36}$$

The value of h_1 is set by the required tolerance δ to

$$h_1 := cons_l \, | \omega \nu_\beta(0) |^{-1} \delta^{1/2l} \tag{2.37}$$

where $cons_l$ is a known constant, representing the Gaussian quadrature error constant with l points (see [7]). For the P-SV equations we calculate quantities as in (2.36), (2.37) also with ν_α replacing ν_β and choose for h_j the minimum of the two obtained values. In addition, if $min\,(Re\,\nu_\alpha(z_j), Re\,\nu_\beta(z_j)) < 0.1$ then, having less trust in this mesh selection process, we set $h_j := \frac{1}{2} h_j$.

From (2.37) it can be seen that h_1 may be taken larger for larger l and that this advantage of higher order collocation methods is particularly important when δ is very small. On the other hand, lower order (smaller l) methods are cheaper to construct per element. When selecting a mesh for given δ, ω and p, the most appropriate collocation scheme at a point z_j is the one which can cover a fixed size interval $(z_j, \bar{z})$ at the least expense. In practice, we consider only two alternatives: a higher order scheme with l collocation points (usually $l = 3$ or 4) or the midpoint scheme ($l = 1$). Calculating h_j by (2.36), (2.37), both for l collocation points and for the midpoint scheme, we would prefer the latter if

$$h_j(collocation) < h_j(midpoint) \cdot \epsilon_M$$

where ϵ_M is the relative efficiency of the midpoint scheme. Taking into account the number of $A(z)$ evaluations needed, the overhead per element in constructing R_j and S_j, the cost of the solution of the assembled sparse system of algebraic equations and the overhead in selecting the mesh, we use

$$\epsilon_M = \frac{3n^2 + 2/3\, n^3 + l + n^3(l-1)^2(1 + \frac{1}{3}(l-1))}{3n^2 + 2/3\, n^3 + 1} \tag{2.38}$$

with $n = 2$ or 4 for SH or P-SV problems, respectively.

2.5. Turning points and leaky modes

Here we consider the possibility that $\nu_\alpha(z) \approx 0$ or $\nu_\beta(z) \approx 0$, $z_j \leq z < z_{j+1}$. For simplicity of presentation, let us assume that the velocities α and β are real. Then comparing them to the phase velocity p^{-1} (recall (2.7)) we distinguish among 3 cases.

(i) $p^{-1} \approx \beta(z)$ (S wave turning point):

Then $\nu_\beta(z) \approx 0$, $\nu_\alpha^2(z) > 0$. This case, the only one occurring for the SH problem, is handled quite well by the collocation scheme, because $|\omega\nu_\beta|$ is not large, while ν_α is real.

(ii) $\beta(z) < p^{-1} < \alpha(z)$ ("leaky mode"):

This situation occurs between the P and S waves turning points. Then $\nu_\beta^2(z) < 0$, $\nu_\alpha^2(z) > 0$. For large ω we must decouple the S and the P wave components, because the fundamental matrix method of §2.3 is suitable only for the oscillatory S waves while the collocation method is suitable only for the evanescent P waves. Without separating these modes, neither method can be effectively applied. By (2.28), this can be done for

$$\hat{U} = (I + \omega^{-1}P_1)^{-1}\hat{V} \tag{2.39a}$$

which satisfies the diagonal differential system

$$\hat{U}' = (\omega\Lambda(z) + Q_1'(z))\hat{U} \tag{2.39b}$$

(see (2.24)) so each row of $\hat{U}$ can be dealt with separately.

For the P-SV problem we see from (2.6) that the second and fourth rows of $\hat{U}$ can be dealt with as in §2.3, while the first and third rows of $\hat{U}$ can be well-approximated by a collocation scheme. Applying the appropriate discretization to each row of (2.39b), we then transform back to the original variables to obtain (2.19). For details of the resulting scheme, see [6]. The feasibility of using this approximation is determined by requiring that conditions (2.30a,b) (but not (2.30c)) hold at both z_j and z_{j+1} and that there be no turning point there. The element size h_j is the minimum of the step size allowed by the collocation scheme for the P wave and the step size allowed by the body wave approximation for the S wave, for a given accuracy tolerance δ.

(iii) $p^{-1} \approx \alpha(z)$ (P wave turning point):

Then $\nu_\beta^2(z) < 0$, $\nu_\alpha(z) \approx 0$. Here again we wish to decouple P and S waves at high frequencies and use collocation for the P wave and the approximate fundamental matrix for the S wave. However, unlike case (ii) above we cannot use $D = E$ in (2.3), (2.4), because (2.2) does not hold. Instead, the decoupling can be done as described in (2.14). With a slight modification we obtain

$$D^{-1}AD = \begin{bmatrix} 0 & p & 0 & 0 \\ \nu_\alpha^2/p & 0 & 0 & 0 \\ 0 & 0 & \nu_\beta & 0 \\ 0 & 0 & 0 & -\nu_\beta \end{bmatrix} \equiv \hat{A} \equiv \begin{pmatrix} H_\alpha & 0 \\ 0 & \Lambda_s \end{pmatrix} \qquad (2.40)$$

where D^{-1} has terms like ν_β^{-1} but not ν_α^{-1} and so stays well-defined and smooth in the passage through an α turning point.

Let, as before

$$V(z;z_j) = D^{-1}(z)B(z;z_j)D(z_j), \quad U = (I - \omega^{-1}\hat{P})V \qquad z \geq z_j. \quad (2.41)$$

Note that V satisfies (2.3). The matrix $\hat{P}(z)$ is defined, similarly to P_1 of (2.26), (2.27), so that U satisfy

$$U' = (\omega\hat{A} + \tilde{A} + O(\omega^{-1}))U \qquad (2.42)$$

where $\tilde{A}$ has zero 2×2 off-diagonal blocks and a diagonal lower right block. Then we neglect the $O(\omega^{-1})$ term and (2.42) decouple. This allows us again to apply the appropriate scheme to each row of the decoupled system. The remaining construction is similar to case (ii) for leaky modes discussed earlier; for details see [6]. The feasibility of using this approximation is determined by requiring that conditions (2.30a) (but not (2.30b)) hold at both z_j and z_{j+1}.

Note that in (2.40) we are using an unusual transformation. Thus, it is not growing and decaying solution components which are decoupled here. The P wave components are handled together implicitly by the collocation scheme. Rather, the P wave is decoupled from the highly oscillatory S wave.

2.6. An algorithm for selecting mesh and method

For given values of frequency ω, slowness p and error tolerance δ, we wish to construct a mesh (2.15) and on each element of the mesh to specify which numerical scheme is to be used to obtain the solution $\{\mathbf{b}_j\}_{j=1}^{N+1}$ to within the specified accuracy. In the preceding section we have presented for each mesh element five possible schemes:

(1) Collocation with l points ($l > 1$, an input parameter; default value $l = 3$).
(2) The midpoint scheme (Collocation with $l = 1$)
(3) The body wave method (2.29)
(4) The leaky mode method (case (ii) of §2.5).
(5) The P- turning point method (case (iii) of §2.5).

Let

$$0 = d_1 < d_2 < \cdots < d_M = z_> \tag{2.43}$$

be a set of depths which must be included in the mesh (recall requirement (a) on (2.15) in §2.2. Also d_i may be a point where $\mathbf{b}(d_i)$ values are desired). The algorithm for mesh and method selection produces the mesh size N, an array $\{z_1, z_2, ..., z_{N+1}\}$ of mesh points (2.15) and an array $\{method_1, method_2, ..., method_N\}$, where $method_j$ indicates the scheme to be used on (z_j, z_{j+1}), $1 \leq method_j \leq 5$, $1 \leq j \leq N$.

The rationale behind the algorithm is quite simple: At each step, the more efficient between schemes (1) and (2) is used, unless one of the schemes (3)-(5) is both applicable and more efficient. Note that schemes (3), (4) and (5) are checked for feasibility in this order and at most one of them may be found feasible at any step.

Algorithm: Mesh and method selection

```
N:=1;  z_1:=0
FOR i:=1 to M-1 DO   {construct mesh between each two given depths}
  hh:=d_{i+1}-d_i
  WHILE hh>0 DO
     determine h_c = h_j from (2.36) as described in §2.4
     determine h_M similarly for the midpoint scheme
     IF h_M >= hh THEN   {use the midpoint scheme}
        method_N:=2;  h:=hh
     ELSE
        h:=max{h_c, ε_M h_M}
        {check feasibility of using schemes (3), (4), (5)}
        IF h<hh AND for some j, 3<=j<=5, scheme (j) is feasible with step size
              h*>h {as described in §2.3, 2.5; at most one j can qualify}  THEN
              method_N:=j;  h:=h*     {use scheme (j)}
        ELSE  {use collocation}
          IF h=h_c THEN
              method_N:=1;  h:=min{h, hh}
          ELSE
              method_N:=2;  h:=h_M
          END IF
        END IF
     END IF
     z_{N+1}:=z_N+h;  hh:=hh-h;  N:=N+1
  END WHILE
END FOR
```

3. An example

In order to demonstrate how our hybrid method works, we calculate synthetic seismograms for a test velocity structure whose parameters are given in Table 1. The structure is perfectly elastic (α and β are purely real), and β has been derived from α assuming a Poisson solid. All material properties vary linearly with depth except at the points given in Table 1.

The test model has been chosen to have two regions having rather different velocity gradients ($\alpha' = 2s^{-1}$ vs $\alpha' = 0.33s^{-1}$) in order to demonstrate how the computational algorithm of §2.6 chooses mesh points and solution schemes on each mesh subinterval depending upon ω, k, and the velocity gradient. Fig. 2 shows mesh points and solution methods chosen for various values of wavenumber k holding ω fixed. The cases displayed correspond, from top to bottom, to a highly evanescent wave, an S-turning point (case (i) of §2.5), a leaky mode (case (ii) of §2.5), a P-turning point (case (iii) of §2.5) and a situation above both turning points (§2.3). Note the automatic choice of schemes for each mesh subinterval, resulting in a hybrid method tailored for each case. The collocation scheme successfully resolves the S-turning point with a reasonably sparse mesh. In the highly evanescent case, collocation points are densely placed near the surface, where the solution varies fast. Then, for larger depths, the mesh becomes sparse according to (2.36) and, in fact the limit on the step size is eventually dictated by the next point of medium discontinuity, invoking the choice of the cheaper midpoint scheme.

The velocity gradients for $0 < z < 0.5$ are so large in this model that collocation had to be used in all cases for that part of the domain in Fig. 2. When $\nu_\beta^2 < \nu_\alpha^2 < 0$ it may happen that (2.30a) holds but (2.30b) does not, hence earlier use of the body wave approximation for the SH problem than for the P-SV problem is feasible. The use of schemes (3)-(5) for the P-SV problem in $0 < z < 0.5$ is enabled only at a higher frequency.

To show that the effort of constructing the hybrid method with the five schemes is worthwhile, we display in Fig. 3 corresponding results to those of Fig. 2 when using collocation only, with the mesh chosen by (2.36). Clearly, many more mesh points are needed at high frequencies to obtain a solution of a similar quality without using the schemes (3)-(5).

Having demonstrated how individual boundary value problems are solved, the next step is to integrate in the ω-k domain and construct the resulting synthetic seismograms. The sampling of the ω-k domain is controlled by the integral transform methods used; we proceed as in [22] and use a fast Fourier transform and a

Depth, km	α, km/s	β, km/s	ρ, g/cm^3
0	4.00	2.31	2.37
0.50	5.00	2.89	2.54
2.00	5.50	3.18	2.63
>2.00	6.00	3.46	2.71

Table 1. Test velocity model

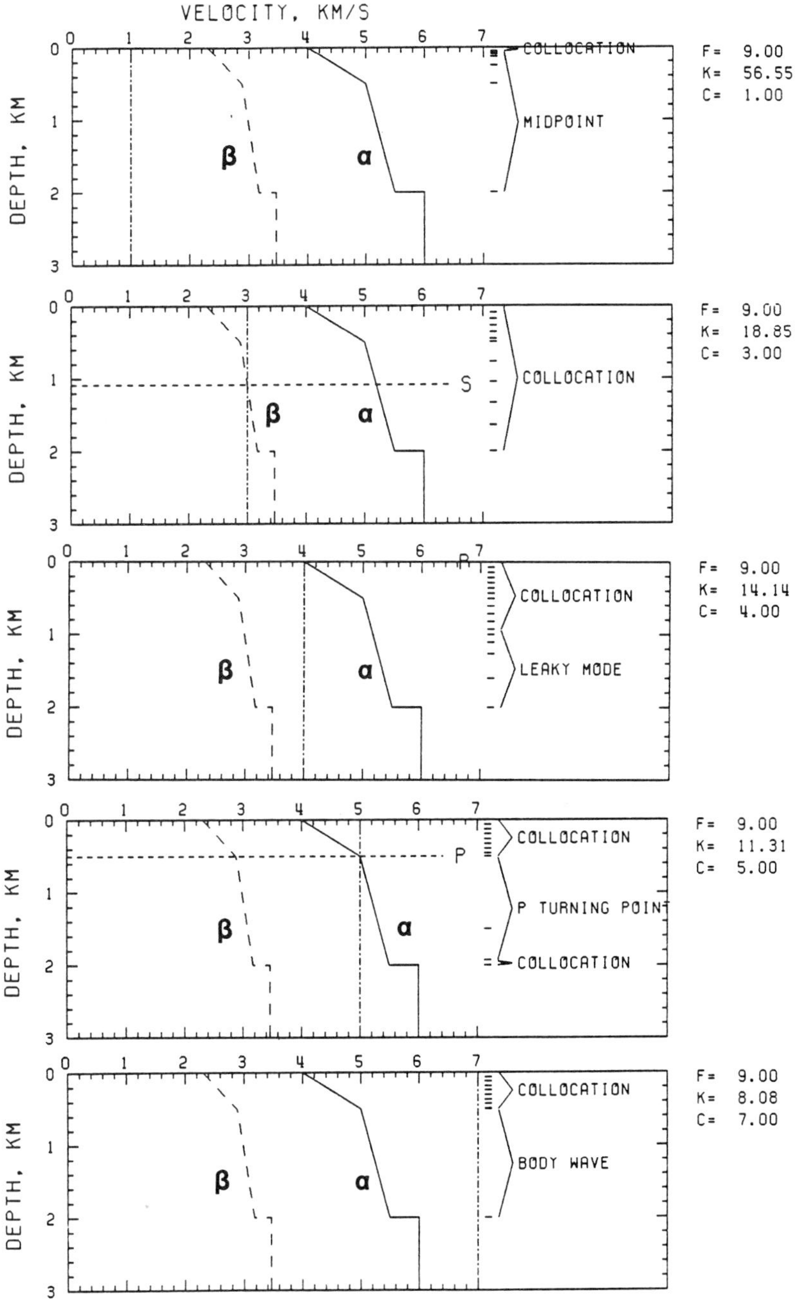

Figure 2. Mesh points and solution schemes chosen by the hybrid method

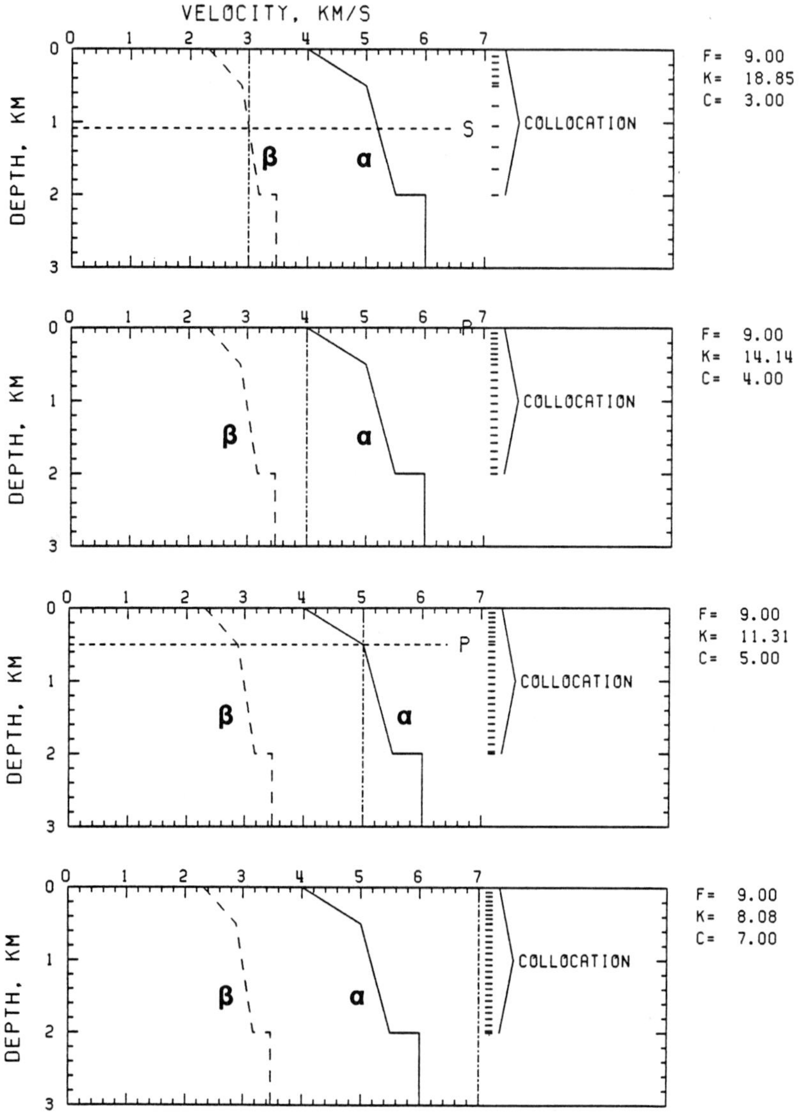

Figure 3. Mesh points for the P-SV problem when using collocation only.

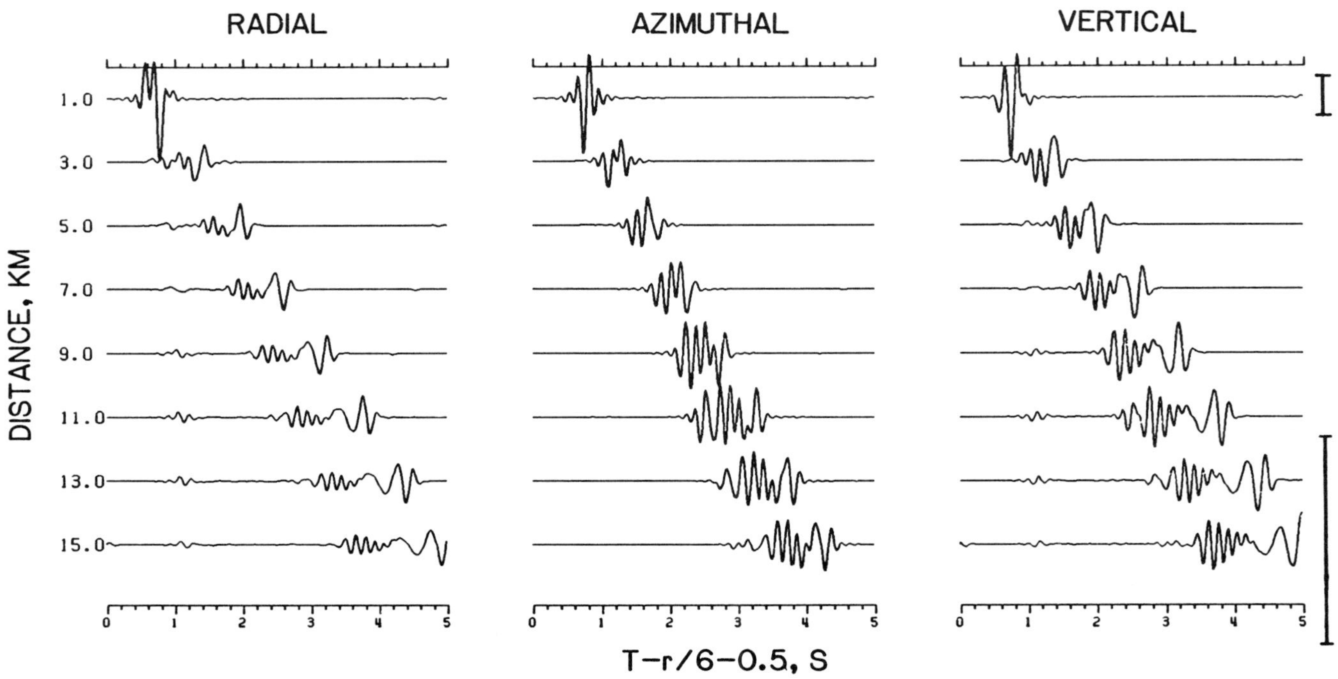

Figure 4. Theoretical seismograms for a point double-couple source

discrete-wavenumber transform. The synthetic seismograms for a 0.5-km-deep point double-couple source are shown in Fig. 4. Full details of this calculation are given in [6] and are beyond the scope of this presentation. Here we make a few comments.

These seismograms are calculated using our hybrid method and are identical to those calculated using collocation only (not shown). The radial and vertical components show a very small initial P wave well separated from a complicated, oscillatory wave train consisting of S body waves and higher-mode surface waves. The fundamental-mode Rayleigh wave is the large long-period pulse terminating the seismogram. For both P and S waves, the headwave travelling in the lower half-space becomes the first arrival at about 12 km epicentral distance. While not very evident in the P waveforms, the emergence of the S headwave is quite conspicuous in the azimuthal seismogram for 15 km distance.

While use of the hybrid method reduces computation time, the amount of the reduction is highly dependent upon the particular problem being solved. The greatest time savings are obtained for high-frequency solutions in velocity models with gentle gradients. For example, in our test problem (Fig. 4), solutions to the boundary value problem were obtained at about 23000 points in the ω–k domain. On a VAX-11/780, computation time was 20500s using collocation only (methods 1 and 2) and 11800s using all methods. Using the same test velocity structure in a higher frequency range would have resulted in (possibly much) greater time savings.

References

[1] Aki, K. & Richards, P.G., 1980. *Quantitative Seismology,* 1, W.H. Freeman, San Francisco.

[2] Alekseev, A.S. & Mikhailenko, B.G., 1980. The solution of the dynamic problem of elastic wave propagation in inhomogeneous media by a combination of partial separation of variables and finite difference methods, *J. Geophys. 48,* 161-172.

[3] Apsel, R.J., and Luco, J.E., 1983. On the Green's functions for a layered half-space, *Bull. Seism. Soc. Am., 73,* 931-951.

[4] Ascher, U. & Bader, G., 1985. Stability of collocation at Gaussian points, *SIAM J. Numer. Anal.,* to appear.

[5] Ascher, U., Christiansen, J. & Russell, R.D., 1977. A collocation solver for mixed order systems of boundary value problems, *Math. Comp. 33,* 659-679.

[6] Ascher, U. & Spudich, P., 1985. A Hybrid collocation method for calculating complete theoretical seismograms in vertically varying media, *Geophys. J. R. astr. Soc.* , to appear.

[7] Ascher, U. & Weiss, R., 1984. Collocation for singular perturbation problems II: linear first order systems without turning points, *Math. Comp. 43,* 157-187.

[8] Chapman, C.H., 1973. The earth flattening approximation in body wave theory, *Geophys. J. R. astr. Soc. 35,* 55-70.

[9] Chapman, C.H., 1974. The turning point of Elastodynamic waves, *Geophys. J.R. astr. Soc. 39,* 613-621.

[10] Chapman, C.H., 1976. Exact and approximate generalized ray theory in vertically inhomogeneous media, *Geophys. J.R. astr. Soc. 46,* 201-234.

[11] Frazer, L.N. and Gettrust, J.F., 1984. On a generalization of Filon's method and the computation of the oscillatory integrals of seismology, *Geophys. J.R. astr. Soc. 76,* 461-481.

[12] Haskell, N.A., 1953. The dispersion of surface waves on multilayered media, *Bull. seism. Soc. Am. 43,* 17-34.

[13] Kennett, B.L.N., 1983. *Seismic Wave Propagation is Stratified Media,* Cambridge University Press.

[14] Kennett, B.L.N. & Illingworth, M.R., 1981. Seismic waves in a stratified half space III - piecewise smooth models, *Geophys. J.R. astr. Soc. 66,* 633-675.

[15] Kennett, B.L.N. & Kerry, N.J., 1979. Seismic waves in a stratified half space, *Geophys. J.R. astr. Soc. 57,* 557-583.

[16] Olson, A.H., Orcutt, J.A. & Frazier, G.A., 1984. The discrete wavenumber/finite element method for synthetic seismograms, *Geophys. J. R. astr. Soc., 77,* 421-460.

[17] Phinney, R.A., 1965. Theoretical calculations of the spectra of first arrivals in layered elastic mediums, *J. Geophys. Res., 70,* 5107-5123.

[18] Richards, P.G., 1971. Elastic wave solutions in stratified media, *Geophysics, 36,* 798-809.

[19] Richards, P.G., & Frasier, C.W., 1976. Scattering of elastic waves from depth-dependent inhomogeneities, *Geophysics, 41,* 441-458.

[20] Scheid, R., 1982. The accurate numerical solution of highly oscillatory ordinary differential equations, Ph.D. Thesis, CalTech.

[21] Schwab, F., and Knopoff, L., 1972. Fast surface wave and free mode computations, in *Methods of Computational Physics, 11,* 87-180, ed. Bolt, B.A., Academic Press, New York.

[22] Spudich, P. & Ascher, U., 1983. Calculation of complete theoretical seismograms in vertically varying media using collocation methods, *Geophys. J. R. astr. Soc. 75,* 101-124.

[23] Takeuchi, H. & Saito, M., 1972, Seismic surface waves, in *Methods of Computational Physics, 11,* 217-295, ed. Bolt, B.A. Academic Press, New York.

ON A NEW BOUNDARY ELEMENT SPECTRAL METHOD

F. K. Hebeker

Abstract: An efficient numerical algorithm for partial differential equations in complicated 3-D geometries is developed in case of viscous fluid flows. The algorithm consists essentially of a combination of a boundary element method (where the resulting linear algebraic system is solved efficiently with a multigrid procedure) and a spectral method to treat the nonhomogeneous part in the differential equations. Our investigations cover an exact mathematical foundation, a rigorous convergence analysis, and some 3-D numerical tests.

1. Introduction

Recently, boundary element methods have been successfully applied to a large variety of problems arising in the engineering sciences (see e.g. Brebbia [3], Wendland [30]). Various additional fields of application may be opened if boundary element methods are combined with fast spectral methods to overcome the crucial problem of nonhomogeneous terms in the differential equations. These methods, called boundary element spectral methods, are briefly described in the present paper.

To be concrete, let us restrict ourselves to the Stokes problems of viscous hydrodynamics, but most of the proposed methods carry over easily to other problems of the applications. We consider the interior Stokes problem

$$\begin{aligned} -\Delta \underline{u} + \nabla p &= \underline{f}, \qquad \operatorname{div} \underline{u} = 0 \quad \text{in } \Omega \\ \underline{u}|_{\partial\Omega} &= \underline{g} \qquad \text{on the boundary } \partial\Omega. \end{aligned} \tag{1.1}$$

Here $\underline{u}$ denotes the velocity vector and p the scalar pressure function of a viscous flow in a smoothly bounded ($\partial\Omega \in C^\infty$, say) 3-D cavity Ω (at constant viscosity $\nu = 1$, say). Correspondingly we consider the exterior Stokes problem

$$-\Delta \underline{u} + \nabla p = \underline{f}, \quad \operatorname{div} \underline{u} = 0 \text{ in } \mathbb{R}^3 \setminus \bar{\Omega}$$
$$\underline{u}_{|\partial\Omega} = \underline{g} \quad \underline{u}_{|\infty} = 0 \text{ at infinity} \tag{1.2}$$

These Stokes problems play a central role in the development of efficient numerical algorithms for the viscous hydrodynamics, for the more general Navier Stokes equations are commonly reduced to problems similar to (1.1) or (1.2) by means of iterative or time-stepping procedures.

The present paper continues some work previously published by the author in [8], [9], [10], [11], [12], [13], [14]. More details and extensions of this paper will be published soon jointly with W. Borchers (cf. [2]).

2. The Boundary Integral Equations

The Stokes differential equations form a generalized elliptic system in the sense of Agmon, Douglis, and Nirenberg. However, a simple fundamental matrix

$$\Gamma_{ij}(x) = -\frac{1}{8\pi}\left(\frac{\delta_{ij}}{|x|} + \frac{x_i x_j}{|x|^3}\right); \qquad i,j = 1,2,3 \tag{2.1}$$

exists, and the ith column $\underline{\Gamma}^{(i)}$ of $\underline{\Gamma}$ solves Stokes equations with pressure function

$$P_i(x) = -x_i/(4\pi|x|^3) + \text{const.} \tag{2.2}$$

A hydrodynamical potential theory, fundamental to any boundary element method, is available (see Ladyzhenskaja [19]). In particular we are given a Green's representation formula for any regular solution of the Stokes differential equations as a sum of a (hydrodynamical) volumn -, simple layer -, and double layer potential.

A special attention should be payed to the volumn potential. In [13] a useful formula of numerical cubature is developed. But, in any case, the numerical evaluation of volumn potentials is an extremely expensive procedure. Hence to circumvent this task the way of using fast spectral solvers has been introduced by the author and subsequently elaborated jointly with W. Borchers. At first we suppress this task by putting $\underline{f} = \underline{0}$ in (1.1) and (1.2), we will return to the general case in Sec. 4.

The interior problem (1.1) is solvable if the data satisfy (see [19])

$$\int_{\partial\Omega}(\underline{g}\cdot\underline{n})do = 0 \tag{2.3}$$

In this case the ansatz

$$\underline{u} = \underline{W\psi}, \quad \text{p analogous} \tag{2.4}$$

in terms of a hydrodynamical double layer potential

$$(\underline{W\phi})_i(x) = \int_{\partial\Omega} \underline{t}'_z\,(\underline{\Gamma}^{(i)}(x-z))\cdot\underline{\phi}(z)do_z, \tag{2.5}$$

where

$$\underline{t}'(\underline{u},p) = p\underline{n} + \Sigma(\nabla u_j)n_j + \frac{\partial u}{\partial n} \tag{2.6}$$

denotes the (adjoint) stress vector and n the normal vector (exterior w.r.t. Ω), leads (by means of the jump relations) to the boundary integral equations system of the second kind (cf. Borchers [2])

$$(\underline{I} + 2\underline{W} + 2\underline{N})\underline{\psi} = 2\underline{g} \quad \text{on } \partial\Omega \tag{2.7}$$

to determine the unknown surface source vector $\underline{\psi}$. Here denotes $\underline{I}$ the identity matrix and N the boundary operator

$$\underline{N\phi}(x) = n(x) \int_{\partial\Omega} (\underline{n}\cdot\underline{\phi})do. \tag{2.8}$$

It turns out [2] that (2.8) has a unique solution $\underline{\psi}$ (when g satisfies (2.3)), and consequently the corresponding potential (2.4) solves the problem (2.1).

The exterior problem (2.2) is unconditionally solvable (if $f = 0$), and the ansatz (Hebeker [12], [13], [14])

$$\underline{u} = (\underline{W} + \eta\underline{V})\underline{\psi}, \quad p \text{ analogous} \tag{2.9}$$

in terms of a combined hydrodynamical double layer- and simple layer potential (η = const., arbitrary) produces the system

$$(\underline{I} - 2\underline{W} - 2\eta\underline{V})\underline{\psi} = -2g \quad \text{on } \partial\Omega. \tag{2.10}$$

It has been previously shown [13], that (2.10) has a unique solution $\underline{\psi}$ if η is chosen positive. Then the corresponding potential (2.9) solves the problem (2.2).

The surface potentials in (2.7) and (2.10) form weakly singular integrals on the boundary, hence these integral equations are well fitted for numerical purposes. For the evaluation of the potentials (2.4) and (2.9) in the space the Gaussian integral [19] should be utilized.

Further, the most valuable and simple formulae of the drag $\underline{D}$ and the moments $\underline{M}$ of a body Ω immersed in a viscous onflow should be notified (cf. [6], [13]):

$$\underline{D} = \eta \int_{\partial\Omega} \underline{\psi} \, do, \quad \underline{M} = \eta \int_{\partial\Omega} \underline{x} \times \underline{\psi} \, do. \tag{2.11}$$

3. The Boundary Element Method

The most simple and efficient way to discretize the boundary integral equations seems to be a collocation-type boundary element method. Let us describe it for the typical case of a body "similiar to a ball", i.e. represented by (normalized) polar coordinates:

$$x \in \partial\Omega : x = F(\theta,\varphi), \ (\theta,\varphi) \in [0,1]^2. \tag{3.1}$$

Note that a large class of bodies including even those with corners and edges are representable in this way.

The method consists of decomposing the parameter space $[0,1]^2$ of the boundary into small quadratic elements Q of mesh size h. As trial functions we use globally continuous and piecewise bilinear polynomials on the parameter space $[0,1]^2$, subordinated to the quadrangulation. Then we are looking for an approximate surface source ψ_h of this kind so that the unknown degrees of freedom are computed from the collocation equations

$$(\underline{I}_h + 2\underline{W}_h + 2\underline{N}_h)\underline{\psi}_h = 2g \quad \text{in } (\theta_i,\varphi_j) \tag{3.2}$$

for the interior problem, or

$$(\underline{I}_h - 2\underline{W}_h - 2n\underline{V}_h)\underline{\psi}_h = -2g \quad \text{in } (\theta_i,\varphi_j). \tag{3.3}$$

for the exterior problem. These equations have to be satisfied at the collocation nodes (θ_i,φ_j) only, namely at all of the grid points of the mesh. Hence we are led to linear algebraic system containing a nonsparse and nonsymmetric, but relatively small and compact system matrix. For details see [12], [13], [14], for a fast multigrid solver see [10], [13].

An important point is the chosen way of numerical quadrature of the surface integrals in (3.2) and (3.3). By choice of the trial functions they are reduced to integrals over the small elements Q, and consequently they are evaluated numerically by the

a) simple midpoint rule;

b) 2×2 Gaussian rule.

Note that, in any case, the quadrature nodes and the collocation nodes form disjoint sets, hence a special treatment of the weakly singular integrals is not always required.

But let us point out that, nevertheless, a special treatment of the singularity proves to be valuable. The crucial case is that the singularity is located at one corner of the element Q. For this we may utilize the formula of Duffy [4], to which we like to draw attention for its simplicity and applicability. For instance, consider the integral

$$F = \int_Q f(x,y)dydx, \quad \text{with } Q = [0,h]^2, \tag{3.4}$$

f affected with a singularity at the origin. Then by splitting Q into two triangels and relabeling,

$$F = \int_0^h \int_0^x (f(x,y) + f(y,x))dydx$$

is obtained. The transformation on triangular coordinates ("blowing-up")

$$x = s, \quad y = \frac{1}{h}st$$

produces an extra factor s in the integral:

$$F = \frac{1}{h} \int_Q \left(f(s,\tfrac{1}{h}st) + f(\tfrac{1}{h}st,s)\right)\cdot s\,ds\,dt. \qquad (3.5)$$

Therefore the numerical quadrature is applied to this last integral rather than to (3.4). We carried out a simple numerical example with $f(x,y) = 1/(x+y)$, using the 2×2 Gaussian rule: applying it to (3.4) directly we are faced to a relative error of about 10%, while by application to (3.5) a result accurate up to a relative error of 0,12% only is produced.

The considerable advantage of this formula for boundary element methods is obvious. Recently, it has been implemented by Atkinson [1], too, and a related but more sophisticated method has been analyzed by Scott and Johnson [25]. Following [25], this "contradicts the common opinion (cf. Wendland [29]) that [accurate] quadrature would require extensive work. It also makes the approximation of the surface by piecewise polynomials, as proposed by Nedelec [20], seem unnecessary". A more sophisticated approach that is applicable even to strongly singular and hypersingular integrals has recently been analyzed by Schwab and Wendland [24].

Finally, we notify a convergence result when applying the simple midpoint rule and neglecting a special treatment of the sinularity. The following estimates have been proved in [13]:

$$\sup_{\partial\Omega} |\psi_h - \psi| = O(h\cdot\log\tfrac{1}{h}), \qquad (3.6)$$

and for the corresponding potentials

$$\sup_{G} |\underline{u}_h - \underline{u}| = O(h\cdot\log\tfrac{1}{h}) \qquad (3.7)$$

holds, uniformly in any domain G with positive distance from $\partial\Omega$. The factor $\log\frac{1}{h}$ is originating from potential theoretic considerations.

4. The case of a Nonhomogeneity

Let us return now to the general case $f \neq 0$ in the problems (1.1) and

(1.2). The usual treatment by means of hydrodynamic volumn potentials proves to be extremely expensive, if a sufficiently high accuracy is required. For an impressive numerical test computation see below Sec. 5.

As an alternative here we propose a spectral method ("boundary element spectral method") seems to be very promising as preliminary numerical results have shown.

Consider the interior problem (1.1). We split this linear problem:

$$\underline{u} = \underline{u}_1 + \underline{u}_2, \qquad p = p_1 + p_2, \tag{4.1}$$

where $(\underline{u}_1, p_1)$ is any solution of the nonhomogeneous Stokes differential equations, but $(\underline{u}_2, p_2)$ solves (1.1) with $\underline{f} = \underline{0}$ and the boundary conditions

$$\underline{u}_2|_{\partial\Omega} = \underline{g} - \underline{u}_1|_{\partial\Omega} \quad . \tag{4.2}$$

The latter is numerically obtained by the boundary element method described in Sec. 2 and 3. Hence we merely have to determine any particular solution (u_1, p_1) of the nonhomogeneous Stokes equations.

Extend $\underline{f}$ (smoothly) to a cube containing Ω so that $\underline{f}|_{\partial C} = \underline{0}$. A complete set of eigensolutions of Stokes equations in a cube with periodicity condition has been constructed recently by Borchers [2]. These eigensolutions $\{\underline{e}_{k,\alpha}\}$ (with constant pressure!) are suitably composed trigonometric functions, hence they allow to use heavily the FFT for the numerical enforcement. Furthermore, in [2] an explicit series representation for the orthogonal projection of any Sobolev space (subjected to the periodicity condition on C) onto its divergence-free subspace is given.

Therefore, assume $\underline{f}$ divergence-free and periodic in C. Let $\underline{f}_0 = |C|^{-1} \int_C \underline{f} dx$. Then we split again:

$$\underline{u}_1 = \underline{u}_3 + \underline{u}_4, \qquad p_1 = p_3 + p_4 \quad . \tag{4.3}$$

Here $(\underline{u}_3, p_3)$ is obtained from the Stokes problem

$$-\Delta\underline{u}_3 + \nabla p_3 = \underline{f} - \underline{f}_0, \quad \text{div } \underline{u}_3 = 0 \quad \text{in } C \tag{4.4}$$

with peridicity condition, and $(\underline{u}_4, p_4)$ solves

$$-\Delta \underline{u}_4 + \nabla p_4 = \underline{f}_0, \text{ div } \underline{u}_4 = 0 \quad \text{in } C. \tag{4.5}$$

It has been shown by Borchers [2] that (4.4) is solvable (since $\int_C (\underline{f} - \underline{f}_0) dx = \underline{0}$), and any solution is given as a spectral series (which allows to use heavily the FFT). On the other hand, it is easy to construct a particular solution of (4.5).

The method indicated in this section easily extends to related problems of the applications.

5. Some Computational Results

Many numerical results in 3-D have been obtained by the author to test the proposed boundary element methods [8], [9], [10], [11], [12], [13], [14]. Subsequently we will give three test examples. Further, some encouraging numerical examples corresponding to the problem (4.4) of the spectral part are contained in [2]. Currently the combination of both software packages to test the full boundary element spectral method is under implementation.

The first example to our boundary element method concerns the test against Stokes' wellknown formula for the viscous drag of a sphere [23] which has been obtained by analytical means. Let the uniform onflow $v_\infty = (1,0,0)'$ be parallel to the x-axis. Then the exact value of the viscous drag of a ball of the radius R = 1 is

$$D = (6\pi, 0, 0)'$$

Our numerical computations have been carried out (without using any symmetry properties of the sphere) on a finite element mesh of the mesh size h = 1/16. The nonsparse system matrix here is of size 867x867, and the numerical computations are carried out with single precision on a PRIME 75o computer. The approximate drag has been computed from formula (2.11). Then by applying the different formulae of numerical quadrature a), b) the following comparison is obtained (only the x-component of the drag is given):

	computed drag	exact drag	relative error
quadrature a)	19.45	18.85	3.2 %
quadrature b)	18.99	18.85	o.7 %

In the second example we compute the flow around an irregular body, represented in polar coordinates by

$$r \leq 1 + \varepsilon.\sin^3 2\pi\theta.\cos 6\pi\theta + \varepsilon.\sin 2\pi\theta$$

using normalized polar coordinates $o \leq \theta,\phi \leq 1$. In case of $\varepsilon = o$ the body degenerates to a ball. When ε increases, some conical corners turn out at the "poles" $\theta = o$ or $\theta = 1$, and the body looses all of its symmetry properties. The body is illustrated in Fig. 1, where its cross-section in the x,z-plane is plotted in case of $\varepsilon = o.3$.

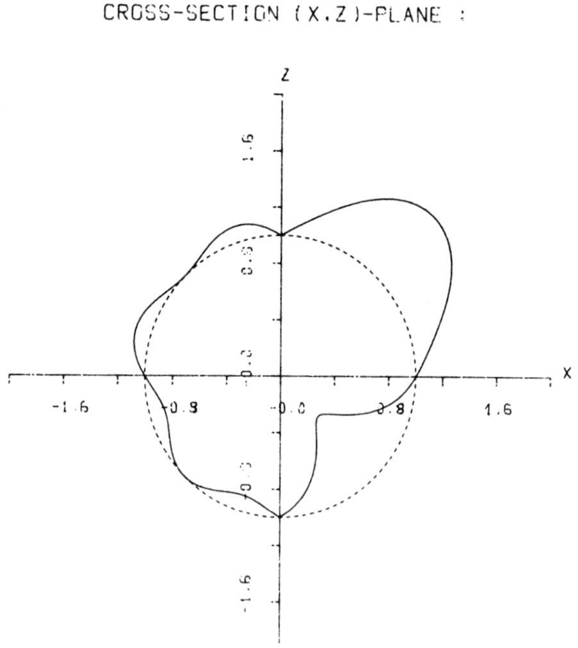

FIG. 1 : Irregular Body, Cross-Section x,z-Plane.

Moreover, in case of $\varepsilon = o.5$ a spine arises ($\theta = o.75$ in Fig. 1), hence any mathematical theory must fail there. In this case our boundary element method is tested against an artificial solution of the problem (1.2). We use those parameters of our first example and apply the formula a) of numerical quadrature. Here the mean relative error of the computed velocity field is given as a function of ε :

ε	o.o	o.1	o.2	o.3	o.4	o.5
relative error %	5.5	5.6	6.1	7.5	9.2	16.9 (!)

Particularly , the spine arising at θ = o.5 proves to destroy the numerical procedure, hence confirming the presently available mathematical theory (see Wendland [27],[29] in case of Laplace and Helmholtz equations).

In the third example our method is applied to solve the initial boundary value problem of the nonstationary Navier Stokes equations of homogeneous fluids, where we consider the flow included in a ball of radius R = 1 . By time differencing methods (implicit for the viscous term, but explicit for the convective term) we obtain for each time step a stationary problem similiar to (1.1), with the differential equations replaced by (time step δt)

$$\frac{1}{\delta t}u - \Delta u + \nabla p = f \quad , \quad \operatorname{div} u = o \qquad \text{in} \quad \Omega .$$

A simple fundamental solution corresponding to this is available, and a tested numerical algorithm is due to Varnhorn [26], who extended our computer programs. The results are given in Fig. 2 , where the parameters h = 1/12 and δt = 1/1o have been used. Here the computational work took many hours of computing time, mainly due to the frequent evaluation of the volumn potentials.

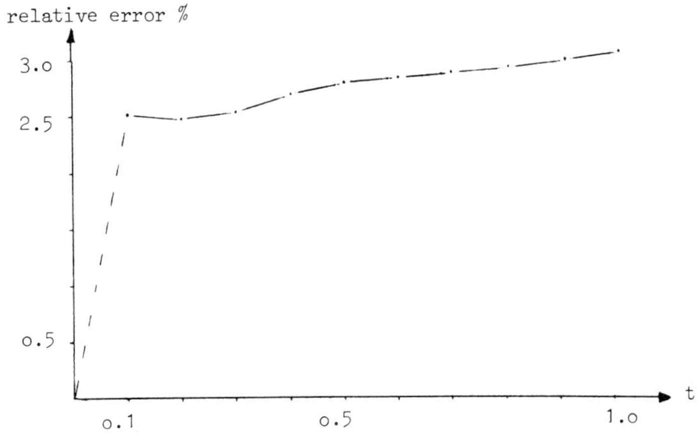

FIG. 2 : Nonstationary Navier Stokes Equations, Moderate Reynolds Number.

Recently, also a plot program has been created in order to visualize the viscous onflow in the case of a ball. Any projection of the velocity field onto a plane can be plotted. A first result is shown in Fig. 3 , where the uniform onflow is parallel to the (positive) x-axis, and the plane is fixed by one of its points (o,o,o.24) and its normal vector n = (o,o,1)' . And the flow configuration is observed from the point of view (5,-5,5) .

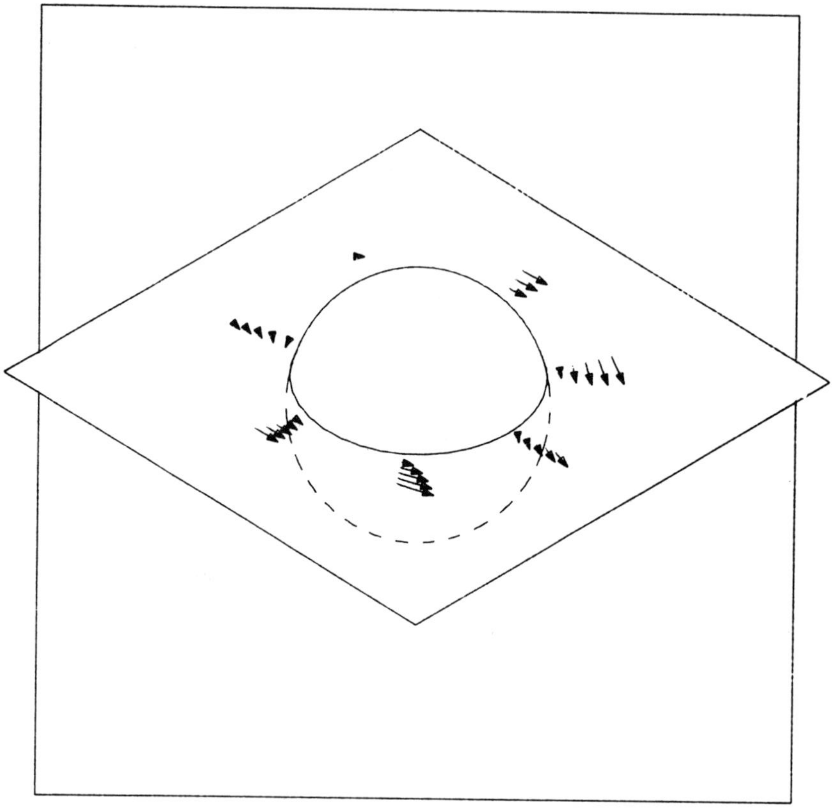

FIG. 3 : Visualization of 3-D Flow, Case of a Ball.

Acknowledgement

Some financial support due to the Deutsche Forschungsgemeinschaft is gratefully acknowledged.

References

[1] Atkinson K.E., Solving Integral Equations on Surfaces in Space. G. Hämmerlin (ed.), *Numerical Solution of Integral Equations.* Basel 1985, 2o-43.

[2] Borchers W., Eine Fourier-Spektralmethode für das Stokes-Resolventenproblem. Submitted for publication, Univ. of Paderborn 1985, 13 pp.

[3] Brebbia C.A. and Maier G. (eds.), *Boundary Elements VII.* 2 vols., Berlin 1985.

[4] Duffy M.G., Quadrature over a Pyramid or Cube of Integrands with a Singularity at a Vertex. *SIAM J. Numer. Anal.* 19 (1982), 126o-1262.

[5] Fischer T.M., An Integral Equation Procedure for the Exterior 3-D Slow Viscous Flow. *Integral Equ. Oper. Th.* 5 (1982), 49o-5o5.

[6] Fischer T.M. and Rosenberger R., A Boundary Integral Method for the Numerical Computation of the Forces Exerted on a Sphere in Viscous Incompressible Flows Near a Plane Wall. Preprint, Univ. of Darmstadt 1985, 3o pp.

[7] Hackbusch W., Die schnelle Auflösung der Fredholm'schen Integralgleichung zweiter Art. *Beitr. Numer. Math.* 9 (1982), 47-62.

[8] Hebeker F.K., A Theorem of Faxén and the Boundary Integral Method for 3-D Viscous Incompressible Flows. Techn. Rep., Univ. of Paderborn 1982, 14 pp.

[9] Hebeker F.K., A Boundary Integral Approach to Compute the 3-D Oseen Flow Past a Moving Body. M. Pandolfi and R. Piva (eds.), *Numerical Methods in Fluid Dynamics.* Braunschweig 1984, 124-13o.

[1o] Hebeker F.K., On a Multigrid Method for the Integral Equations of 3-D Stokes Flow. W. Hackbusch (ed.), *Efficient Solution of Elliptic Systems.* Braunschweig 1984, 67-73.

[11] Hebeker F.K., Zur Randelemente-Methode in der 3-D viskosen Strömungsmechanik. Habilitation Thesis, Univ. of Paderborn 1984, 127 pp.

[12] Hebeker F.K., A Boundary Element Method for Stokes Equations in 3-D Exterior Domains. J.R. Whiteman (ed.), *The Mathematics of Finite Elements and Applications.* London 1985, 257-263.

[13] Hebeker F.K., Efficient Boundary Element Methods for 3-D Exterior Viscous Flows. Submitted for publication, Univ. of Paderborn 1985, 35 pp.

[14] Hebeker F.K., Efficient Boundary Element Methods for 3-D Viscous Flows. C.A. Brebbia and G. Maier (eds.), *Boundary Elements VII.* Berlin 1985, 9:37-44.

[15] Hsiao G.C. and Kopp P. and Wendland W.L., A Collocation Method for some Integral Equations of the First Kind. *Computing* 25 (198o), 89-13o.

[16] Hsiao G.C. and Kopp P. and Wendland W.L., Some Applications of a Galerkin Collocation Method for Boundary Integral Equations of the First Kind. *Math. Meth. Appl. Sci.* 6 (1984), 28o-325.

[17] Hsiao G.C. and Kreß R., On an Integral Equation for the 2-D Exterior Stokes Problem. *Appl. Numer. Math.* 1 (1985), 77-93.

[18] Hsiao G.C. and Wendland W.L. and Fischer T.M., Singular Perturbations for the Exterior 3-D Slow Viscous Flow Problem. *J. Math. Anal. Appl.*, in press.

[19] Ladyzhenskaja O.A., *The Mathematical Theory of Viscous Incompressible Flow.* New York 1969.

[2o] Nedelec J.C., Approximation des Equations Integrales en Mechanique et en Physique. Ecole Polytechnique Palaiseau 1977, 127 pp.

[21] Novak Z., Use of the Multigrid Method for Laplacean Problems in 3-D. W. Hackbusch and U. Trottenberg (eds.), *Multigrid Methods.* Berlin 1982, 576-598.

[22] Quarteroni A. and Maday Y., Spectral and Pseudospectral Approximations of the Navier Stokes Equations. *SIAM J. Numer. Anal.* 19 (1982), 761-78o.

[23] Schlichting H., *Grenzschicht-Theorie.* 8th ed., Karlsruhe 1982.

[24] Schwab Ch. and Wendland W.L., 3-D BEM and Numerical Integration. C.A. Brebbia and G. Maier (eds.), *Boundary Elements VII.* Berlin 1985, 13:85-1o1.

[25] Scott L.R. and Johnson C., An Analysis of Quadrature Errors in second-kind Boundary Integral Equation Methods. Preprint, Univ. of Michigan (Ann Arbor), 47 pp.

[26] Varnhorn W., Zur Numerik der Gleichungen von Navier-Stokes. Doctoral Thesis, Univ. of Paderborn, to appear.

[27] Wendland W.L., Die Behandlung von Randwertaufgaben im $\mathbb{R}^3$ mit Hilfe von Einfach- und Doppelschichtpotentialen. *Numer. Math.* 11 (1968), 38o-4o4.

[28] Wendland W.L., On the Asymptotic Convergence of some Boundary Element Methods. J.R. Whiteman (ed.), *The Mathematics of Finite Elements and Applications.* London 1982, 281-312.

[29] Wendland W.L., Boundary Element Methods and Their Asymptotic Convergence. P. Filippi (ed.), *Theoretical Acoustics and Numerical Techniques.* CISM Courses and Lectures 277 , Wien 1983, 135-216.

[3o] Wendland W.L., On some Mathematical Aspects of Boundary Element Methods for Elliptic Problems. J.R. Whiteman (ed.),

The Mathematics of Finite Elements and Applications V. London 1985, 193-227.

[31] Zhu J., A Boundary Integral Equation Method for the Stationary Stokes Problem in 3-D. C.A. Brebbia (ed.), *Boundary Element Methods*. Berlin 1983, 283-292.

[32] Temam R., *Navier Stokes Equations and Nonlinear Functional Analysis*. SIAM, Philadelphia 1983.

PART III

HYPERBOLIC PDE PROBLEMS

A HIGH ORDER ESSENTIALLY NON-OSCILLATORY SHOCK CAPTURING METHOD

Bjorn Engquist, Ami Harten, Stanley Osher

Abstract

A special class of shock capturing methods for the approximation of hyperbolic conservation laws is presented. This class of methods produce essentially non-oscillatory solutions. This means that a Gibbs phenomenon at discontinuities is avoided and the variation of the numerical approximation may only grow due to the truncation error in the smooth part of the solution. The schemes have thus many of the desirable properties of total variation diminishing schemes, but they have the advantage that any order of accuracy can be achieved.

1. Introduction

In this paper we shall present the use of essentially non-oscillatory high order accurate schemes for the numerical approximation of weak solutions of conservation laws.

$$\begin{aligned} &u_t + f(u)_x = 0 \\ &u(x,0) = u_0(x). \end{aligned} \tag{1}$$

Here $u(x,t)$ is a state vector and the vector valued function f of u corresponds to a flux.

A computational difficulty with nonlinear conservation laws of this type is the occurence of discontinuities in the solution. A typical example is shocks in aerodynamics applications. Standard finite difference or finite element methods that are based on local approximations of smooth solutions often run into difficulties. Numerical oscillations are created close to the discontinuities.

In many so called high resolution shock capturing schemes upwind differencing is used in order to produce numerical solutions with good approximations at discontinuities. For an important class of discontinuities the upwind differencing has the effect that spatial derivatives are replaced by one sided or one sided biased differencing for which the computational stencil point away from the discontinuity. The errors in the approximation at a discontinuity does then not pollute the approximation in the smooth part of the solution [4].

For upwind schemes the useful property that differences are often pointing away from discontinuities and are thus applied to the smooth part of the solution is a by-product of the upwind design principle. For the non-oscillatory or essentially non-oscillatory schemes this property is a central part in the design [3], [6]. The computational stencil is automatically adjusted to give good approximations close to discontinuities. The form of the stencil is given by a non-oscillatory interpolation step in the algorithm. Furthermore these new schemes do not require limiters that reduce the order of accuracy at critical points [3]. High order upwind schemes need limiters [2].

An important feature in the new schemes is that the computational stencil is all the time adapted to the solution. A striking result of this adaptivity is the stability of the scheme. The algorithm is computationally stable even if local linear analysis indicates instabilities.

2. The Structure of the schemes.

The essentially non-oscillatory schemes are finite difference methods in conservation form. We assume that the initial value problem (1) is well-posed and denote its evolution operator by $E(t)$

$$u(\cdot, t) = E(t) \cdot u_0. \tag{2}$$

Let $\bar{w}(x)$ denote the sliding average of a function $w(x)$

$$\bar{w}(x) \equiv \frac{1}{h} \int_{-h/2}^{h/2} w(x+y)dy \equiv (A_h \cdot w)(x) \tag{3}$$

The sliding average in x of a weak solution of (1), $\bar{u}(x,t)$, satisfies

$$\frac{\partial}{\partial t}\bar{u}(x,t) + \frac{1}{h}[f(u(x+h/2,t)) - f(u(x-h/2,t))] = 0. \tag{4}$$

Integrating this relation from t to $t+\tau$, we get

$$\bar{u}(x,t+\tau) = \bar{u}(x,t) - \lambda[\hat{f}(x+h/2,t;u) - \hat{f}(x-h/2,t;u)] \tag{5}$$

where $\lambda = \tau/h$ and

$$\hat{f}(x,t;w) = \frac{1}{\tau}\int_0^\tau f(w(x,t+\eta))d\eta.$$

Let $\{I_j \times [t_n, t_{n+1}]\}$ where $I_j = [x_{j-\frac{1}{2}}, x_{j+\frac{1}{2}}]$, $x_\alpha = \alpha h$, $t_n = n\tau$, be a partition of $\mathbf{R} \times \mathbf{R}^+$. Writing relation (5) at $x = x_j, t = t_n$ we get

$$\bar{u}_j^{n+1} = \bar{u}_j^n - \lambda[\hat{f}(x_{j+\frac{1}{2}}, t_n; u) - \hat{f}(x_{j-\frac{1}{2}}, t_n; u)]. \tag{6}$$

Here

$$\bar{u}_j^n = \bar{u}(x_j, t_n) = \frac{1}{h}\int_{I_j} u(x, t_n)dx \tag{7}$$

is the "cell-average" of u at time t_n.

The class of numerical schemes described in this paper generalizes Godunov's scheme and its second order extensions to any finite order of accuracy [1], [2], [7], [8]. These schemes can be written in standard conservation form

$$v_j^{n+1} = v_j^n - \lambda(\bar{f}_{j+\frac{1}{2}} - \bar{f}_{j-\frac{1}{2}}) \equiv (\bar{E}_h(\tau) \cdot v^n)_j. \tag{8}$$

Here $\bar{E}_h(\tau)$ denotes the numerical solution operator and $\bar{f}_{j+\frac{1}{2}}$, the numerical flux, denotes a function of $2k$ variables

$$\bar{f}_{j+\frac{1}{2}} = \bar{f}(v_{j-k+1}^n, ..., v_{j+k}^n),$$

which is consistent with the flux $f(u)$ in (1.1) in the sense that $\bar{f}(u, u, ..., u) = f(u)$.

When $f(u)$ is a nonlinear function of u, the approximation of $\hat{f}(x_{j+\frac{1}{2}}, t_n; u)$ to $O(h^r)$ requires knowledge of pointwise values of the solution to the same order of accuracy. In order to design an accurate numerical flux , we must extract high order accurate pointwise information from the given $\{v_j^n\}$, which are approximations to $\{\bar{u}_j^n\}$, the cell averages (7) of the solution. Solving this reconstruction problem to arbitrarily high-order of accuracy r, without introducing $O(1)$ Gibbs-like spurious oscillations at points of discontinuity, is the most important step in the design of our new schemes.

Given $\bar{w}_j = \bar{w}(x_j)$, cell averages of a piecewise smooth function $w(x)$, we construct $R(x; \bar{w})$, a piecewise polynomial function of x of uniform polynomial degree $(r-1)$ that satisfies:

(i) At all points x for which there is a neighborhood where w is smooth

$$R(x; \bar{w}) = w(x) + e(x)h^r + O(h^{r+1}).$$

(ii) conservation in the sense of

$$\bar{R}(x_j; \bar{w}) = \bar{w}_j$$

here $\bar{R}$ denotes the sliding average (3) of R.

(iii) It is essentially non-oscillatory

$$TV(R(\cdot; \bar{w})) \leq TV(w) + O(h^r),$$

where TV denotes total-variation in x.

The inequality (iii) implies that the reconstruction R is essentially non-oscillatory in the sense that it does not have a Gibbs-like phenomenon of generating $O(1)$ spurious oscillations at points of discontinuity that are proportional to the size of the jump there.

Using the reconstruction above, we can express the abstract form of our new schemes by

$$\bar{E}_h(\tau) \cdot \bar{w} \equiv A_h \cdot E(\tau) \cdot R(\cdot; \bar{w}). \tag{9}$$

Here A_h is the cell-averaging operator on the RHS of (3); $E(t)$ is the exact evolution operator (2) and w is any piecewise smooth function of x.

In practice $E(\tau)$ has to be approximated. Since this step is common to all generalized Godunov schemes we shall not discuss it further but instead concentrate on the new step R in the following sections. The generalized Riemann solver $E(\tau)$ can be approximated in many different ways [1], [2], [3] [8].

3. Reconstruction.

In this section we present a brief description of the reconstruction $R(x; w)$ to be used in (9).For this purpose we introduce $H_m(x; w)$, a piecewise polynomial function of x that interpolates w at the points $\{x_j\}$, i.e.

$$\begin{aligned} &H_m(x_j; w) = w(x_j), \\ &H_m(x; w) \equiv q_{m,j+\frac{1}{2}}(x; w) \qquad \text{for } x_j \leq x \leq x_{j+1}, \end{aligned} \tag{10}$$

where $q_{m,j+\frac{1}{2}}$ is a polynomial in x of degree m.

We take $q_{m,j+\frac{1}{2}}$ to be the m-th degree polynomial that interpolates $w(x)$ at the $(m+1)$ successive points $\{x_i\}$, $i_m(j) \leq i \leq i_m(j) + m$, that include x_j and x_{j+1}.

We turn now to describe reconstruction via primitive function. It is also possible to do reconstruction via deconvolution, [3].

Given cell averages $\bar{w}_j$ of a piecewise smooth function w

(11) $$\bar{w}_j = \frac{1}{h_j}\int_{x_{j-\frac{1}{2}}}^{x_{j+\frac{1}{2}}} w(y)dy, \quad h_j = x_{j+\frac{1}{2}} - x_{j-\frac{1}{2}},$$

we can immediately evaluate the point-values of the primitive function $W(x)$

(12) $$W(x) = \int_{x_0}^{x} w(y)dy$$

by

(13) $$W(x_{j+\frac{1}{2}}) = \sum_{i=i_0}^{j} h_j \bar{w}_j.$$

Since

$$w(x) \equiv \frac{d}{dx}W(x)$$

we apply interpolation to the point values (13) of the primitive function $W(x)$ (12) and then obtain an approximation to $u(x)$ by defining

(14) $$R(x;\bar{w}) = \frac{d}{dx}H_r(x;W).$$

We note that this procedure does not require uniformity of the mesh.

The primitive function $W(x)$ is by one derivative smoother than $w(x)$, therefore it follows from the Theorem below that wherever $W(x)$ is smooth

$$\frac{d^k}{dx^k}H_r(x;W) = \frac{d^k}{dx^k}W(x) + O(h^{r+1-k}) :$$

thus we get from the definition (14) that

$$\frac{d^l}{dx^l}R(x;\bar{w}) = \frac{d^l}{dx^l}w(x) + O(h^{r-l}),$$

which implies (i) in section 2 for $l = 0$.

The conservation property of the reconstruction (ii) follows immediately from the definition (14):

$$\frac{1}{h_j}\int_{x_{j-\frac{1}{2}}}^{x_{j+\frac{1}{2}}} R(x;\bar{w})dx = \frac{1}{h_j}[H_r(x_{j+\frac{1}{2}};W) - H_r(x_{j-\frac{1}{2}};W)]$$
$$= \frac{1}{h_j}[W(x_{j+\frac{1}{2}}) - W(x_{j-\frac{1}{2}})] = \bar{w}_j.$$

The non-oscillatory nature of the reconstruction (iii) follows primarily from the non-oscillatory nature of the interpolation Theorem below.

4. Essentially non-oscillatory interpolation. Let us first define the piecewise polynomial $H_m(x;w)$, which interpolates $w(x)$ at the points $\{x_j\}$. In the interval $x_j \leq x \leq x_{j+1}$ H_m is given by the polynomial $q_{m,j+\frac{1}{2}}$ of degree m

(15) $$H_m(x;w) \equiv q_{m,j+\frac{1}{2}}(x;w) \text{ for } x_j \leq x \leq x_{j+1}.$$

We define $q_{m,j+\frac{1}{2}}$ to be the unique m-th degree polynomial in x that interpolates $w(x)$ at the $m+1$ points $\{x_i\}$, $i_m(j) \leq i \leq i_m(j)+m$ such that x_j and x_{j+1} is included, i.e.

(16) $$q_{m,j+\frac{1}{2}}(x_i;w) = w(x_i) \qquad \text{for } i_m(j) \leq i \leq i_m(j)+m.$$

(17) $$1-m \leq i_m(j) - j \leq 0.$$

Clearly there are exactly m such polynomials corresponding to the m different choices of $i_m(j)$ subject to (17). This freedom is used to assign to (x_j, x_{j+1}) a stencil of $(m+1)$ points so that $w(x)$ is "smoothest" in $(x_{i_m(j)}, x_{i_m(j)+m})$ in some asymptotic sense.

The information about smoothness of $w(x)$ is extracted from a table of divided difference of w. The latter can be defined recursively by

(18) $$w[x_i] = w(x_i)$$

(19) $$w[x_i,...,x_{i+k}] = (w[x_{i+1},...,x_{i+k}] - w[x_i,...,x_{i+k-1}])/(x_{i+k}-x_i).$$

It is well known that if w is C_∞ in $[x_i, x_{i+k}]$ then

(20) $$w[x_i,...,x_{i+k}] = \frac{1}{k!}\frac{d^k}{dx^k}w(\xi_{i,k}), \; x_i \leq \xi_{i,k} \leq x_{i+k}.$$

However if w has a jump discontinuity in the p-th derivative in this interval, $0 \leq p \leq k$, then

(21) $$w[x_i,...,x_{i+k}] = O(h^{-k+p}[w^{(p)}]);$$

here $[w^{(p)}]$ denotes the jump in the p-th derivative. (20) shows that $|w[x_i, ..., x_{i+k}]|$ provides an asymptotic measure of the smoothness of w in (x_i, x_{i+k}), in the sense that if w is smooth in (x_{i_1}, x_{i_1+k}) but is discontinuous in (x_{i_2}, x_{i_2+k}), then for h sufficiently small $|w[x_{i_1}, ..., x_{i_1+k}]| < |w[x_{i_2}, ..., x_{i_2+k}]|$. Hence the problem of choosing a stencil of points for which w is "smoothest" is basically the same as that of finding an interval in which w has the "smallest divided differences."

The following recursive algorithm is used to evaluate $i_m(j)$. We start by setting

$$i_1(j) = j, \tag{22}$$

i.e. $q_{1,j+\frac{1}{2}}$ is the first-degree polynomial interpolating w at x_j and x_{j+1}. Let us assume that we have already defined $i_k(j)$, i.e. $q_{k,j+\frac{1}{2}}$ is the k-th degree polynomial interpolating w at

$$x_{i_k(j)}, ..., x_{i_k(j)+k}.$$

We consider now as candidates for $q_{k+1,j+\frac{1}{2}}$ the two $(k+1)$-th degree polynomials obtained by adding to the above stencil the neighboring point to the left or the one to the right; this corresponds to setting $i_{k+1}(j) = i_k(j)-1$ or $i_{k+1}(j) = i_k(j)$, respectively. We choose the one that gives a $(k+1)$-th order divided difference that is smaller in absolute value, i.e.

$$i_{k+1}(j) = \begin{cases} i_k(j) - 1 & \text{if } |w[x_{i_k(j)-1}, ..., x_{i_k(j)+k}]| < |w[x_{i_k(j)}, ..., x_{i_k(j)+k+1}]| \\ i_k(j) & \text{otherwise.} \end{cases}$$

We shall now prove the basic property that this interpolation procedure is essentially non-oscillatory.

Theorem

For any piecewise smooth $w(x)$

$$TV(H_m(\cdot; w)) \leq TV(w) + O(h^{m+1}) \tag{23}$$

and if w is smooth at x

$$\frac{d^k}{dx^k} H_m(x; w) = \frac{d^k}{dx^k} w(x) + O(h^{m+1-k}, \quad 0 \leq k \leq m \tag{24}$$

Proof.

Consider the interval $x_j \le x \le x_{j+1}$ and study the two cases:

(i) w is smooth in $[x_j,\ x_{j+1}]$

(ii) w has a jump discontinuity in $[x_j,\ x_{j+1}]$.

Case (i):

If w is smooth over the full interval of interpolation $[x_{jl}, x_{jr}]$, $jl = i_m(j)$, $jr = i_m(j)+m$, standard interpolation results imply $H_m(x,w) = w(x)+O(h^{m+1})$ and both (23) and (24) follows. Otherwise, for h small enough there exist an interval containing $m+1$ consecutive points such that all divided differences $w[\ ,\ ,\ ]$ involving points in this interval are bounded independently of h. The definition of E_m guarantees that, for h small enough there will be no discontinuity of w in the interval of interpolation. This follows from the explicit form of the divided difference. If $[x_{j1}, x_{j2}]$ contains one discontinuity then

$$w[x_{j1}, ..., x_{j2}] = \frac{w(x_{j1}) - w(x_{j2})}{r!h^r} + O(h^{-r+1})$$

where $r = j2 - j1 + 1$. For h sufficiently small such intervals will always have a divided difference which is larger in absolute value than the divided difference over an interval without a discontinuity. Thus (23) and (24) follows as when $w(x)$ is smooth.

Case (ii):

We may assume that h is small enough such that $w(x)$ has only one discontinuity in $[x_{jl}, x_{jr}]$ and that the discontinuity is in $[x_j,\ x_{j+1}]$. For a given interval of interpolation we write $w(x)$ as a sum of a Lipschitz continuous function $v(x)$ and a piecewise constant function $z(x)$ with a single jump in $[x_j, x_{j+1}]$.

In $[x_j, x_{j+1}]$ we then have

$$H_m(x;w) = H_m(x;v) + H_m(x;z) \tag{25}$$

where

$$H_m(x;v) = \sum_{v=jl}^{jr} v[x_v, ..., x_{jr}] \prod_{\mu=v+1}^{jr} (x - x_\mu)$$

and

$$|v[x_v, ..., x_{jr}]| \le C\ h^{v-jr+1}$$

This implies that

(26) $$\left|\frac{d}{dx}\, H_m(x;v)\right| \le C$$

By Rolle's Theorem the interpolant $H_m(x;z)$ of the piecewise constant function must have an extremum in every interval (x_v, x_{v+1}) for $v \neq j,\ jl \le j \le jr$. This gives a total of $m-1$ extrema and since the interpolant is of degree m, it must be monotone in $[x_j, x_{j+1}]$.

Thus for $h = 1$ we have

$$\max_{x_j \le x \le x_{j+1}} \left|\frac{d}{dx} H_m(x;z)\right| \ge C > 0$$

and after scaling for a general h

(27) $$\max_{x_j \le x \le x_{j+1}} \left|\frac{d}{dx} H_m(x;z)\right| \ge \frac{c}{h}$$

Thus (26), (27) imply that $H_m(x;w)$ is monotone in $[x_j, x_{j+1}]$. On this interval

$$TV(Hm(x;w)) = |w(x_{j+1}) - w(x_j)| = TV(w)$$

which implies (23) and the theorem is proven.

The $O(h^{n+1})$ term in the theorem above can potentially create weak oscillations. We have seen that, in specially designed test cases, but not in realistic calculations. The practical need for a strictly non-oscillatory method is thus perhaps not so great.

It is easy to derive second order schemes which are non-oscillatory [5]. Higher order of approximation are also possible by limiting the divided differences w [ ] such that the first order divided difference dominates the higher order differences in the sum for the Newton interpolant q. The resulting q is then monotone if w is monotone. At a minimum or maximum of w the second order divided difference should dominate the higher order ones.

5. Numerical examples

In this section we shall present the application of the essentially non-oscillatory scheme to the Euler equations of gas dynamics for a polytopic gas:

$$u_t + f(u)_x = 0$$

$$u = (\rho, m, E)^T \tag{28}$$

$$f(u) = qu + (0, P, qP)^T$$

$$P = (\gamma - 1)(E - \frac{1}{2}\rho q^2).$$

Here ρ, q, P and E are the density, velocity, pressure and total energy, respectively; $m = \rho q$ is the momentum and γ is the ratio of specific heats.

The reconstruction R is done following the description above for each characteristic quantity separately. The characteristic quantities are defined from the system matrix, the Jacobian $\frac{\partial f}{\partial u}$.

For the evolution step E in the algorithm (9) the differential equation and a local Taylor expansion is used to define the values of u at $x_{j+\frac{1}{2}}$ and $x_{j-\frac{1}{2}}$ in formula (6). Details of the algorithm is given in [3].

The approximation to the solution of a Riemann problem is displayed on the following page. The solution contains a shock, a contact discontinuity and a rarefaction wave.

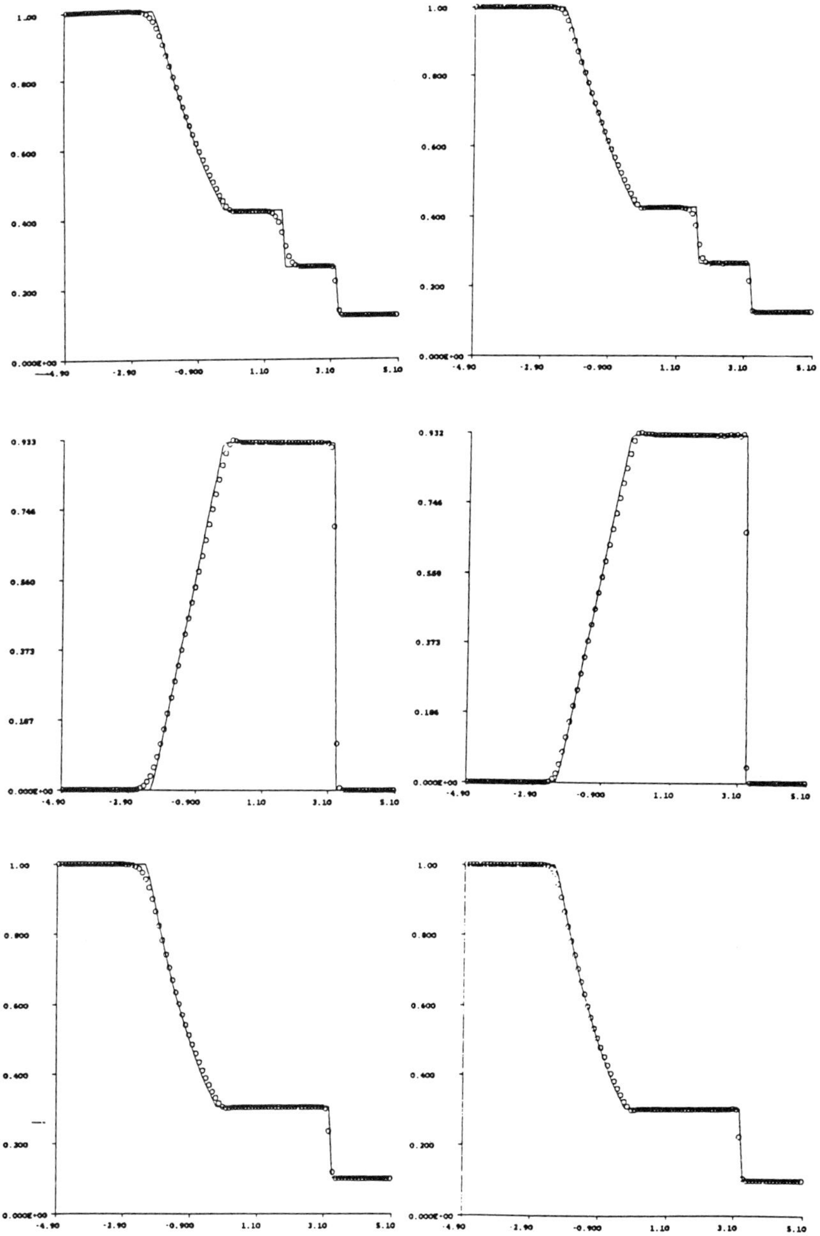

Approximation to Sod's problem. From above to below: density, velocity and pressure. Second order (left), fourth order (right).

Bibliography

[1] S.K. Godunov, "A finite difference method for the numerical computation of discontinuous solutions of the equations to fluid dynamics, " *Mat. Sb., 47* (1959), pp. 271-290.

[2] A. Harten, "High resolution schemes for hyperbolic conservation laws," *J. Comp. Phys.*, V. 49 (1983), pp. 357-393.

[3] A. Harten, B. Engquist, S. Osher, S. Chakravarthy, "Uniformly high order accurate essentially non-oscillatory schemes III," *ICASE Report No. 86-22* (1986).

[4] A. Harten, P.D. Lax and B. van Leer, "On upstream differencing and Godunov-type schemes for hyperbolic conservation laws," *SIAM* Rev., 25 (1983), pp. 35-61.

[5] A. Harten and S. Osher, "Uniformly high-order accurate non-oscillatory schemes, I.," *MRC Technical Summary Report #2823,* May 1985, to appear in SINUM.

[6] A. Harten, S. Osher, B. Engquist and S.R. Chakravarthy, "Some results on uniformly high-order accurate essentially non-oscillatory schemes," to appear in J. App. Num. Math.

[7] B. van Leer, "Towards the ultimate conservative difference schemes V. A second order sequel to Godunov's method," *J. Comp. Phys.*, V. 32, (1979) pp. 101-136.

[8] P.L. Roe, "Approximate Riemann solvers, parameter vectors, and difference schemes," *J. Comp. Phys.*, V. 43 (1981), pp. 357-372.

Research supported by NSF Grant No. DMS85-03294, ARO Grant No. DAAG29-85-K-0190, NASA Consortium Agreement No. NCA2-IR390-403, and NASA Langley Grant No. NAG1-270.

VORTEX DYNAMICS STUDIED BY LARGE-SCALE SOLUTIONS TO THE EULER EQUATIONS

Arthur Rizzi

1. ABSTRACT

Numerical solutions to the Euler equations have been found to exhibit the realistic behaviour of some vortex flows, but the fundamental question surrounding these is the way by which the vorticity initially is created in the flow. One explanation lies in the nonlinear occurrence of vortex sheets admitted as a solution to the Euler equations. The sheets may arise from the build up and breaking of simple shear waves, and seems associated with singular points of the surface velocity field where the flow separates from the body. In order to gain insight into this situation a numerical method to solve either the compressible or incompressible Euler equation in three dimensions is described. The computational procedure is of finite-volume type and highly vectorized, achieving a rate of 125 M flops on the CYBER 205, and allows the use of very dense meshes necessary to resolve the local features like the singular points. Several flow problems designed to test the vortex-sheet hypothesis are solved upon a mesh with over 600,000 grid cells. These solutions, even when the body is smooth, present numerical observations of such vortex sheets and singular points which support the hypothesis. One such computation runs in about 2 hours of CPU time on the CYBER 205 supercomputer.

2. INTRODUCTION

Vortex flows are among the most baffling to understand and the most challenging to simulate by numerical methods.

For the most part they are not steady flows, and very often not stable. Therefore it is advisable to look first at a generic subclass that under certain conditions is steady and stable, so-called edge-vortex flows (Fig. 1). Such a flow develops when a stream meets a flat delta-shaped plate at some incidence angle. For sufficiently large Reynolds number the flow separates in a vortex sheet all along the edges of the plate, and under the influence of its own vorticity, the sheet coils up to form a spiral vortex. The essential parameters governing the stability of this fundamentally three-dimensional situation are the angle of incidence α and the apex angle of the delta Λ.

The past few years have brought a number of computer simulations of such flows based on the Euler equations for either a compressible[1-3] or incompressible fluid,[4,5] which have been at least qualitatively realistic. The new feature of such inviscid simulations is the capturing of the vortex sheet as a part of the weak solution, in the same sense as shock waves have been captured. Hoeijmakers and Rizzi[4] have assessed the validity of such a result by carrying out the first comparison between a vortex sheet captured in a numerical Euler solution and one fitted as a discontinuity surface to a numerical solution of the Laplace equation for incompressible flow. But the fundamental question of what is the mechanism responsible for inviscid vortex separation remains open. Explanations suggested so far have pointed to macro-scale features of either the differential equations like a shock[6] or a geometrical singularity that forces the solution to develop a vortex sheet in order to remain bounded[7], or a feature present only in the discrete equations, like the truncation error which may be a source of vorticity[2].

In this paper I argue that the origin of the mechanism lies in the differential equations on the micro-scale of shear waves, which, if invoked by the boundary geometry, can grow nonlinearly into a vortex sheet. Several computed examples are presented to support the hypothesis starting with a 2D one exemplifying how truncation error causes separation, to 3D ones that indicate vortex sheets form even in the absence of a geometrical singularity. In order to minimize the truncation error, a fine grid of 614,000 grid cells is used for the latter examples.

3. INVISCID FLOW SEPARATION

The three common models for fluid flow are the potential equation, the Euler equations, and the Navier-Stokes equations. The common thinking holds that only the viscous model, the Navier-Stokes equations, can represent vortex separation because the vorticity can only be created in the boundary layer (also strong shocks). This viewpoint is founded on Kelvin's theorem which states that in the absence of viscous forces the circulation around a closed fluid circuit remains constant. The corrollary then is if it is initially zero somewhere, it must remain zero there after. But as Meyer[8] cautions us, we must be discerning when we apply this conclusion to the model of the Euler equations. Because it holds for a fluid circuit, Meyer points out (Fig. 2) that the theorem does not rule out the possibility of vortex-sheet separation for the Euler equations since such a sheet is a stream surface and a fluid circuit cannot traverse it. Unlike the potential equation, the Euler equations do admit the vortex-sheet discontinuity as a solution, just as they do shock waves. Vorticity exists only within the discontinuity, but may seep out due to diffusion in the numerical solution. The main hypothesis here then is that separation is fundamentally different for the Euler equations than for the potential equation. The central issue in this case comes down to explaining the origin of the vortex sheet. Does it exist in the differential model or is it only a feature of the discrete solution? In either case it is essentially a nonlinear phenomenon.

4. FORMATION OF VORTEX SHEET

Study of the underlying eigensystem of the differential system may give some clues on how a finite amplitude wave like a vortex sheet may be built up from simple infinitesmal waves. For simplicity let's consider the incompressible Euler equations in the context of the hyperbolic system given by the artificial compressibility method[9]. This system has three characteristic speeds $U \pm a$ and U where $U = \alpha u + \beta v$ is the velocity in the direction normal to the wave front (α, β). The first two are pseudo-acoustic fronts which are irrotational. The third is a simple velocity wave front with disturbances given by

$$\begin{pmatrix} \delta p \\ \delta u \\ \delta v \end{pmatrix} = k \; U \begin{pmatrix} 0 \\ -\beta \\ \alpha \end{pmatrix} \qquad (1)$$

which is a transverse, or shear wave, because the change produced is along the front itself and not in its direction of travel (Fig. 3). If we think of the front as the surface $|\vec{V}|$ = constant, then the wave rotates the velocity vector, setting up an infinitesimal shear. This is a rotational disturbance, not allowed in the potential model. It is not clear, however, how such a disturbance might build up into a finite-amplitude discontinuity. This family of characteristics does not overtake one another in the way that acoustic waves coalesce into shocks, but since the system is nonlinear, we cannot rule out the possibility of such growth, particularly near solid boundaries where the velocity is large and the waves may be focussed. In addition there may be anomalous growth due to the truncation errors in a numerical solution.

But it should be emphasized that numerical errors do not determine the strength of the resulting sheet. Koeck[2], for example, has suggested that the errors in the total pressure, which should be zero outside of the sheet in the diffferential model, may be the responsible agent for the formation of the sheet. More recently Powell et al.[10] have explained how such losses in total pressure are a consequence of capturing the sheet in a numerical solution, and not a cause for its development. The argument is based on the realization that the jump in velocity across the sheet must be supported by a profile of several discrete values. These points in transition from one side to the other must necessarily be in error (Fig. 4). Their numerical experiments also demonstrate that the level of loss is a function of the strength of the sheet i.e. $(\vec{V}_2-\vec{V}_1)\times\vec{n}$, and not particularly sensitive to the numerical dissipation.

5. STREAMLINE SINGULAR POINTS

In two-dimensional flow the condition for continuity prevents streamlines from running together. Separation necessarily involves a stagnation point. The addition of the third direction relieves this constraint, so we must investigate the types of singular points that can occur on the surface of a body where streamlines meet one another. Following Smith's analysis[11], let ξ and η be local coordinates on the body surface with origin at the singular point where the velocity components are both zero $u(0,0)=v(0,0)=0$. We want to determine the integral curves of the equation $(d\xi)/(u(\xi,\eta))= (d\eta)/(v(\xi,\eta))$ passing through the origin. Let's assume that u and v are well behaved at the origin in the sense

$$\begin{pmatrix} u \\ v \end{pmatrix} = \begin{bmatrix} \alpha & \beta \\ \gamma & \delta \end{bmatrix} \begin{pmatrix} \xi \\ \eta \end{pmatrix} + \dots \qquad (2)$$

with $\alpha\delta \neq \beta\gamma$. Smith's analysis therefore is linear, but singular points do arise for nonlinear systems where the Jacobian is zero. In the linear case he finds the six curves presented in Fig. 5, of which only the first four, the nodes and saddle-points, can occur in the streamline pattern of irrotational flow. The remaining two, the spiral point and the center point, requires the flow to be rotational. It seems worthwhile to ask whether, in connection with the emergence of a vortex sheet, does a spiral or center point arise in the numerical solution of the Euler equations. In order to gain insight into the matter, I carry out a number of numerical experiments to try to observe such features. The calculations are performed on a large scale with a fine mesh of 614,000 cells in three dimensions so that the resolution is as high, and the truncation as low, as possible with today's advanced supercomputers.

6. FINITE-VOLUME METHOD

The first step in carrying out a numerical solution is to discretize the flowfield by creating a mesh that spans the entire region. Eriksson's method of transfinite interpolation[12] constructs a boundary-conforming O-O grid around a 70 deg. swept delta wing (Fig. 6). An equivalent form of the compressible Euler equations can be expressed as an integral balance of the conservation laws

$$\frac{\partial}{\partial t} \iiint q \, dvol + \iint \underset{\sim}{H}(q) \cdot \underset{\sim}{n} \, dS = 0 \qquad (3)$$

where q is the vector with elements $[\rho, \rho u, \rho v, \rho w]$ for density and Cartesian components of momentum with reference to the fixed coordinate system x,y,z. The flux quantity $\underset{\sim}{H}(q) \cdot \underset{\sim}{n} = [q\underset{\sim}{V} + (0, \underset{\sim}{e}_x, \underset{\sim}{e}_y, \underset{\sim}{e}_z)p] \cdot \underset{\sim}{n}$ represents the net flux of q transported across, plus the pressure p acting on, the surface S surrounding the volume of fluid. The mesh segments the flowfield into very many small six-sided cells, in each of which the integral balance (3) must hold. The finite-volume method then discretizes (3) by assuming that q is a cell-averaged quantity located in the center of the cell, and the discretized flux term $[\underset{\sim}{H} \cdot \underset{\sim}{S}]_{ijk} = [\underset{\sim}{H}(\mu_I q_{ijk}) \cdot \underset{\sim}{S}_I + \underset{\sim}{H}(\mu_J q_{ijk}) \cdot \underset{\sim}{S}_J + \underset{\sim}{H}(\mu_K q_{ijk}) \cdot \underset{\sim}{S}_K]$, where μ is the averaging operator, is defined only at the cell faces by averaging the values on each side (see Fig. 7). With these definitions and adding the artificial viscosity model T to make the system slightly dissipative in the sense of Kreiss, we obtain the semi-discrete finite-volume form for cell ijk

$$\frac{d}{dt} q_{ijk} + [\delta_I(\underset{\sim}{H} \cdot \underset{\sim}{S}_I) + \delta_J(\underset{\sim}{H} \cdot \underset{\sim}{S}_J) + \delta_K(\underset{\sim}{H} \cdot \underset{\sim}{S}_K)]_{ijk} \cdot \underset{\sim}{n} = T \qquad (4)$$

where $\delta_I(\underset{\sim}{H}\cdot\underset{\sim}{S}_I) = (\underset{\sim}{H}\cdot\underset{\sim}{S}_I)_{i+1/2} - (H\cdot S_I)_{i-1/2} = \delta_I f_I$ is the centered difference operator. To this the boundary conditions for the particular application must be specified. They occur at the six bounding faces of the computational domain and when included, we can write Eq.(4) in the semi-discrete form

$$\frac{d}{dt} q_{ijk} + FD = 0 \tag{5}$$

where FD now represents the total flux difference, i.e. the differences of the convective and dissipative fluxes including the boundary conditions. A more detailed description of the method is given in Ref. 13. With the spatial indices suppressed we integrate this last equation with the two-level three-stage scheme in parameter $\theta=1/2$ or 1

$$\begin{aligned} q^{(0)} &= q^n \\ q^{(1)} &= q^n + \Delta t\ FD(q^{(0)}) \\ q^{(2)} &= q^n + \Delta t\left[(1-\theta)FD(q^{(0)}) + \theta\ FD(q^{(1)})\right] \\ q^{(3)} &= q^n + \Delta t\left[(1-\theta)FD(q^{(0)}) + \theta\ FD(q^{(2)})\right] \\ q^{n+1} &= q^{(3)} \end{aligned} \tag{6}$$

that steps the solution forward in time until a steady state is reached. The number of cells in a suitable grid around a wing can range from 50 thousand to several millions and the computational burden to solve Eq.(6) repeatedly for say 1000 time steps is large and requires a supercomputer with the greatest speed and largest memory.

Algorithm (6) is explicit, i.e. the state q^n for each point known at time t is updated pointwise to the new state q^{n+1} by simple algebraic equations involving no simultaneous solution of equations or recursive relations. Therefore the updating of different points in the field are independent of each other and can be carried out concurrently. And it is this concurrency that has led to the reexamination of explicit algorithms for large computational problems with the aim to maximize the obtainable vector processing rate.

7. DATA STRUCTURE AND METHODOLOGY FOR VECTOR PROCESSING

A three dimensional structure suitable for vector processing is correctly visualized as consisting of a collection of adjacent pencils of memory cells with suitable boundary conditions. This leads to the natural grouping for the formation of differences in J by means of differencing adjacent pencils. The natural differencing in the K-direction thus becomes the difference of a plane of adjacent pencils. Differences in the I-direction are formed by use of single elements offset one against the next within the pencil. See Fig. 8 for these relationships which allow for the complete vectorization of all of the internal difference expressions. In the I-direction the contiguous group is the individual storage cell, in the J-direction it is the pencil of all I cells for a given J and K, and in the K-direction it is the plane of all I*J cells for a given K. Because of the ordering of that data by the grid transformation, the total vector difference throughout the interior domain can then be formed by subtracting the respective group from its forward neighbor simply by off-setting the starting location of the flux vector for each of the three directions. The inherent concurrency of the algorithm now becomes apparent. In this way all of the work in updating interior points is exclusively vector operations without any data motion, because of the ordering that results from the structured grid. The problem remaining is how to interleave the boundary conditions into this long vector. They require different operations and interrupt the vector processing of the interior values. One approach, in conjunction with the data-structure design, that overcomes this interruption of vector processing leads to the following key features of the procedure: 1) separate storage arrays are assigned for the dependent variables q, flux component F, and flux differences FD, 2) dimension one extra group in each direction for q(I+1,J+1,K+1) and F(I+1,J+1,K+1) in order to insert the forward boundary values in q and to provide intermediate scratch space during the flux computation which is then overwritten with the rear boundary condition on F. Four variables are updated at each boundary cell. These boundary conditions are set on every stage of each iteration. In each direction the starting boundary value is established first on the property q, and the ending value on the flux F. Now following this procedure we can difference the entire three-dimensional flux field F together with its boundary conditions with one vector statement involving vectors whose length can be as large as the total number of nodal points in the grid. The variable q is then eventually updated as a function of the differences in F (i.e. FD) according to Eq.(6). Figure 9 attempts to illustrate this algorithm schematically for differences in the K-direction which can be computed with full vectorization and no data motion. To begin, the right boundary (i.e. last plane, K+1) values are loaded in q(1,1,K+1;M) in one vector statement where M= (I+1)* (J+1) is the number of elements in the plane. This value is picked up by the averaging operator μ when computing the leading or outward (i.e. K+1) flux for the Kth plane of cells. The left boundary (first plane) then must be set, and because there is scratch storage already there, I choose to enforce this directly on the flux F(1,1,1;M) entering the first plane of cells. Now all differences in the K-direction are ready to be taken in one vector instruction simply by offsetting the starting location of the flux vector

FD(1,1,1;N) = F(1,1,2;N) - F(1,1,1;N)

Diagrammatically the data flow in the algorithm may be viewed as a series of steps starting with q and pyramiding up to the flux difference FD over the whole K-direction.

8. COMPUTED EXAMPLES OF VORTEX FLOWS

In this section we see the results computed with either the compressible or incompressible Euler model of flow around bodies in a variety of geometrical shapes. The first is a circular cylinder, the next two are delta wings, one flat with a sharp leading edge and the other 6% thick with a rounded edge, and the last is an elongated spheroid. These were chosen as good tests to bring out the effects of geometry and the addition of the third dimension.

Transverse Circular Cylinder

Incompressible flow past a transverse circular cylinder is two dimensional and is described by the analytical solution to the Laplace equation. Singular points cannot occur and, since the flow is steady, neither may vortex sheets remain in the flow. The only mechanism left, therefore, that we have discussed to account for vortical features, is numerical truncation error. Figure 10a demonstrates how, when these errors are large due to a poor mesh distribution and an inappropriate second-order artificial viscosity, the flow, as it begins to recompress on the lee-side of the cylinder, looses total pressure and cannot reach the near-most stagnation point. Instead it stagnates somewhat before that point, separates, and then because of continuity, fills in behind forming a recirculation region. Completely bogus, this is simply a feature of poor numerical accuracy. When the mesh is improved and a more suitable fourth-order artificial viscosity is used, a solution is obtained (Fig. 10b) closely approximating the exact analytical one. It seems then, unlike an airfoil, the circular geometry does not excite any transient vortex sheets. The physical mechanism for generating sheets, however, are much stronger in three dimensions, to which we now turn.

Two Delta Wings

Since I want to explore the vortex-sheet generating mechanism, it's very important to hold the truncation errors to a minimum. An effective means is to concentrate computational cells locally to the region where the gradients are greatest, and to use meshes with as many total number of cells as current computer memories allow. In my case using the CYBER 205 with 8 M words of 32-bit length, the largest O-O mesh contains 160 cells around the half span, 80 along the chord, and 48 outward from the wing surface. At the freestream flow conditions M_∞=0.7, α=10 deg. Fig. 11a shows the results computed for a flat 70-deg. swept delta wing of zero thickness and with a sharp leading edge. The concentric isobar patterns over the wing are typical of shed vortex flow. The vortex begins in a flow singularity at the apex of the wing and remains highly conical over the centire wing. The effects of flow compressibility do not appear to be significant here. But the results in Fig. 11b demonstrate the dramatic effect of changing the wing cross-section to one of 6% thickness and a round leading edge. Much reduced in size the vortex pattern now is nonconical. It begins somewhere past the 50% chord, not at the apex, and the actions of compressibility are larger. Spanwise views of the v-w velocity vectors indicate that the flow is indeed separating in a vortex sheet from a geometrically smooth wing surface. A nonpotential feature, it strongly suggests that, if the body shape has a bounded but sufficiently high curvature, a vortex sheet forms in the flow, perhaps by the build-up and eventual breaking of shear waves.

Prolate Spheroid

Do three-dimensional vortex phenomena appear only if the body is shaped like a wing, and if they are found for non-wing bodies, are they the result of truncation error as in the case of the cylinder? The fine-grid solution of incompressible flow past a 9 1/2:1 prolate spheriod at 10 deg. incidence using a mesh of 144×60×72 cells offers some insight. The side view of the isobars and surface velocity vectors (Fig. 12) indicate attached flow on the fore region of the body. By midbody the leeside flow begins to meet the stream from the windside creating a pinching effect that may lead to vortex-sheet separation. But near the tail where the tendency to separate is greatest, the velocity vectors in the body surface show a feature very much like a spiral singularity. It is very well supported in this mesh, and therefore probably is a real aspect of the Euler equations, and another example of the formation of a vortex sheet. In any case it is not a potential phenomenon, but just what relevance it has to a real flow must be determined in a further investigation.

9 CLOSING COMMENTS

The paper addresses the issue of vortex phenomena modelled by the Euler equatons. Instead of the usual explanations of a transient shock wave or numerical truncation error being responsible for the creation of vortex sheets, I argue that it may be the nonlinear build-up of shear waves encompassed in this system of equations. Three numerical experiments are presented to show that, when the grid is sufficiently fine and the truncation error is small, vortex sheets are observed to form at the body. The formation of the sheets seems to be associated with singular points in the surface velocity.

10 REFERENCES

1 Rizzi, A., Eriksson, L.E., Schmidt, W. and Hitzel, S.: "Numerical Solutions of the Euler Equations Simulating Vortex Flows Around Wings", AGARD-CP-342, 1983.

2 Koeck, C.: "Computation of Three Dimensional Flow Using the Euler Equations and a Multiple-Grid Schemes", Int. J. Num. Meth. Fluids, Vol. 5, 1985, pp. 483-500.

3 Raj, P. and Sikora, J.S.: "Free-Vortex Flows Recent Encounters with an Euler Code", AIAA Paper 84-0135, Jan 1984.

4 Hoeijmakers, H.W.M. and Rizzi, A.: "Vortex-Fitted Potential Solution Compared with Vortex-Captured Euler Solution for Delta Wing with Leading-Edge Vortex Separation", AIAA Paper 84-2144, 1984.

5 Krause, E., Shi, X.G. and Hartwich, P.M.: "Computation of Leading-Edge Vortices", AIAA Paper 83-1907, 1983.

6 Salas, M.D.: "Foundations for the Numerical Solution of the Euler Equations", in Advances in Computational Transonics, ed. W.G. Habashi, Pineridge Press, 1985.

7 Schmidt, W. and Jameson, A.: "Euler Solvers as an Analysis Tool for Aircraft Aerodynamis", in Advances in Computational Transonic, ed. W.G. Habashi, Pineridge Press, 1985.

8 Meyer, R.E.: "Introduction to Mathematical Fluid Dynamics, Wiley-Interscience, New York, 1971, p. 75ff.

9 Rizzi, A. and Eriksson L.E.: "Computation of Inviscid Incompressible Flow with Rotation", J. Fluid Mech., Vol. 153, April 1985, pp. 275-312.

10 Powell, K., Murman, E., Perez, E, and Baron, J.: "Total Pressure Loss in Vortical Solutions of the Concial Euler Equations", AIAA Paper 85-1701, 1985.

11 Smith, J.H.B.: "Remarks on the Structure of Conical Flow", in Progress in Aerospace Sciences, ed. D. Kuchemann, Pergamon Press, Oxford, 1972.

12 Eriksson, L.E.: "Generation of Boundary-Conforming Grids around Wing-Body Configurations using Transfinite Interpolation", AIAA J., Vol. 20, Oct 1982, pp. 1313-1320.

13 Rizzi, A. and Eriksson, L.E.: "Computation of Flow Around Wings Based on the Euler Equations", J. Fluid Mech., Vol. 148, Nov 1984, pp. 45-71.

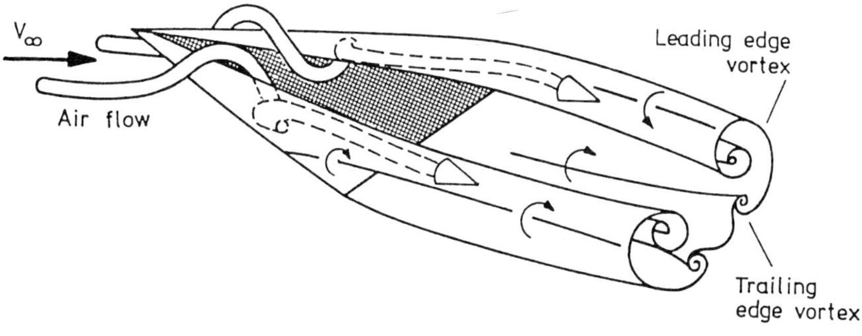

Fig. 1 Schematic of rolling up of shed vortex sheets into stable vortex structures.

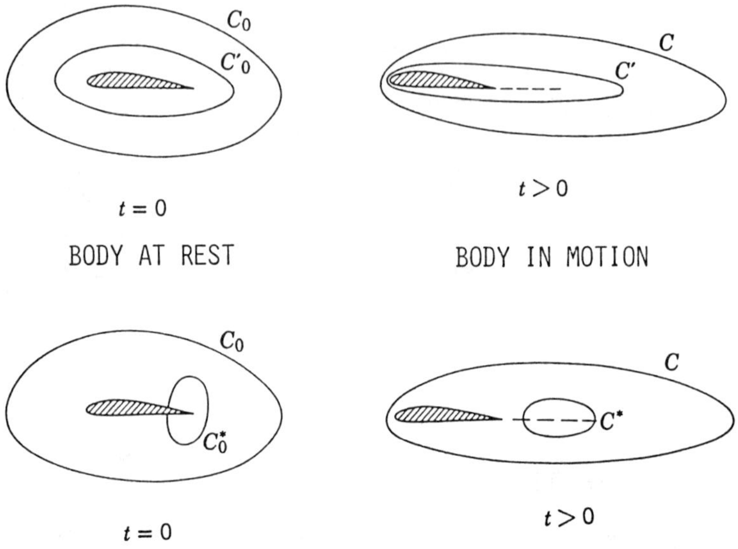

Fig. 2 The closed fluid circuits C and C´ have zero circulation as do their images C_0 and C_0' at t=0. The circuit C* has nonzero circulation but its image shows it is not a closed fluid circuit (Ref. 8).

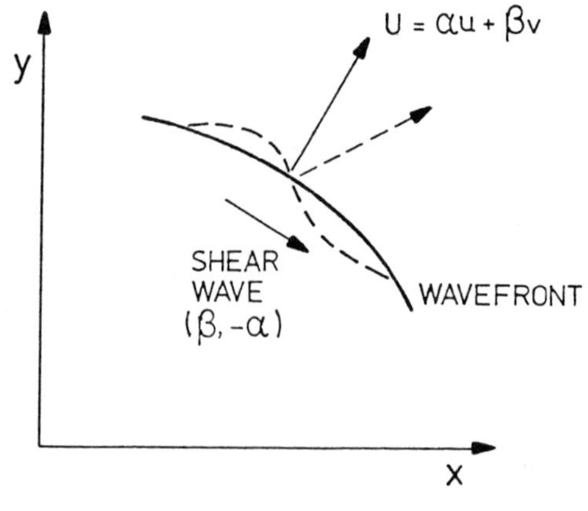

Fig. 3 A shear wave is a disturbance travelling along a wavefront and rotates the propagating velocity vector.

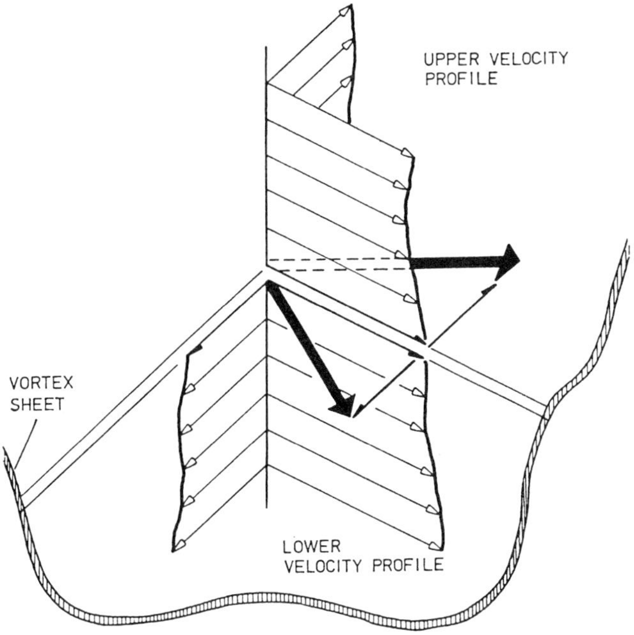

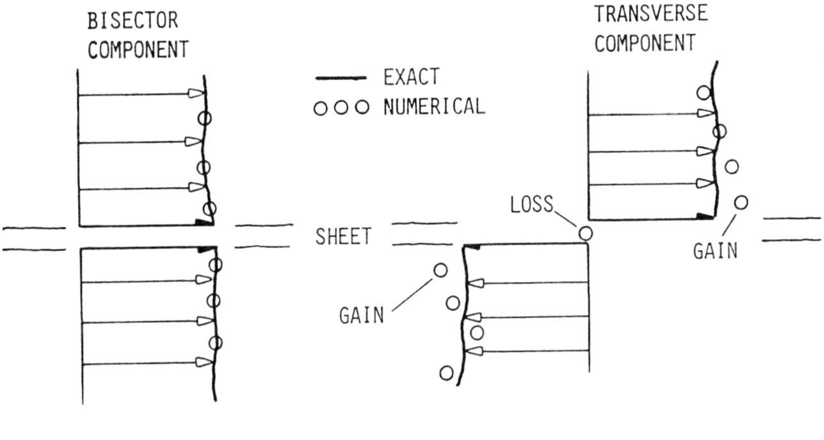

Fig. 4 A vortex sheet has a discontinuity in velocity vector. Any numerical solution that supports the jump in the transverse component may show either a loss or gain in total pressure.

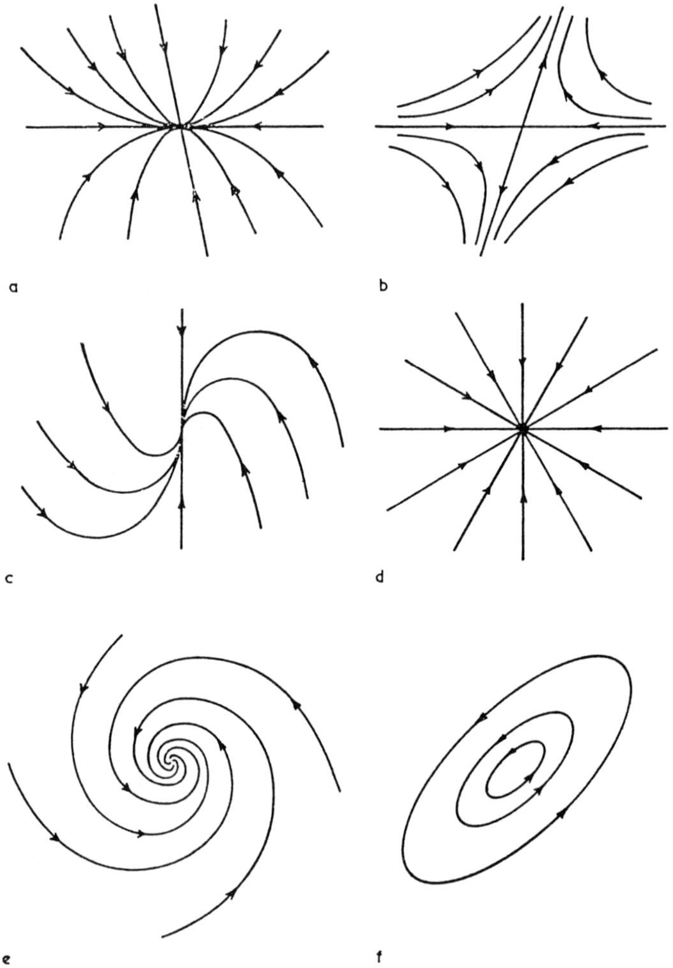

Fig. 5 Singular points of surface velocity field for which $J \neq 0$. (a) Improper node, (b) Saddle-point, (c) Impoper node, (d) Proper node, (e) Spiral point, (f) Centre. Points a to d occur for irrotational velocity, e and f for rotational valocity. (Ref. 11).

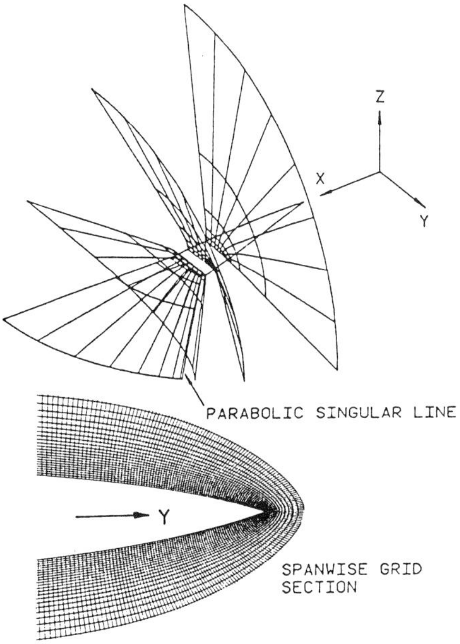

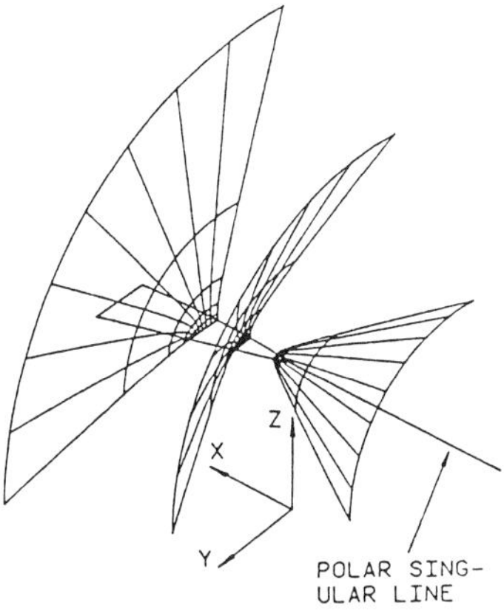

Fig. 6 (a)

Grid generated around a delta-shaped small aspect ratio wing has an O-O topology. The polar singular line produces a dense and nearly conical distribution of points at the apex which is needed to resolve the rapidly varying flow there. This mesh is well-suited for computing the flow around wings of combat aircraft.

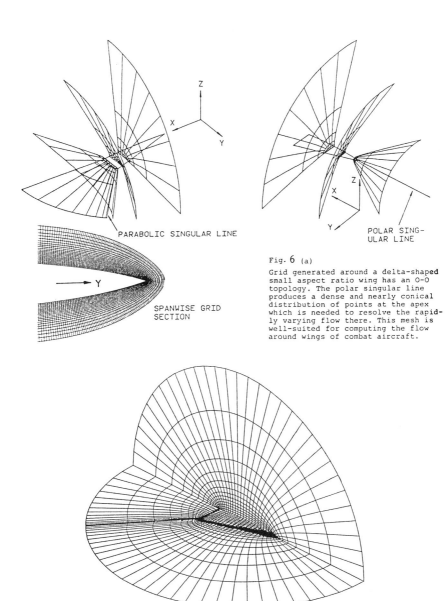

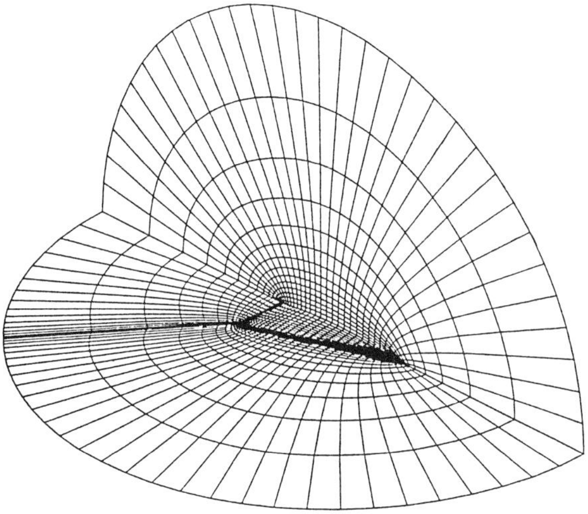

Fig. 6 (b) three-dimensional view of the delta wing mesh

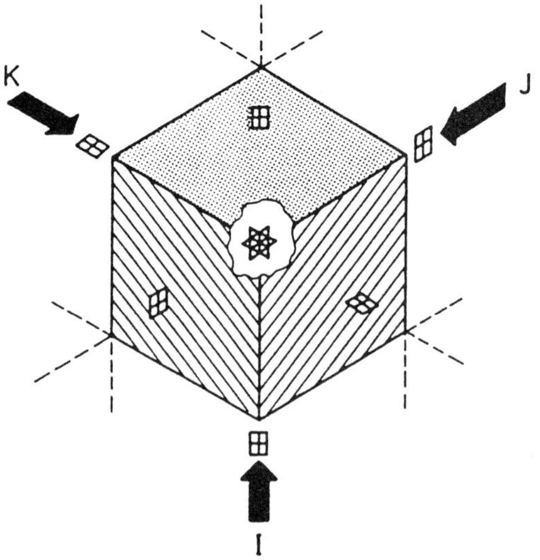

a) the cell averaged variable q is positioned at the center of the cell

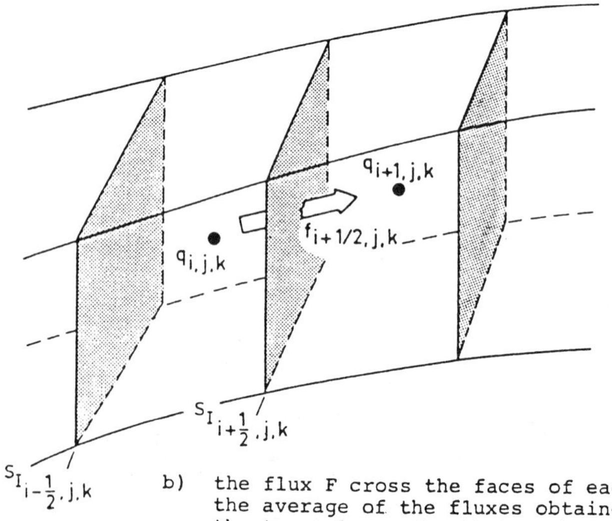

b) the flux F cross the faces of each cell is the average of the fluxes obtained from the two values of q that each face straddles.

Fig. 7

Hexahedrons discretize the domain with the finite volume concept

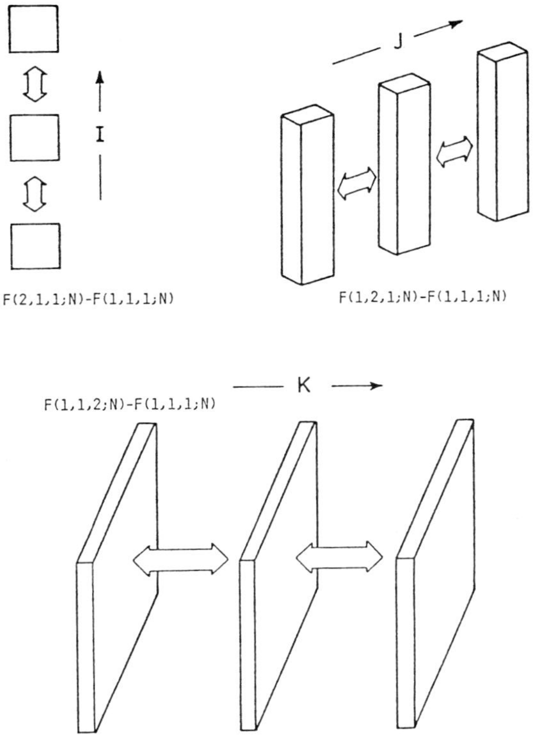

Fig. 8 Because of the grid structure the flux F can be differenced as a vector by offsetting its starting location.

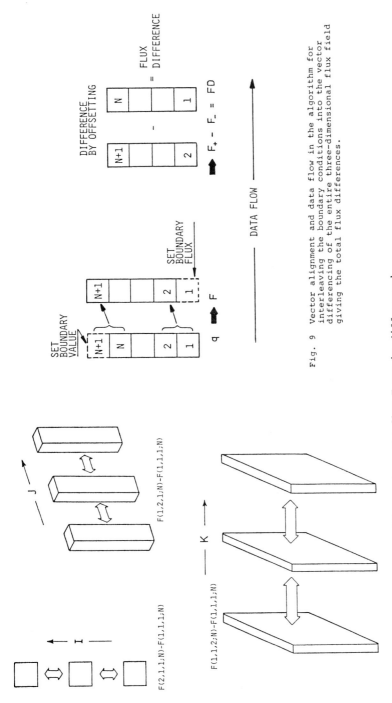

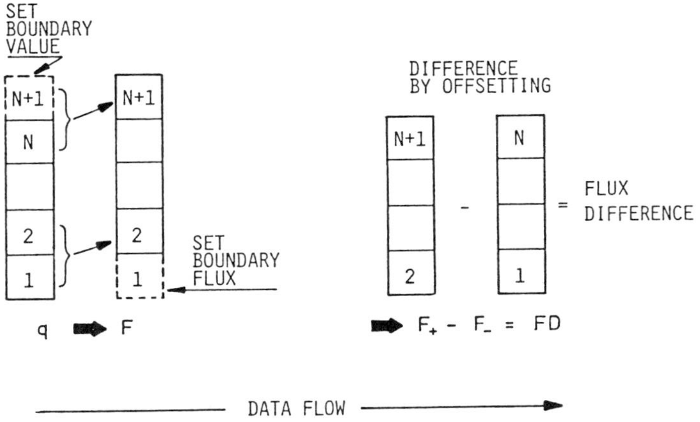

Fig. 9 Vector alignment and data flow in the algorithm for interleaving the boundary conditions into the vector differencing of the entire three-dimensional flux field giving the total flux differences.

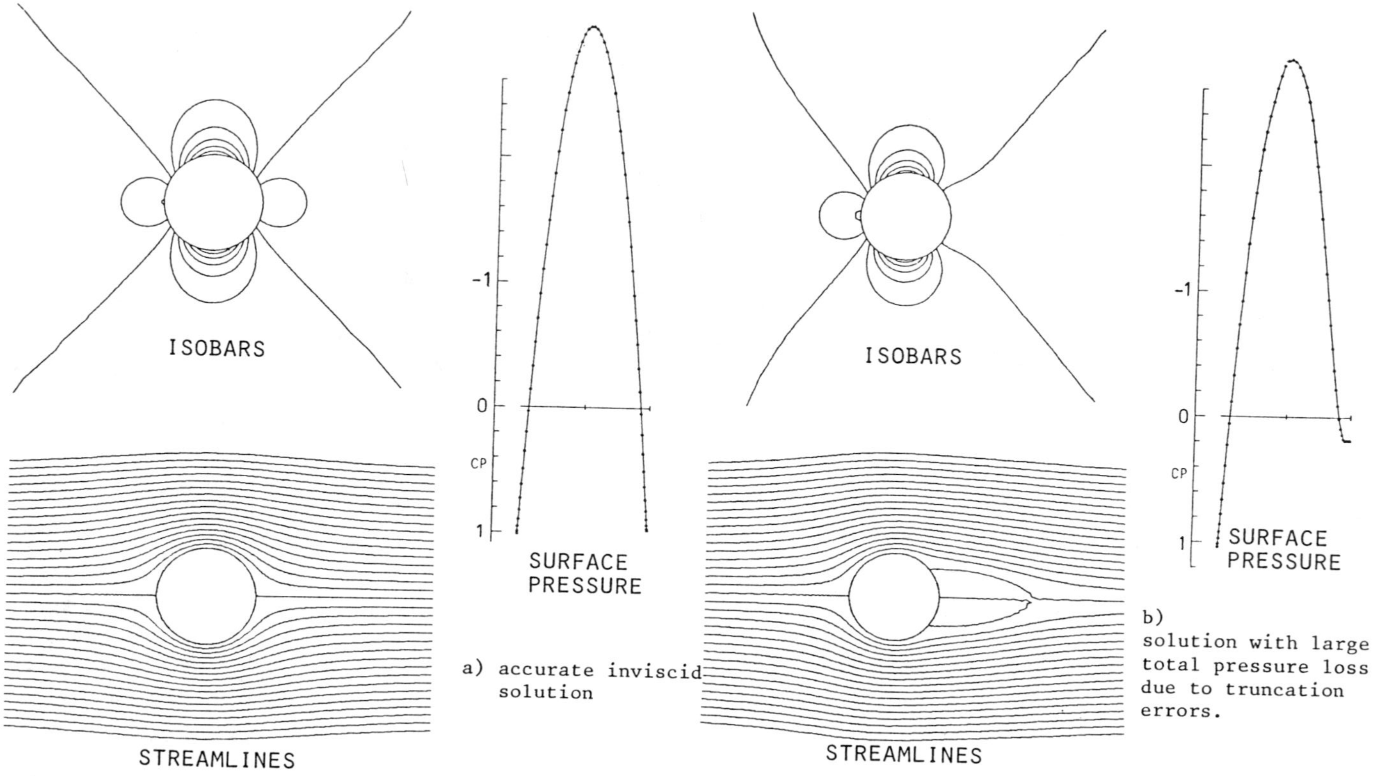

Fig. 10 Comparison of accurate inviscid isobars, streamlines and pressure distribution with ones having large total pressure loss and artifical separation for incompressible flow past a cylinder.

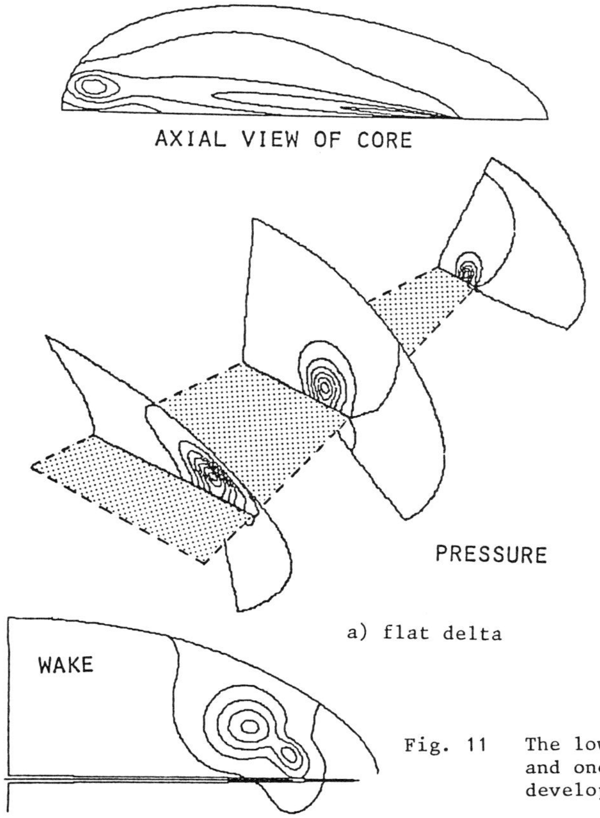

a) flat delta

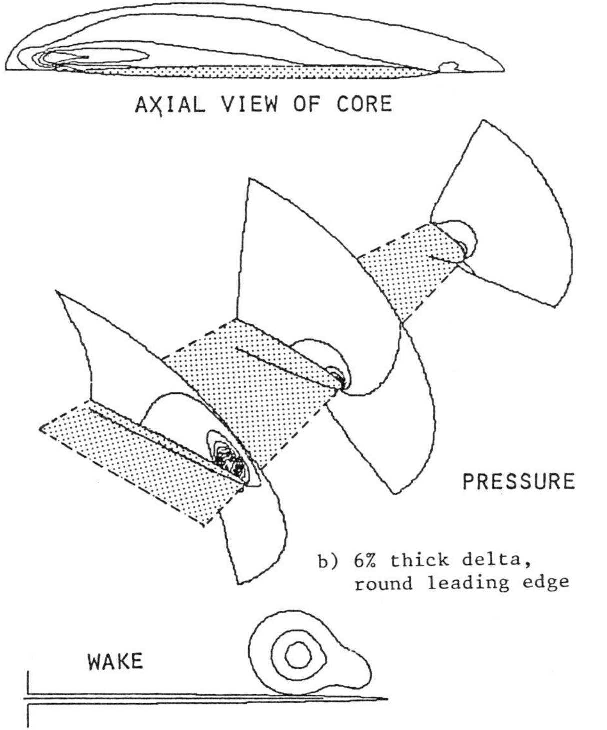

b) 6% thick delta, round leading edge

Fig. 11 The low pressure regions in the flow computed around a flat delta wing and one with thickness and rounded leading edges show that a vortex developes in both cases. $M_\infty = 0.7$ $\alpha = 10$ deg.

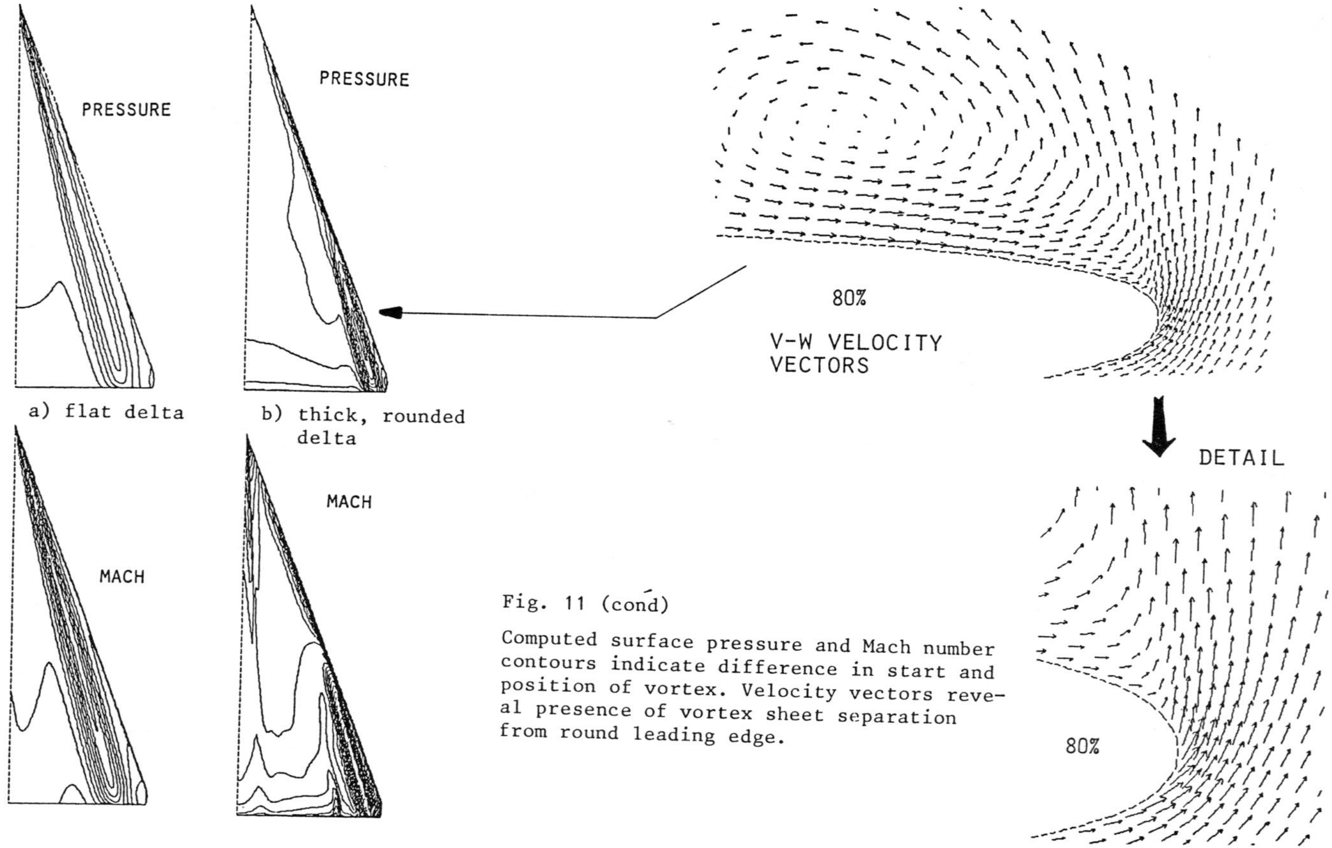

Fig. 11 (cond)

Computed surface pressure and Mach number contours indicate difference in start and position of vortex. Velocity vectors reveal presence of vortex sheet separation from round leading edge.

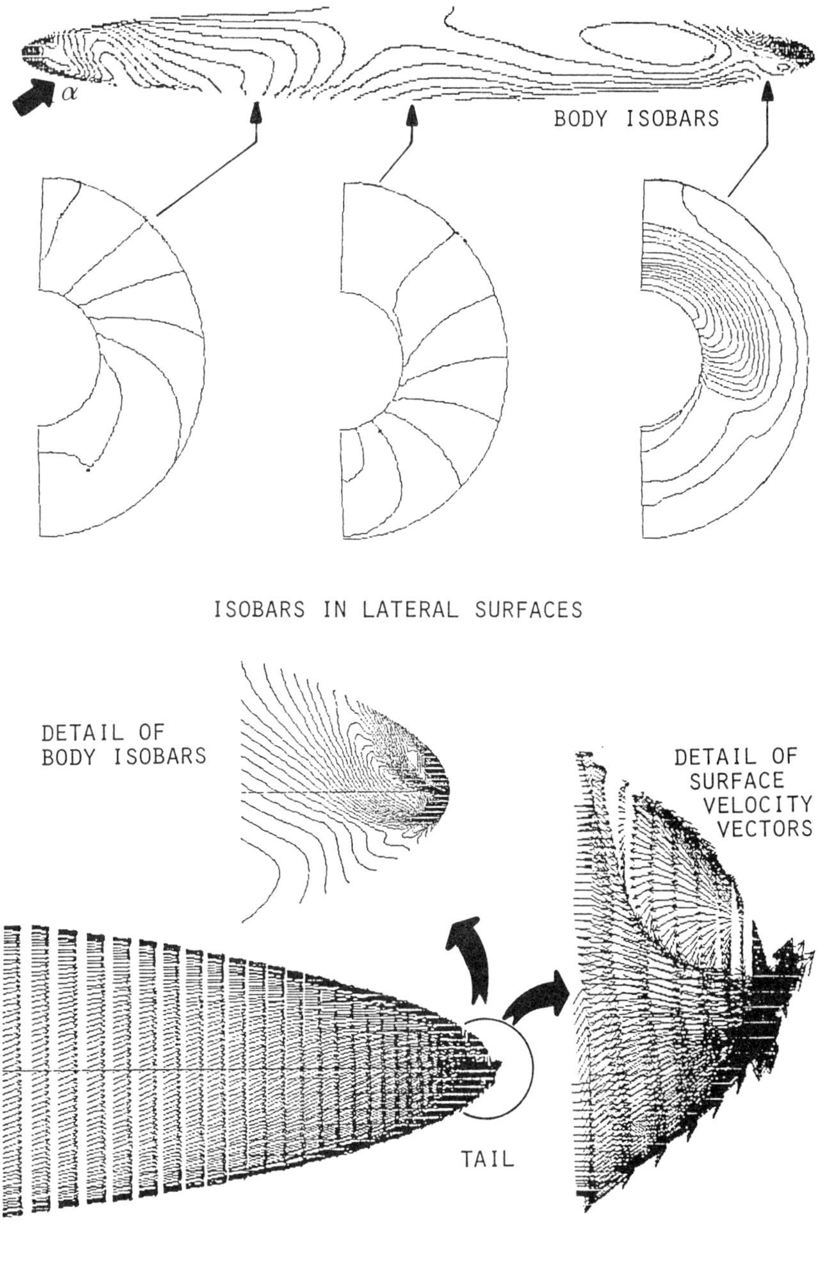

Fig. 12 Isobars and velocity vectors computed for incompressible flows around 9½:1 prolate spheroid. $\alpha = 10$ deg.

PART IV

INVERSE PROBLEMS

NUMERICAL BACKPROJECTION IN THE INVERSE 3D RADON TRANSFORM

A. K. Louis and R. M. Lewitt

Abstract: In this paper we study the error caused by numerically approximating the integral over the sphere in three dimensions as part of Radon's inversion formula. Using special functions we demonstrate that, if suitable filters in the first reconstruction step - differentiation of order two - are used, essential errors are produced only outside the region of interest. The directions needed in measuring the data fulfil also another requirement, they guarantee optimal resolution in the reconstruction.

1 INTRODUCTION

In the simultaneous methods of magnetic resonance (MR) imaging the measured data pertain to planar integrals of a three-dimensional distribution of atomic nucleii in the human body. If a field gradient is applied in a direction ω the integrals over parallel planes perpendicular to ω are measured. In mathematical terms this is formulated in the following way. The searched-for density is a real-valued function f defined on a subset Ω of $\mathbb{R}^3$. Being interested only in functions with compact support we assume that, after a suitable normalization, the domain Ω is contained in the unit ball V(0,1) in $\mathbb{R}^3$. With S^2 we denote the directions in $\mathbb{R}^3$; i.e., the vectors of unit length. Then the RADON transform of f is defined as the real-valued function Rf defined on the unit cylinder $Z = \mathbb{R} \times S^2$ by

$$Rf(s,\omega) = \int_{\mathbb{R}^3} f(x)\delta(s-x\cdot\omega)dx \tag{1.1}$$

where δ denotes the delta distribution. In the cw-mode of MR imaging the data are samples of the RADON transform of the distribution of protons. If the more efficient pulsed techniques are used the data are the FOURIER transform of Rf with respect to the first variable; i.e.,

$$(Rf)^{\wedge}(\sigma,\omega) = \int_{\mathbb{R}} Rf(s,\omega)e^{-2\pi is\sigma}dx. \tag{1.2}$$

The retrieval of the searched-for density distribution f from (1.1) or (1.2) consists of two steps. First the data have to be "filtered" with respect to the first variable, see (2.4). Then an averaging process over the directions takes place, see (2.5). The usage of these formulae necessitates the knowledge of the Radon transform on the whole cylinder, a requirement which is never fulfilled in practice, where only a finite number of samples can be measured. This means that the inversion formula has to be approximated numerically.

In the following we study the effect of replacing the exact integration in (2.5) by a numerical integration formula which then prescribes the directions in which the data have to be taken. The effect of the discretization in s-direction is contained in the filter function F, see (3.1). Both the data in the cw-mode and in the pulsed techniques can be subsumed this way because the inverse discrete Fourier transform can be written as

$$\sum_{\ell=-N/2}^{N/2} \hat{g}_\ell e^{i\ell\sigma} = g * K_N(\sigma)$$

with

$$K_N(s) = U_{N+1}(\cos \tfrac{s}{2})$$

where $*$ denotes the convolution and U_n is the Chebyshev polynomial of the second kind.
The numerical backprojection produces perturbations in the reconstruction which are called artifacts. These effects are studied in § 4 and the integration formula is chosen such that these artifacts are minimal. It so happens that the so deriv‿d directions also fulfil another optimality criterion given in Lou s [5], [6], which guarantees that for a minimal number of measurements the resolution is maximal. Here we interpret resolution of size d in the way that an object of diameter d is reliably recovered in the reconstruction.

2 THE RADON TRANSFORM

A simple and very fast to implement inversion formula for the Radon transform (1.1) is based on a fundamental relation between the Radon transform and the Fourier transform. Denoting with

$$\hat{f}(\xi) = \int_{\mathbf{R}^3} f(x) e^{-2\pi i x \xi} dx$$

the Fourier transform of f, then we have the following relation

$$\hat{f}(\sigma \cdot \omega) = (Rf)^\wedge(\sigma, \omega), \qquad \sigma \in \mathbf{R}, \ \omega \in S^2, \tag{2.1}$$

with $(Rf)^\wedge$ given in (1.2). This is the so-called projection theorem, see e.g. Ludwig [10]. Applying the inverse Fourier transform in (2.1) leads to

$$R^{-1} = R^* K \tag{2.2}$$

where the operator K acts on the first variable,

$$(Kg)^\wedge(\sigma, \omega) = \sigma^2 \hat{g}(\sigma, \omega), \tag{2.3}$$

which can be written as

$$Kg(s, \omega) = -\frac{1}{4\pi^2} \frac{\partial^2}{\partial s^2} g(s, \omega). \tag{2.4}$$

The operator R^* is the so-called backprojection,

$$R^* q(x) = \frac{1}{2} \int_{S^2} q(x \cdot \omega, \omega) d\omega. \tag{2.5}$$

In the following we are concerned with the approximation of this operation.

3 EXACT RECONSTRUCTION FROM FILTERED DATA

In the following we assume that the data are filtered; i.e., instead of the exact data we use

$$\hat{g}_F(\sigma,\omega) = \hat{F}(\sigma)\ (Rf)^\wedge(\sigma,\omega) \tag{3.1}$$

where $\hat{F}$ is real and symmetric. This filter F can contain the interpolation process necessary when only a finite number of data per directions are given, see e.g. Marr - Chen - Lauterbur [11], it can contain the effect of numerical differentiation, see Louis [7] and also the discrete inverse Fourier transform, see § 1. Besides that the filter serves of course as a regularization which avoids oscillations in the reconstruction due to high frequency components in the data. For this purpose either a low-pass filter with suitable cut-off frequency is applied, which cancels completely the high frequencies, or a filter is used which dies off fast enough; e.g. a sinc^2-filter.

In this chapter we first study the effect of filtering the data on the reconstruction when the exact inversion formula (2.2) is applied on the data defined in (3.1).

The Radon transform and its inverse are linear operators, therefore it suffices to consider as input data f the delta distribution δ_ξ concentrated at a point $\xi \in \mathbf{R}^3$, then the result of the inversion is the so-called point response function.

The Fourier transform of the delta distribution δ is equal to 1, therefore the Fourier transform of the shifted version δ_ξ is $e^{-2\pi i x\xi}$.

Using the projection theorem (2.1) the Fourier transform of the data is then

$$(R\delta_\xi)^\wedge(\sigma,\omega) = e^{-2\pi i\sigma\omega\xi}. \tag{3.2}$$

The application of the operator K defined in (2.3) on the filtered data according to (3.1) leads to

$$q_F(s,\omega) = Kg_F(s,\omega)$$

$$= \int_{\mathbf{R}} \sigma^2 \hat{F}(\sigma) e^{2\pi i\sigma(s-\omega\cdot\xi)} d\sigma \ .$$

Finally we use the backprojection to get

$$\delta_{\xi,F}(x) = \frac{1}{2} \int_{S^2} q_F(x\cdot\omega,\omega)d\omega$$

$$= \int_{S^2} \int_0^\infty \sigma^2 \hat{F}(\sigma) e^{2\pi i\sigma\omega(x-\xi)} d\sigma \, d\omega. \tag{3.3}$$

Interchanging the order of integration and replacing $x-\xi = \sigma'\omega'$, $\sigma' \in \mathbf{R}$, $\omega' \in S^2$, the integral over the sphere can be computed to be

$$\int_{S^2} e^{2\pi i\sigma\sigma'\omega\omega'} d\omega = 2\pi \int_{-1}^{1} e^{2\pi i\sigma\sigma' t} dt$$

$$= \frac{2}{\sigma\sigma'} \sin (2\pi\sigma\sigma')$$

$$= 4\pi \; j_0(2\pi\sigma\sigma') \tag{3.4}$$

where we have used in the first step the Funk-Hecke theorem, see e.g. Erdelyi et al. [2] and in the last step the spherical Bessel function j, which is defined as

$$j_\nu(z) = (\frac{\pi}{2z})^{1/2} \; J_{\nu+1/2}(z)$$

see e.g. Abramowitz-Stegun [1], p. 437. Here J_ν denotes the Bessel function of the first kind and especially

$$j_0(z) = \mathrm{sinc}(z/\pi) = \frac{\sin z}{z} \ .$$

Inserting (3.4) in (3.3) we finally get

$$\delta_{\xi,F}(x) = 4\pi \int_0^\infty \sigma^2 \hat{F}(\sigma) \; j_0(2\pi\sigma|x-\xi|)d\sigma \ . \tag{3.5}$$

If we use the low-pass filter; e.g. $\hat{F}(\sigma) = 1$ for $|\sigma| \leq \tau$ and 0 otherwise, we get

$$\delta_{\xi,F}(\xi) = \frac{4\pi}{3} \tau^3.$$

4 NUMERICAL BACKPROJECTION

In this section we study the error caused by numerically approximating the integral in the backprojection (2.5). Introducing the usual spherical coordinates θ and φ to label $\omega \in S^2$ so that

$$\omega = \omega(\theta,\varphi) = (\sin\theta\cos\varphi, \sin\theta\sin\varphi, \cos\theta)^T$$

with $\theta \in [0,\pi[$, $\varphi \in [0,2\pi[$, the integration measure $d\omega$ becomes $d\varphi \sin\theta \, d\theta$. On the circle; i.e., integration with respect to φ, we use the Riemannian sum

$$\int_0^{2\pi} h(\varphi)\, d\varphi \doteq \frac{2\pi}{N} \sum_{\nu=0}^{N-1} h(\varphi_\nu)$$

where

$$\varphi_\nu = \nu \frac{2\pi}{N}, \quad \nu = 0,\ldots,N-1, \tag{4.1}$$

which in this case is the same as the most accurate Gaussian integration rule. Not so straight forward is the selection of the integration formula in the θ-direction. Using the change of coordinates $t = \cos\theta$; i.e.,

$$\int_0^{\pi} g(\cos\theta)\sin\theta\, d\theta = \int_{-1}^{1} g(t)dt, \tag{4.2}$$

we can use formulas known for the unit interval; i.e., we approximate by

$$\int_0^{\pi} g(\cos\theta)\sin\theta\, d\theta \doteq \sum_{\mu=1}^{M} w_\mu \sin\theta_\mu \cdot g(\cos\theta_\mu).$$

In order to avoid redundancy in the data because of

$$s\omega(\theta,\varphi) = -s\omega(\pi-\theta,\varphi+\pi) \ , \tag{4.3}$$

we use in the following an odd N in (4.1).
We denote by

$$\omega_{\mu\nu} = \omega(\theta_\mu,\varphi_\nu),\ \mu = 1,\dots,M,\quad \nu = 0,\dots,N-1 \tag{4.4}$$

the integration points on the sphere. When we replace the exact integration in (3.3) by our numerical integration formula we get

$$\begin{aligned}\delta_{\xi,F,N}(x) &= \frac{\pi}{N}\sum_{\mu=1}^{M} w_\mu \sin\theta_\mu \sum_{\nu=0}^{N-1} q_F(x\cdot\omega_{\mu\nu},\omega_{\mu\nu}) \\ &= \frac{\pi}{N}\sum_{\mu=1}^{M} w_\mu \sin\theta_\mu \sum_{\nu=0}^{N-1} \frac{1}{2}\int_{-\infty}^{\infty} \sigma^2 \hat{F}(\sigma) e^{2\pi i\sigma\omega_{\mu\nu}(x-\xi)} d\sigma\ .\end{aligned} \tag{4.5}$$

In order to recognize the effect of the integration with respect to θ and φ we expand the plane wave

$$e^{2\pi i\sigma\omega(x-\xi)} = e^{2\pi i\sigma\sigma'\omega\omega'}$$

in terms of spherical harmonics. Again we use $x-\xi = \sigma'\omega'$ where

$$\omega' = \omega(\theta',\varphi') \tag{4.6}$$

and get

$$\begin{aligned}e^{-2\pi i\sigma\sigma'\omega\omega'} &= \sum_{n=0}^{\infty} (2n+1)(-i)^n j_n(2\pi\sigma\sigma') \\ &\cdot \sum_{m=0}^{n} \varepsilon_m \frac{(n-m)!}{(n+m)!} \cos[m(\varphi-\varphi')] P_n^m(\cos\theta) P_n^m(\cos\theta')\end{aligned} \tag{4.7}$$

where $\varepsilon_0 = 1$ and $\varepsilon_m = 2$ for $m > 0$, see Morse-Feshbach [12] formula (11.3.46). Here P_n^m denotes the Legendre functions. Inserting (4.7) into (4.5) we get

$$\begin{aligned}\delta_{\xi,F,N}(x) &= \sum_{n=0}^{\infty} (-i)^n \psi_n(\sigma') \sum_{m=0}^{n} \varepsilon_m \frac{(n-m)!}{(n+m)!} P_n^m(\cos\theta') \\ &\cdot \frac{\pi}{N}\sum_{\nu=0}^{N-1} \cos[m(\varphi_\nu-\varphi')] \sum_{\mu=1}^{M} w_\mu P_n^m(\cos\theta_\mu) \sin\theta_\mu\end{aligned} \tag{4.8}$$

where

$$\psi_n(\sigma') = \frac{1}{2}(2n+1)\int_{-\infty}^{\infty} \sigma^2 \hat{F}(\sigma)\, j_n(2\pi\sigma\sigma')d\sigma. \tag{4.9}$$

Comparing this with the result of the exact integration in (3.5) we realize that, if the integration formulas are exact for $m = n = 0$, the term with $n = 0$ is the result of exact integration. All the other terms are error terms.
As mentioned in the beginning of Section 3 the filter function $\hat{F}$ is real and symmetric, hence

$$\psi_n \equiv 0 \qquad \text{for } n \text{ odd}. \tag{4.10}$$

With the choice (4.1) for the integration with respect to φ we can compute

$$\frac{\pi}{N} \sum_{\nu=0}^{N-1} \cos\,[m(\varphi_\nu - \varphi')] = \begin{cases} 0 & \text{for} \quad m \neq 0 \bmod N \\ \pi \cos kN\varphi' & \text{for} \quad m = kN, \quad k \in \mathbb{N}_0 . \end{cases}$$

This cancels all the terms where $m \neq 0 \bmod N$ and we get

$$\delta_{\xi,F,N}(x) = \pi \sum_{n=0}^{\infty} (-1)^n \psi_{2n}(\sigma') \sum_{k=0}^{[2n/N]} \varepsilon_{kN} \frac{(2n-kN)!}{(2n+kN)!} P_{2n}^{kN}(\cos\theta') \cdot \cos(kN\varphi')$$
$$\cdot \sum_{\mu=1}^{M} w_\mu \sin\theta_\mu \, P_{2n}^{kN}(\cos\theta_\mu).$$

Here $[x]$ denotes the integral part of x. The last sum approximates

$$\int_{-1}^{1} P_{2n}^{kN}(t)dt$$

which is 0 for $n > 0$ and $k = 0$ due to the orthogonality of the Legendre polynomials. For odd k this integral is also 0 because then P_{2n}^{kN} is an odd function. We thus use a symmetric integration formula for the θ-integration; i.e., if θ_μ is an integration point then also $\pi-\theta_\mu$ with the same weight. This cancels all the terms with odd k. We cannot avoid backprojecting error when the $k = 2$ term starts contributing. The first term of this type is P_{2N}^{2N}, followed by P_{2N+2}^{2N}, P_{2N+4}^{2N} etc. Hence we select the integration formula such that it is exact for the even polynomials of degree less than 2N.
Putting everything together we get

$$\delta_{\xi,F,N}(x) = \delta_{\xi,F}(x)$$

$$+ 2\pi \sum_{n=N}^{\infty} (-1)^n \psi_{2n}(\sigma') \sum_{k=0}^{[n/N]} \varepsilon_{2kN} \frac{(2n-2kN)!}{(2n+2kN)!} \tag{4.11}$$

$$\cdot \cos(2kN\varphi')\, P_{2n}^{2kN}(\cos\theta') \sum_{\mu=1}^{M} w_\mu P_{2n}^{2kN}(\cos\theta_\mu) \sin\theta_\mu .$$

The final decision concerns the integration formula with respect to θ. In order to minimize the number of measurements; i.e., the integration points on the sphere, for maximal accuracy we select the Gaussian formula on the interval [-1,1], see Abramowitz - Stegun [1], Formula 25.4.29. Then $\theta_\mu = \cos^{-1} x_\mu$ for $\mu = 1,\dots,M$, where the x_μ are the zeroes of the Legendre polynomial of degree M. This formula is exact for polynomials of degree 2M-1. As discussed above it makes no sense to use M with $2M-1 > 2N-2$ and this leads to

$$M = N. \tag{4.12}$$

Using the symmetry of the Radon transform it is sufficient to take

$$N(N+1)/2$$

directions. This coincides with the number of directions proposed in Louis [6], where the resolution is maximized for a minimal number of directions. Besides that, the directions considered here also satisfy the optimality criterion (7) given in Louis [6].

To study the behaviour of the error we notice that the Bessel functions are small for small argument. More precisely we have for $0 \leqq x \leqq 1$

$$0 \le J_n(xn) \le (2\pi n)^{-1/2}(1-x^2)^{-1/4}\exp\left(-\tfrac{n}{3}(1-x^2)^{3/2}\right)$$

see for example Formula (4.6) in Louis [8].

If we use a low-pass filter with cut-off frequency $\tau = \frac{N}{2\pi}$ we see that because of

$$\sigma' = |x-\xi| \leqq 2 \qquad \text{for } x,\xi \in V(0,1)$$

we integrate for determining $\psi(\sigma')$ over σ with

$$|2\pi\sigma'\sigma| \leqq 2\pi|\sigma'|\tau \leqq 2N,$$

and here the Bessel functions with index greater or equal 2N are almost zero. That means that the artifacts emanating from an object in the reconstruction region starts contributing essential errors only outside the region of interest.

ACKNOWLEDGEMENT

The research of the first author was supported by the Deutsche Forschungsgemeinschaft under project number Lo 310/1-1 and by grant HL28438 from the National Heart, Lung and Blood Institute.
The work of the second author was supported by National Cancer Institute under grant CA31843 and by National Heart, Lung and Blood Institute under grants HL28438 and HL4664.

REFERENCES

[1] Abramowitz, M., Stegun, I.A.: Handbook of mathematical formulas, New York: Dover, 1965

[2] Erdelyi, A., Magnuus, W., Oberhettinger, F., Tricomi, F.: Tables of integral transforms, Vol. II, New York: McGraw Hill, 1954

[3] Gradshteyn, I.S., Ryzhik, I.M.: Tables of integrals, series and products, New York: Academic Press, 1980

[4] Lewitt, R.M.: Reconstruction algorithms: Transform methods, Proc. IEEE, 71, 390-408, 1983

[5] Louis, A.K.: Orthogonal function series expansions and the null space of the Radon transform, SIAM J. Math. Anal. 15, 621-633, 1984

[6] Louis, A.K.: Optimal sampling in nuclear magnetic resonance tomography, Comput. Assist. Tomogr. 6, 334-340, 1982

[7] Louis, A.K.: Approximate inversion of the 3D Radon transform, Math. Meth. Appl. Sci. 5, 176-185, 1982

[8] Louis, A.K.: Nonuniqueness in inverse Radon problems: The frequency distribution of the ghosts. Math. Z. 185, 429-440, 1984

[9] Louis, A.K., Natterer, F.: Mathematical problems of computerized tomography, Proc. IEEE 71, 379-389, 1983

[10] Ludwig, D.: The Radon transform on Euclidean spaces, Comm. Pure Appl. Math. 19, 49-81, 1966

[11] Marr, R.B., Chen, C.N., Lauterbur, P.C.: On two approaches to 3D reconstruction in NMR zeugmatography. In Herman, G.T., Natterer, F. (eds.): Mathematical aspects of computerized tomography, Berlin: Springer LNMI 8, 1981

[12] Morse, P.M., Feshbach, H.: Methods of theoretical physics II, New York, McGraw Hill, 1953

[13] Shepp, L.A.: Computerized tomography and nuclear magnetic resonance, J. Comput. Assist. Tomogr. 4, 94-107, 1980

A DIRECT ALGEBRAIC ALGORITHM IN COMPUTERIZED TOMOGRAPHY

Hermann Kruse, Frank Natterer

1. Introduction

In computerized tomography an x-ray source is moved around a body, the initial intensity I_0 of the x-rays is known, the intensity I after having passed the body is measured. A simple physical model yields the equation

$$I = I_0 \cdot \exp\left(- \int_L f(x)dx\right) , \tag{1.1}$$

with L being the way of the x-ray through the body and f(x) the attenuation coefficient of the tissue at the point x which is to be reconstructed by these values. See Herman [8] for the principles of computerized tomography.

The mathematical problem in this procedure is connected with the Radon transform:

<u>DEFINITION:</u> If $f \in \mathcal{S}(\mathbb{R}^2)$, $s \in \mathbb{R}$, $\Theta = \begin{pmatrix} \cos\varphi \\ \sin\varphi \end{pmatrix}$, $\varphi \in [0,2\pi]$, then

$$\mathcal{R}f(s,\Theta) = \int_{-\infty}^{\infty} f(s \cdot \Theta + t \cdot \Theta^{\perp})\, dt$$

is the Radon transform of the function f.

From (1.1) we immediately see that

$$- \ell n \left(\frac{I}{I_0}\right) = \int_L f(x)\, dx \quad , \tag{1.2}$$

hence the problem in computerized tomography is just the inversion of the Radon transform.

The standard method for doing this in computerized tomography is "filtered backprojection" which is nothing else than the numerical realization of the inversion formula

$$f(x) = (2\pi)^{-3/2} \int_0^{\pi} \int_{-\infty}^{\infty} |\sigma| (\mathbb{R}f)^\wedge(\sigma,\Theta)\, e^{i\sigma\Theta\cdot x}\, d\sigma\, d\varphi \,, \qquad (1.3)$$

with $(\mathbb{R}f)^\wedge$ denoting the Fourier transform of $\mathbb{R}f$ with respect to the first argument. (See Herman [8], Louis, Natterer [14], Natterer [16] for details.)

Those methods which discretize the integral equation in some way and solve the resulting system of linear equations are called "algebraic" algorithms.

The most famous and simplest algebraic method is ART (= algebraic reconstruction technique), which was already used in the first commercial scanner built by G.N. Hounsfield [9]:
One simply covers the plane by a square grid and assumes that f is constant on each pixel. Then the integral equation becomes a system of linear equations which has to be solved.

Typical for algebraic methods is the large size of the linear system which therefore usually is solved by an iterative method; ART uses Kaczmarz's method, see [6], [7], [10].

We shall present an algebraic method leading to a linear system which can be solved very efficiently by direct methods.

2. Special problems

We shall always assume that $\mathrm{supp}\,(f) \subset D := \{x \in \mathbb{R}^2 \mid |x| \leq 1\}$. Then we distinguish the following four cases (compare Louis, Natterer [14], Natterer [16]):

i) The complete problem: The data $g(s,\Theta)$ are given for all values $\varphi \in [0,2\pi]$, $s \in [-1,1]$.

This is the standard case.

ii) The interior problem: $g(s,\Theta)$ is given only for $\varphi \in [0,2\pi]$ and $|s| \leq a < 1$. This problem occurs e.g. if one is interested only in the region $\{|x| \leq a\}$ and does not want to waste radiation in the exterior of this region.

iii) The exterior problem: $g(s,\Theta)$ is given for $\varphi \in [0,2\pi]$, $0 < a \leq |s| \leq 1$. Such problems arise e.g. if there are metallic implants inside the body.

iv) The limited angle problem: $g(s,\Theta)$ is given for $0 \leq \varphi \leq \phi < \pi$, $s \in [-1,1]$. Limited angle problems occur e.g. in nondestructive testing or electron microscopy.

3. The idea of the algorithm

First of all we of course get only a finite number of data, second the detectors of a real scanner have finite width, so it is justified to assume that the data are integrals over strips instead of lines (see figure 1).

More general we assume that we receive $p \cdot q$ measurements g_{ij} at equally spaced angles $\varphi_i = i \cdot \Delta\varphi$

$$g_{ij} = \int_{-1}^{1} \omega_j(s) Rf(s,\Theta_i)\, ds =: R_{ij} f \tag{3.1}$$

$$i = 0,\dots,p-1\ ,\quad j = 1,\dots,q\ ,\quad \Theta_i = \begin{pmatrix} \cos\varphi_i \\ \sin\varphi_i \end{pmatrix},$$

with certain weights $\omega_j(s)$, e.g.

$$\overline{\omega}_j(s) := \begin{cases} 1 & |s-s_j| \leq d/2 \\ 0 & \text{else} \end{cases}, \quad s_j \in [-1,1],\ d > 0\ . \tag{3.2}$$

It is easily seen that

$$\begin{aligned} g_{ij} &= \int_{-1}^{1} \omega_j(s) \int f(s\Theta_i + t\Theta_i^{\perp})\, dt\, ds \\ &= \int_D \omega_j(x \cdot \Theta_i) f(x)\, dx \qquad (3.3) \\ &= \langle \kappa_{ij}, f \rangle_{L_2(D)} \end{aligned}$$

with $\kappa_{ij}(x) = \omega_j(x \cdot \Theta_i)$ and $<\cdot,\cdot>_{L_2(D)}$ denoting the inner product in $L_2(D)$, see Buonocore et al. [4].

For the special weight functions $\bar{\omega}_j$ defined by (3.2) κ_{ij} is just the characteristic function of the strip $L_{ij} = \{x \in \mathbb{R}^2 \mid |x \cdot \Theta_i - s_j| \le d/2\}$.

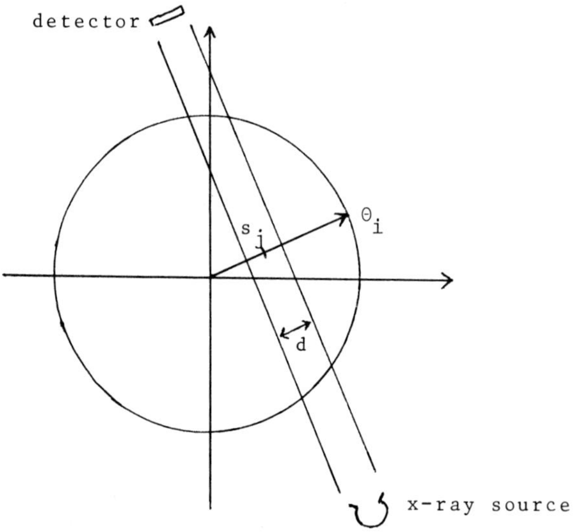

Figure 1

Let us consider now the system of integral equations

$$R_{ij} f = g_{ij} \qquad , \ i = 0,\dots,p-1 \ , \ j = 1,\dots,q \ .$$

We arrange q equations into a group and get

$$R_i f = g_i \quad \text{with} \quad R_i = \begin{pmatrix} R_{i1} \\ \vdots \\ R_{iq} \end{pmatrix} , \ g_i = \begin{pmatrix} g_{i1} \\ \vdots \\ g_{iq} \end{pmatrix},$$

and this is written as

$$Rf = g \qquad \text{with} \qquad R = \begin{pmatrix} R_0 \\ \vdots \\ R_{p-1} \end{pmatrix} \quad , \; g = \begin{pmatrix} g_0 \\ \vdots \\ g_{p-1} \end{pmatrix} .$$

In order to allow for measurement errors we apply the Tikhonov - Phillips method to this underdetermined system and consider the problem

$$\text{minimize } \{ \| Rf - g \|^2_{\mathbb{R}^{p \cdot q}} + \gamma \| f \|^2_{L_2(D)} \mid f \in L_2(D) \} \tag{3.4}$$

with $\gamma > 0$ being a regularization parameter.

The solution of this minimization problem is

$$f_\gamma = R^*(R R^* + \gamma I)^{-1} g \tag{3.5}$$

with $R^* h := \sum_{i,j} \kappa_{ij} h_{ij} \quad , \; h \in \mathbb{R}^{p \cdot q} \quad ,$

being the adjoint of the operator $R : L_2(D) \to \mathbb{R}^{p \cdot q}$ defined by $R_{ij} f = \langle \kappa_{ij}, f \rangle_{L_2(D)}$.

Defining $h := (R R^* + \gamma I)^{-1} g$ we are able to gain the solution f within two steps:

i) Compute the solution of $(R R^* + \gamma I) h = g$.

ii) Compute $f_\gamma = R^* h$.

The crucial point obviously is step i) which is a $(p \cdot q) \times (p \cdot q)$ - system of linear equations, at first view far from being solved directly for large p and q; so let us have a second view:

The operator $R R^*$ is a $p \times p$ - matrix consisting of $q \times q$-blocks $(R_i R_j^*)$, whose (k, ℓ) - th element is

$$(R_i R_j^*)_{k\ell} = \langle \kappa_{ik}, \kappa_{j\ell} \rangle_{L_2(D)} \quad . \tag{3.6}$$

Let U be a rotation with angle $\Delta\varphi$ around the origin, then

$$\begin{aligned}\kappa_{j\ell}(x) &= \omega_\ell(x \cdot \Theta_j) \\ &= \omega_\ell(x \cdot U^j \Theta_0) \\ &= \omega_\ell(U^{-j} x \cdot \Theta_0) \\ &= \kappa_{0\ell} \circ U^{-j}(x) \quad .\end{aligned} \tag{3.7}$$

It immediately follows that

$$\begin{aligned}(R_i R_j^*)_{k\ell} &= \langle \kappa_{0k} \circ U^{-i}, \kappa_{0\ell} \circ U^{-j} \rangle_{L_2(D)} \\ &= \langle \kappa_{0k}, \kappa_{0\ell} \circ U^{i-j} \rangle_{L_2(D)} \quad .\end{aligned} \tag{3.8}$$

Hence the block $(R_i R_j^*) =: S_{j-i}$ only depends on the difference $j - i$. Note that $S_{-j} = S_j^T$.

With this notation our system $(RR^* + \gamma I)h = g$ assumes the form

$$\begin{pmatrix} S_0 + \gamma I & S_1 & S_2 & \cdots & S_{p-1} \\ S_{-1} & S_0 + \gamma I & S_1 & \cdots & S_{p-2} \\ \vdots & & & & \vdots \\ S_{1-p} & \cdot & \cdot & S_{-1} & S_0 + \gamma I \end{pmatrix} \begin{pmatrix} h_0 \\ \vdots \\ h_{p-1} \end{pmatrix} = \begin{pmatrix} g_0 \\ \vdots \\ g_{p-1} \end{pmatrix} \quad . \tag{3.9}$$

Obviously $(RR^* + \gamma I)$ is a block - Toeplitz - matrix!

There are several numerical methods to solve equations of this kind, see e.g. Bareiss [1], Trench [17] or Bunch [3] who gives a nice survey on this area.

Here we only want to sketch the method of Bitmead and Anderson [2] basing on the work of Kailath, Morf et al. [5], [11], [12]:

$(RR^* + \gamma I)$ is invertible for $\gamma > 0$ and its inverse can be decomposed into

$$(RR^* + \gamma I)^{-1} = L_1 U_1 + L_2 U_2 \tag{3.10}$$

with upper resp. lower triangular block - Toeplitz - matrices U_i, L_i. The calculation of L_i, U_i can be done with $O(p \cdot q^2 \cdot \log^2 p)$ operations, the storage required is $O(p \cdot q^2)$. Is $(RR^* + \gamma I)^{-1}$ decomposed in this way once and forever, then for each right hand side g the system (3.9) can be solved by computing $h = L_1 U_1 g + L_2 U_2 g$. This can be done by four convolutions of length p on vectors of dimensions q by means of FFT - techniques with $O(p \cdot q^2 + pq \log p)$ operations.

Let us now confine ourselves to the special case $\Delta\varphi = \frac{2\pi}{p}$, i.e. we exclude the limited angle problem.

Then we have $U^p = I$ and

$$\begin{aligned}(S_{p-j})_{k\ell} &= \langle \kappa_{0k}, \kappa_{0\ell} \circ U^{j-p} \rangle_{L_2(D)} \\ & \langle \kappa_{0k}, \kappa_{0\ell} \circ U^{j} \rangle_{L_2(D)} \\ &= (S_{-j})_{k\ell} \quad .\end{aligned} \tag{3.11}$$

Hence in this case $(RR^* + \gamma I)$ turns out to be a block - cyclic convolution and it is well known that applying the Fourier transform to both sides of equation (3.9) then yields p systems of size $(q \times q)$, namely

$$(\hat{S}_k + \gamma I)\hat{h}_k = \hat{g}_k \quad , \quad k = 0,\ldots,p-1 \quad . \tag{3.12}$$

4. The algorithm

For the latter case we end up with the following algorithm:

Step 1: i) Compute the matrices S_j and apply the discrete Fourier - transform:

$$\hat{S}_k = \sum_{j=0}^{p-1} S_j \, e^{-2\pi ijk/p} \quad , \quad k = 0,\ldots,p-1 \quad .$$

ii) Compute $(\hat{S}_k + \gamma I)^{-1}$ and store them!

Step 2: i) Compute the Fourier transform of the data

$$\hat{g}_k = \sum_{j=0}^{p-1} g_j \, e^{-2\pi ijk/p} \qquad k = 0,\ldots,p-1 \quad .$$

ii) Calculate $\hat{h}_k = (\hat{S}_k + \gamma I)^{-1} \hat{g}_k$.

iii) Apply the inverse Fourier transform:

$$h_j = \frac{1}{p} \sum_{j=0}^{p-1} \hat{h}_k \, e^{2\pi ikj/p} \qquad j = 0,\ldots,p-1 \quad .$$

iv) Do the so - called "backprojection":

$$f_\gamma(x) = \sum_{j,\ell : x \in L_{j\ell}} h_{j\ell} \quad .$$

Note the following essential fact:

As soon as the parameters p,q and a suitable γ are fixed for a certain CT - scanner, step 1 must be executed only once.

Afterwards all the pictures can be reconstructed by applying only step 2.

If we choose our special weight functions

$$\bar{\omega}_\ell(s) = \begin{cases} 1 & |s - s_\ell| \le d/2 \\ 0 & \text{else} \end{cases}$$

i.e. $g_{j\ell} = \int_{L_{j\ell}} f(x)dx$ with $L_{j\ell} = \{x \mid |x \cdot \Theta_j - s_\ell| \le d/2\}$,

then the (k,ℓ) - th element of S_j is defined by

$$(S_j)_{k\ell} = \langle \kappa_{0k} , \kappa_{j\ell} \rangle_{L_2(D)}$$

$$= \text{area } \{D \cap L_{0k} \cap L_{j\ell}\}$$

which can easily be computed. These weights have been used in our implementation.

5. Numerical results

The best way to demonstrate numerical results in computerized tomography is to present some pictures, see figures 2, 3. We have done reconstructions with the filtered backprojection method (bottom left), with ART (bottom right) and with the direct algebraic method (top right). On top left there is the original object we have tried to reconstruct.

Figure 2 is an example dealing with the exterior problem. We have used $p = 512$, $q = 102$, $a = 0.7$.

Figure 3 is the example for an interior problem, here the parameters are $p = 512$, $q = 98$, $a = 0.4$.

There are bad artifacts in the filtered backprojection reconstruction but only few artifacts in the images generated by ART and by the direct method.

The second step of the direct method has taken about $2^{\underline{30}}$ min for each of the reconstructions, filtered backprojection needed about $2^{\underline{00}}$ min, ART about $2^{\underline{30}}$ min each iteration, we did 15 iterations, thus the total time ART needed is $37^{\underline{30}}$ min!

Remarks: The algorithm has first been suggested by Lent [13]. A detailed analysis and an implementation has been given by Natterer [15], see also Buonocore et al. [4].

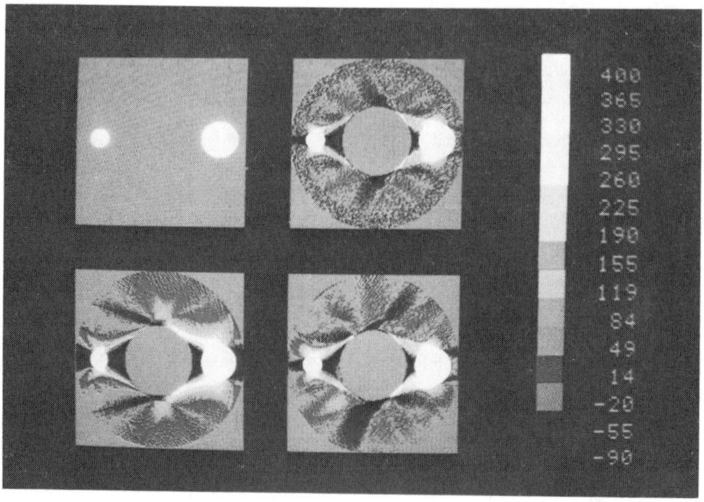

Figure 2

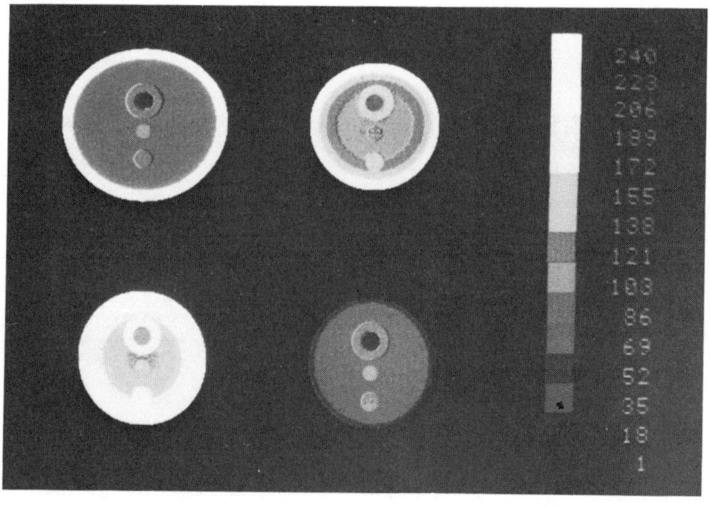

Figure 3

6. References

[1] Bareiss, E.H.: Numerical solution of linear equations with Toeplitz and vector Toeplitz matrices, Numer. Math. 13, pp. 404-424 (1969).

[2] Bitmead, R.R., Anderson, B.D.O.: Asymptotically fast solution of Toeplitz and related systems of equations, Linear Algebra and Appl. 34, pp. 103-116 (1980).

[3] Bunch, J.R.: Stability of methods for solving Toeplitz systems of equations, SIAM J. Sci. Stat. Comput. 6, pp. 349-364 (1985).

[4] Buonocore, M.H., Brody, W.R., Macovski, A.: A natural pixel decomposition for two-dimensional image reconstruction, IEEE Trans. Biomed. Eng. BME-28, No. 2 (1981).

[5] Freidlander, B., Kailath, T., Morf, M., Ljung, L.: New inversion formulas for matrices classified in terms of their distance from Toeplitz metrices, Linear Algebra and Appl. 27, pp. 31-60 (1979).

[6] Gordon, R., Bender, R., Herman, G.T.: Algebraic Reconstruction Techniques (ART) for three dimensional electron microscopy and x-ray photography, J. Theor. Biol. 29, pp.471-481 (1970).

[7] Guenther, R.B., Kerber, C.W., Killian, E.K., Smith, K.T. Wagner, S.L.: Reconstruction of objects from radiographs and the location of brain tumors, Proc. Nat. Acad. Sci. USA 71, pp. 4884-4886 (1974).

[8] Herman, G.: Image reconstruction from projections, the fundamentals of computerized tomography, Academic Press 1980.

[9] Hounsfield, G.N.: Computerized transverse axial scanning tomography: Part I, description of the system, Br. J. Radiol. 46, pp. 1016-1022 (1973).

[10] Kaczmarz, S.: Angenäherte Auflösung von Systemen linearer Gleichungen, Bull. Int. Acad. Pol. Sci. Lett. A, pp. 355-357 (1937).

[11] Kailath, T., Vieira, A., Morf, M.: Inverses of Toeplitz operators, innovations and orthogonal polynomials, SIAM Rev. 20, pp. 106-119 (1978).

[12] Kailath, T., Levy, B., Ljung, L., Morf, M.: The factorization and representation of operators in the algebra generated by Toeplitz operators, SIAM J. Appl. Math. 37, pp. 467-484 (1979).

[13] Lent, A.: Seminar talk at the Biodynamic Research Unit, Mayo Clinic, 1975.

[14] Louis, A.K., Natterer, F.: Mathematical problems in computerized tomography,
Proc. IEEE, Vol. 71, pp. 379-389 (1983).

[15] Natterer, F.: Efficient implementation of 'optimal' algorithms in computerized tomography,
Math. Meth. in the Appl. Sci. 2, pp. 545-555 (1980).

[16] Natterer, F.: Einige Beiträge der Mathematik zur Computer-Tomographie,
ZAMM 64, T252-T260 (1984).

[17] Trench, W.: An algorithm for the inversion of finite Toeplitz matrices,
SIAM J. Appl. Math. 12, pp. 515-522 (1964).

A TWO-GRID-APPROACH TO IDENTIFICATION AND CONTROL PROBLEMS FOR PARTIAL DIFFERENTIAL EQUATIONS

Volkmar Friedrich and Bernd Hofmann

Abstract

Both parameter identification and control problems for partial differential equations (PDE) can be treated by minimization of functionals the values of which depend on the unknown parameters or control functions. Each evaluation of these functionals requires an approximate solution of a boundary value or an initial boundary value problem (IBVP) for the PDE. To get an acceptable trade-off between accuracy and the amount of computation, we preferably use a coarse-grid-approximation of the IBVP for minimizing these functionals. Only at a few points of the parameter or control function space a fine-grid-approximation of the IBVP has to be calculated in order to reduce the systematic error created by the coarse grid.

This approach will be illustrated by the hand of parabolic equations, but it seems to be quite general even for large scale identification and control problems.

1. Introduction

Let us consider the IBVP for the parabolic equation in the time interval $[0,t_E]$

$$\frac{\partial T}{\partial t} = \text{div}\,(\lambda(T,x)\ \text{grad}\ T) \qquad t>0,\ x\in\Omega \tag{1}$$

$$\frac{\partial T}{\partial n} = \varkappa(T,x) \qquad x\in\partial\Omega,\quad T(x,0) = \varphi(x) \qquad x\in\Omega \tag{1'}$$

In an identification problem we might be interested in learning somewhat about the functions $\lambda(T,x)$ and/or $\varkappa(T,x)$ from the measured temperature $T(x_i,t_i)$ in a certain number of points $(x_i,t_i)\in\Omega\times[0,t_E]$. The measured data z_i, in general, include some measurement error.

A control problem for the same equation usually requires to calculate a heat flux Ψ through the boundary

$$\frac{\partial T}{\partial n} = \Psi(x,t,T), \tag{1''}$$

which will guarantee a prescribed temperature profile in time and in space with a certain level of accuracy. In a discretized version, the temperature $T(x_i,t_i)$ at a given number of points $(x_i,t_i) \in \Omega \times [0,t_E]$ should be near to given values z_i. Note that constraints on T are not of interest in this paper.

We suppose the unknown functions $\lambda, \varkappa$ or Ψ to be dependent on a finite number of parameters $u_1,\ldots,u_n$ and use a vector notation $u = (u_1 \ldots, u_n)$. The solution of the IBVP for the equation (1) depends on this u in a nonlinear manner. We denote by A the, in general, nonlinear mapping of u into those values of the solution $T(x_i,t_i)$ which can be compared with the measured or prescribed data z_i $(i = 1,\ldots m)$. Then, a vector u with

$$\|Au - z\| \leq \eta \tag{2}$$

can be accepted, where the error level η corresponds to the model and measurement errors in identification problems or to the accuracy required in a control problem.

If we solve the IBVP by a FDM or FEM approach, we use an approximation A_h of A and an additional approximation error η_h will appear. In this case, for the identification problem, we have to accept not only the vectors u fulfilling the inequality (2), but all elements u satisfying the inequality

$$\|A_h u - z\| \leq \eta + \eta_h \tag{3}$$

A characteristic feature of both types of problems considered here is their ill-posedness, i.e. clearly distinct parameter vectors u may correspond to equal or almost equal values of the measurable quantities $T(x_i,t_i)$. The greater the number of parameters, i.e. the finer the structure of the unknown parameter or control function, the more this ill-posedness makes it difficult to solve these problems numerically by minimizing a functional of residual norm square type

$$\|Au - z\|^2. \tag{4}$$

To overcome this difficulty, we restrict ourselves to a set D of vectors u, for which the parameter or control functions λ, $\varkappa$ or Ψ are physically reasonable (e.g. nonnegative or monotonous functions) and use regularization techniques developed by Tikhonov [6], Philips [5] and others.

2. Regularization for the identification problem

To overcome the ill-posedness arising from the minimization of the functional (4), we can solve one of the nonlinear optimization problems of constrained least square type

$$\min_{u\in D,\ \Omega(u)\leq C} \|Au - z\| \tag{5}$$

$$\min_{u\in D,\ \|Au-z\|\leq \eta} \Omega(u) \tag{6}$$

Here, Ω is a nonnegative stabilizing functional which attains small values if the associated parameter vector $u\in D$ will be preferred as a solution of the identification problem. If otherwise u can almost be excluded, the value $\Omega(u)$ will get large. Often, Ω can be chosen to suppress higher oscillations in the parameter functions λ and $\varkappa$ which, in general, cannot be reconstructed from noisy data z. The option between the problems (5) and (6) will always be a subjective one, since our knowledge of the error level η (and η_h too) is incomplete. Moreover, the choice of the functional Ω and of the constant C is also subjective.

For this reason we usually consider the whole set $\bar{U}$ of regularized solutions u_α, i.e., the set of minimizers u_α $(\alpha>0)$ of the regularized functional

$$F(u) = \|Au - z\|^2 + \alpha\Omega(u) \tag{7}$$

on D. Note that for a linear operator A, a convex set D and a convex functional Ω the solutions of the problems (5) and (6) for all values $\eta>\inf_{u\in D} \|Au-z\|$ and $C>\inf_{u\in D} \Omega(u)$ belong to this set $\bar{U}$. For a nonlinear operator A on the contrary pathological situations are possible, in which not all solutions of (5) or (6) belong to the set $\bar{U}$. But we will not focus our attention on this exceptional case here (see [4]).

In some situations a statistical approach to these problems is useful. If we want to estimate a vector u by the observed data $z \approx Au$, the unknown vector u may be treated as a realization of a random vector by the Bayesian approach or as a nonrandom vector. For a linear operator A the best linear Bayesian estimator can be considered as a regularized solution with optimally chosen Ω and α. For a nonrandom vector u nonbiased estimators cannot be used due to the ill-posedness of our problem here. If we consider the set of minimizers u_α of the functional (7) the cross-validation principle [8] and its generalizations [3] for estimating a nonrandom vector u are an effective way to select a good regularization parameter α. Unfortunately, this technique fails to work in our case for two reasons. First of all, this technique works well only if a systematic model or approximation error can be neglected in comparison with the random measurement error, otherwise it can completely fail. Secondly the numerical realization of the cross-validation technique for a nonlinear operator A is much more expensive than for the linear case.

Consequently, we compute the minimizers of (7) for some values of the regularization parameter α in order to select an appropriate u_α by some heuristic criteria (see also § 5) or by an interpreter of the mathematical model.

3. Regularization for the control problem

Here the functional Ω is to prefer some kinds of control or it can be chosen as a cost functional. The problems (5) and (6) are connected with the functional (7) in the same way as for the identification problem and the parameter α will be chosen to approximate the solution of the problems (5) or (6), resp. In the discrete case, there is another numerical reason for using an appropriate functional Ω. If in some subregions of D the operator A_h does not significantly depend on u, the values of $\|A_h u-z\|$, perturbed by different discretization and round-off errors for different vectors u, could mislead gradient methods for minimizing (7).

4. Minimization of the functional and the two-grid-approach

Let A_h be an approximation of the operator A. For $\Omega(u) = (G(u-\bar{u}),u-\bar{u})$ with an appropriately chosen matrix $G = LL^T > 0$ and a vector $\bar{u}$ and an additional constraint set $D = \{u | g_k(u) \le 0,$ $k = 1,\dots,s\}$ the minimization of the discretized version of (7)

$$\| A_h u - z \|^2 + \alpha \cdot (G(u-\bar{u}), u-\bar{u}) \tag{8}$$

can be handled by the penalty function method with a penalty term $\beta \|p(u)\|^2$, where $p(u) = \{p_k(u)\}_{k=1}^{s}$, $p_k(u) = \max\{g_k(u), 0\}$.

The practical advantages of this approach are the following: We can use standard software for solving overdetermined nonlinear systems $W(u) = 0$, if we rewrite the functional (8) as $\min \|W(u)\|^2$ with

$$W(u) = \begin{Bmatrix} A_h u - z \\ \sqrt{\alpha} L^T (u-\bar{u}) \\ \sqrt{\beta} p(u) \end{Bmatrix}$$

- linear and nonlinear constraints can be dealt with in a unified manner,
- we don't need a high penalty accuracy of u_α^h, since we have only empirical information forming D, G, $\bar{u}$. Consequently, we have not to choose a high penalty level and can generally avoid numerical difficulties, which otherwise would require special shift techniques for high penalty levels.

In standard software package derivative-free modified Gauß-Newton-methods are used.

The most time-consuming part of our minimization procedure is the evaluation of $A_h u$, i.e., the approximate solution of the IBVP (1).

Let us denote by h and H different discretization parameters (in space and/or in time) and refer to them as to a fine and a coarse grid. Then, associated discretization errors for these grids are

$$\eta_h = Au - A_h u \quad \text{and} \quad \eta_H = Au - A_H u \tag{9}$$

For a coarse grid the minimization time would be small, but we would lose accuracy, when substituting the minimizer u_α^h by u_α^H.
To diminish the influence of the coarse grid discretization error η_H we can try to estimate the essential part $\tilde{\eta}_H$ of it. For h much smaller than H the error η_h could be neglected. Thus, we have

$$\eta_H = Au - A_H u \approx A_h u - A_H u = \tilde{\eta}_H \tag{10}$$

Otherwise, extrapolation techniques can be used to estimate $\tilde{\eta}_H$ under the assumption that the structure of the discretization error $\eta_h = Ch^q + o(h^q)$ is known. In our examples below we will use the relation (10).

Our suggestion is to accelerate the minimization of (8) by substituting this fine-grid-functional by the following coarse-grid-functional

$$T_H(u,\tilde{z}) = \|A_H u - \tilde{z}\|^2 + \alpha \cdot (G(u-\bar{u}), u-\bar{u}) + \beta \|p(u)\|^2 \tag{11}$$

with grid-corrected data $\tilde{z} = z - \tilde{\eta}_H$. This suggestion can be motivated by the observation that for most of our problems the difference $A_h u - A_H u$ has changed only slowly with $u \in D$.
The principle scheme of the two-grid-algorithm [2]:

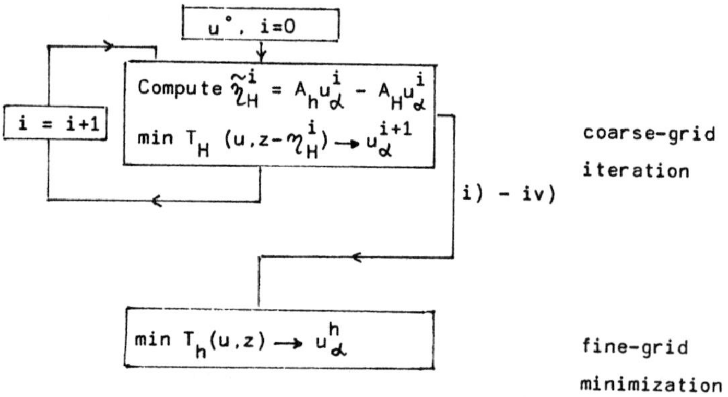

Each minimization step starts with the result u_α^i of the preceding iteration step.

The coarse-grid iteration should be terminated if one of the following situations will appear:

i) $\|\tilde{\eta}_H^i - \tilde{\eta}_H^{i-1}\| \leq \varepsilon_1$

ii) $\|u_\alpha^{i+1} - u_\alpha^i\| \leq \varepsilon_2$

iii) $T_h(u_\alpha^{i+1}, z) > T_h(u_\alpha^{i-1}, z) - \varepsilon_3$

iv) i exceeds a given maximal number of coarse-grid-iteration steps

} Coarse-grid iterations do not work efficiently (iii, iv)

The first two inequalities i) and ii) indicate the convergence of the coarse-grid iteration process. Fine-grid iterations at this stage can be used to improve the result of the coarse-grid iteration.

5. The choice of α and a three-grid approach to identification problems

In the two-grid approach described above the regularization parameter α was supposed to be known. For identification problems however, the minimizer u_α intrinsically depends on this parameter and there are several heuristic strategies for choosing this parameter such as Morozov's discrepancy principle [6], the stochastic smoothing principle by Wahba and Wold [8], the criterion of quasioptimality by Tikhonov and Glasko (see [7]) or Wahba's cross-validation criterion [3]. Each of these criteria requires the (approximate) minimization over α of an additional functional depending on u_α. To avoid complete two-grid iterations for all considered values of the parameter α, it seems to be preferable to use the Tikhonov-Glasko criterion, which recommends to choose α according to

$$\min_{\substack{\alpha > 0 \\ \|Au_\alpha - z\| \leq \eta}} \left\| \alpha \cdot \frac{du_\alpha}{d\alpha} \right\| \tag{12}$$

Using a sequence $\alpha_i = 2^{-i} \cdot \alpha_0$ and a grid $\hat{H}$ and taking into account the possible existence of more than one local minimum of the additional functional (12), we replace (12) by the following criterion: Choose a quasioptimal α_{qo} as the largest value α_i, for which the inequalities

$$\|u^{\hat{H}}_{\alpha_{i+2}} - u^{\hat{H}}_{\alpha_{i+1}}\| > \|u^{\hat{H}}_{\alpha_{i+1}} - u^{\hat{H}}_{\alpha_i}\| < \|u^{\hat{H}}_{\alpha_i} - u^{\hat{H}}_{\alpha_{i-1}}\| \tag{13}$$

and $\|A_{\hat{H}} u^{\hat{H}}_{\alpha_i} - z\| \leq \eta + \bar{\eta}_{\hat{H}}$ with $\bar{\eta}_{\hat{H}}$ an upper bound for $\|\eta_{\hat{H}}\|$

hold. From experiments we have learned, that the value α_{qo} according to (13) is well-adapted to the position of the leading minimum of (12), even if we use a rather coarse grid with $\hat{H} > H$. Furthermore, the quality of the bound $\bar{\eta}_{\hat{H}}$ much less effects the position of α_{qo} than the parameter choice according to Morozov's discrepancy principle.

6. Examples

Now we are going to present both an identification problem for the heat equation and a control problem from rheology. The two-grid approach introduced above is applied to the numerical solution of these two different nonlinear inverse problems.

First we are dealt with an initial-boundary value problem to the onedimensional heat equation in polar coordinates as follows:

$$\varrho(T)\frac{\partial T(r,t)}{\partial t} = \frac{1}{r}\frac{\partial}{\partial r}\left(r\cdot\lambda(T)\frac{\partial T(r,t)}{\partial r}\right),\quad 0<r<R,\ 0<t\leq t_E$$

$$T(r,0) = \varphi(r),\quad 0\leq r\leq R, \tag{14}$$

$$\frac{\partial T(0,t)}{\partial r} = 0,\quad -\lambda(T)\frac{\partial T(r,t)}{\partial r} = \eta(T)(T(R,t)-\bar{\varphi}(t)),\quad 0<t\leq t_E$$

The physical background of the above given problem is formed by considering a heating process for a long, cylindric metal bar of radius R. At an initial time $t = 0$ the bar possesses a radial temperature profile $\varphi(r)$, $0\leq r\leq R$. During the time interval $t\in(0,t_E]$ the solid is heated in a gas stream of temperature $\bar{\varphi}(t)$, where boundary conditions of the third kind with heat conductivity $\lambda(T)$ and heat transfer function $\eta(T)$ are imposed. The quasilinear heat equation allows us to express the heat conduction process in mathematical terms. Here, $\varrho(T)$ is the product of density and heat capacity.

We deal here with the inverse problem of identifying $\lambda(T)$ similar to that studied by Eriksson and Dahlquist [1], although in applications we are also often challenged to identify the heat transfer function η. Thus we assume to know the temperature dependent functions ϱ and η and the initial profile φ and the gas stream temperature $\bar{\varphi}$. It is assumed that the maximum principle yields the lower and upper limitations $0\leq T(t,r)\leq \bar{T}$ for the whole temperature field in $(r,t)\in[0,R]\times[0,t_E]$. We want to identify $\lambda(T)$ from noisy measurements of the final temperature profile $T(r,t_E)$, $0\leq r\leq R$.

Now we agree upon the discretization of the unknown function $\lambda(T)$ by choosing the values $u_i = \lambda(T_i)$, $0\leq T_1<T_2<\dots<T_n\leq\bar{T}$ as parameters. The components z_j of the measurement vector z correspond to the values $T(r_j,t_E)$ of the final temperature at the observation points $0\leq r_1<r_2<\dots<r_m\leq R$. For simplicity, equally spaced grids in $[0,\bar{T}]$ and $[0,R]$ were used to win the numerical results below. Given u the initial-boundary value problem (14) is approximately solved by a finite difference method using both a coarse grid and a fine grid. We denote by $A_H u$ and $A_h u$, respectively, the associated finite difference solutions with the coarse and with the fine grid. The nonlinear operators A_H and A_h transform the conductivity values into resulting final temperature values. For the numerical experiment, D is prescribed by lower and upper bounds for any component u_i.

Furthermore, Ω is the discrete fashion of a smoothing term
$q_0 \|\lambda(T)\|^2_{L_2[0,\bar{T}]} + q_1 \|\lambda'(T)\|^2_{L_2[0,\bar{T}]}$.

In the following we give some numerical results concerning an experiment with $m = 50$ and $n = 11$. The operator A_H uses 10 time steps, whereas A_h is related to 100 time steps. Thus, $A_h u$ approximates the exact solution of (14), for given u very well. On the other hand, computing $A_H u$ requires a ten times smaller expense in comparison with the computation of $A_h u$.

For the results of Table 1, we directly minimize the functional

$$F_h(u,\alpha) = \|A_h u - z\|^2 + \alpha\Omega(u) \tag{15}$$

$\alpha = 10^{-2}$, by using a derivative-free Gauss-Newton method and penalty functions for handling the set D.
The vector u^i_α represents the i-th Gauss-Newton iterate, u^h_α the exact minimizer of $F_h(u,\alpha)$ and the cost unit is equal to the expense of one computed vector $A_h u$.

For Table 2, using the same initial guess we exploit the above introduced two-level algorithm.
The first two iterates u^i_α, i=1,2 are the minimizers of the functional:

$$F^i_H(u,\alpha) = \|A_H u - z + \tilde{\eta}^i_H\|^2 + \alpha\Omega(u). \tag{16}$$

Table 1:

i	$\frac{\|u^i_\alpha - u^h_\alpha\|}{\|u^h_\alpha\|}$	cost
0	1.36	-
1	0.32	17.0
2	0.31	33.0
3	0.19	48.0
4	0.023	60.0
5	0.0011	72.0

Table 2:

i	$\frac{\|u^i_\alpha - u^h_\alpha\|}{\|u^h_\alpha\|}$	cost
0	1.36	-
1	0.171	18.0
2	0.025	33.2
3	0.0046	42.2
4	0.0010	57.2

As we see, the cost of any such minimization is almost equal to the cost of one direct Gauss-Newton step. However, the accuracy of results thus obtained is better. If we continue to minimize $F_h(u,\alpha)$ by direct Gauss-Newton steps u^3_α and u^4_α, starting with initial vector u^2_α, the precision of iterates gets improved. Thus, the two-grid method helps to overcome the starting phase of minimization in an efficient manner.

A comparison of the modified Tikhonov-Glasko criterion with Morozov's discrepancy principle in table 3 shows that the first is much more sensitive to find approximately the best possible regularization parameter.

Table 3 Properties of the parameter u to be identified			$\|Au-z\|/\sqrt{m}$	$\Omega(u).10^{-6}$
			11.99	2.24
Properties of regularized u^h_α				
α	$\frac{\|u^h_\alpha-u\|}{\|u\|}$		$\|A_h u^h_\alpha-z\|/\sqrt{m}$	$\Omega(u^h_\alpha)\cdot 10^{-6}$
10^{-1}	0.315		24.89	1.26
10^{-2}	0.202		12.67	1.75
10^{-3}	0.126		12.00	1.97
6.10^{-4}	0.108		11.94	2.05
4.10^{-4}	0.097	α_{qo}	11.90	2.15
10^{-4}	0.114		11.78	3.05
10^{-5}	0.510		11.57	10.8
10^{-6}	0.531		11.56	12.0
10^{-7}	0.532		11.55	12.1

The quasioptimality criterion even provides the appropriate value $\alpha_{qo} = 4.10^{-4}$ if only four time steps are used to cover the time interval $[0,1]$. Thus obtaining the quasioptimal parameter α_{qo} is rather cheep.

Now we draw the reader's attention to a control problem from rheology. We imagine a homogeneous isotropic viscous fluid (e.g. a melt of glass or polymer) with properties which are dependent on temperature and pressure. This fluid is assumed to flow through a vertical straight circular tube of length one. The fluid adheres to the wall and the heat transfer between fluid and wall we shall describe by boundary conditions of the first kind. The variables $v_r(r,l)$, $v_l(r,l)$, $p(l)$ and $T(r,l)$ denote the quantities radial component of the fluid velocity, axial component of the fluid velocity, pressure and temperature, respectively. These functions are spatially varying with respect to the radial coordinate $0 \leq r \leq 1$ ($r=0$ centre of the tube, $r=1$ wall of the tube) and with respect to the axial coordinate $0 \leq l \leq 1$.

Then the direct problem is determined by an initial-boundary value problem to a system of partial differential equations:

$$\frac{dp}{dl} = \frac{1}{c_1} \cdot \frac{1}{r} \frac{\partial}{\partial r} \left(r \vartheta \frac{\partial v_1}{\partial r}\right), \ 0 \leqslant l \leqslant 1, \ 0 \leqslant r \leqslant 1,$$

$$\int_0^1 \varrho \cdot v_1 \cdot r \cdot dr = \frac{1}{2}, \ 0 \leqslant l \leqslant 1,$$

$$v_1 \frac{\partial \varrho}{\partial l} + \varrho \left(\frac{\partial v_r}{\partial r} + \frac{v_r}{r} + \frac{\partial v_1}{\partial l}\right) = 0, \ 0 \leqslant l \leqslant 1, \ 0 \leqslant r \leqslant 1, \qquad (17)$$

$$\varrho \cdot c \cdot c_1 \cdot c_2 \cdot \left(v_r \frac{\partial T}{\partial r} + v_1 \frac{\partial T}{\partial l}\right) = \frac{1}{r} \frac{\partial}{\partial r} \left(r \cdot \lambda \frac{\partial T}{\partial r}\right), \ 0 \leqslant l \leqslant 1, \ 0 \leqslant r \leqslant 1,$$

$$p(0) = p_0, \ T(r,0) = \varphi(r), \ v_r(r,0) = 0, \ 0 \leqslant r \leqslant 1,$$

$$v_r(1,l) = v_1(1,l) = 0, \ 0 \leqslant l \leqslant 1.$$

In the above system the viscosity is designated by ϑ, the heat conductivity by the symbol λ, the density by ϱ and the heat capacity by c. The values c_1 and c_2 are rheological constants. All material functions and constants are supposed to be known. We choose a grid $0 \leqslant l_1 < l_2 < \ldots$ $\ldots < l_n \leqslant 1$ for the axial direction and a grid $0 \leqslant r_1 < r_2 < \ldots < r_m \leqslant 1$ for the radial direction. Furthermore, define $u_i = T(1,l_i)$ and $z_j = T(r_j,1)$. The control problem under consideration is aimed at finding boundary controls along the tube such that a prescribed final radial temperature profile at the bottom of the tube (l=1) is achieved or approximated as best as possible. We denote by $A_H u$ and $A_h u$ finite difference solutions of the system (17), for given u. As we already assumed in the identification example, $A_H u$ is based on a ten times smaller number of time steps (here, axial direction steps) compared to $A_h u$. Provided the above given assumptions about Ω and D, we present some numerical results concerning the regularized solution of this inverse problem to which the two-grid approach is applied. The following results are related to an experiment with m=42, n=6 and $\alpha = 10^{-10}$. There are comparable data for the direct Gauss-Newton iteration (see Table 4) and for the two-grid approach (see Table 5), which are based on the same initial guess u^0. All vectors u^i, i=1,2,3 of Table 5 come from the minimization of $F_H^i(u,\alpha)$ and no direct Gauss-Newton steps have been added. In constrast to the tables of the identification example, the accuracy is given now by means of the values $F_h(u,\alpha)$ instead of relarive errors.

Table 4:

i	$F_h(u^i,\alpha)$	cost
0	$4.38.10^{-2}$	-
2	$6.84.10^{-3}$	26.0
4	$8.78.10^{-4}$	54.0
6	$1.77.10^{-4}$	76.0

Table 5:

i	$F_h(u^i,\alpha)$	cost
0	$4.38.10^{-2}$	
1	$6.56.10^{-4}$	23.0
2	$2.41.10^{-4}$	34.5
3	$8.42.10^{-5}$	45.9

As the numerical experiment indicates, there are good chances of reducing the amount of computation for realizing regularized solutions whenever a coarse grid is used in order to perform an essential part of minimization work. The introduced method seems to be appropriate for doing this. However, there are still many open questions concerning stopping criteria for the two-grid iteration. If these questions can be answered in an intellegent way, then the method represents a promising technique for the solution of identification and control problems in an efficient manner.

References

[1] G. Erikson and G. Dahlquist. On an inverse non-linear diffusion problem, in P. Deuflhard and E. Hairer (eds.) Numerical Treatment of Inverse Problems in Differential and Integral Equations, Birkhäuser, 1983, pp. 238-245.

[2] V. Friedrich and B. Hofmann. A predictor-corrector technique for constrained least-squares regularization, Wiss. Information 46, Techn. Hochschule Karl-Marx-Stadt, 1984.

[3] G. H. Golub, M. Heath and G. Wahba. Generalized cross-validation as a method for choosing a good ridge parameter, Technometrics 21 (1979), 215-223.

[4] B. Hofmann. Regularization for Applied Inverse and Ill-Posed Problems, Teubner, Leipzig, 1986.

[5] Phillips, D. L. A technique for the numerical solution of certain integral equations of the first kind, JACM 9(1), 84-87 (1962)

[6] A. N. Tikhonov and V. Y. Arsenin. Methods for the Solution of Ill-Posed Problems (2nd Russian edition), Nauka, Moscow, 1979.

[7] A. N. Tikhonov and V. A. Morozov. Regularization methods for ill-posed problems (Russian), in V. A. Morozov and E. S. Nikolaev (eds.) Numerical Methods and Programming 35, Moscow University

[8] Wahba, G., Wold, S. A complete automatic French curve: fitting spline functions by cross validation. Communications in Statistics 4(1), 1-17 (1975).

PART V

OPTIMIZATION AND OPTIMAL CONTROL PROBLEMS

SOLVING LARGE-SCALE INTEGER OPTIMIZATION PROBLEMS

KURT SPIELBERG and UWE H. SUHL

ABSTRACT

We survey proven state-of-the-art solution techniques for solving large-scale integer optimization problems and describe an experimental software system for the solution of large 0-1 integer optimization problems. This system is built around a large commercial LP-system (MPSX/370) and uses sophisticated data structures for an efficient implementation. Numerical results for difficult and large real-life problems were significantly better than with traditional branch-and-bound algorithms such as implemented in commercial software systems.

1. INTRODUCTION

We consider the mixed zero-one optimization problem

$$\begin{aligned} &\text{Min } c'x + d'y \\ &\quad C\,x + D\,y \begin{Bmatrix} \leqslant \\ = \\ \geqslant \end{Bmatrix} b \qquad\qquad (P) \\ &\quad\ x \geqslant 0,\ y \text{ n-dimensional 0-1 vector} \end{aligned}$$

were c, d, C, D are real vectors and matrices of appropriate dimensions. We use the following conventions: e(j) denotes the j-th component of a vector e, a(.,j) denotes the j-th column vector of a matrix A, pa(.,j) and na(.,j) are the vectors containing the positive and negative components of a(.,j), respectively. More precisely, pa(.,j)(i) = a(i,j) if a(i,j) > 0 and 0 otherwise, na(.,j)(i) = a(i,j) if a(i,j) < 0 and 0 otherwise. We use the term LP for Linear Programming. The LP problem corresponding to (P) is defined as (P) with the integrality requirements dropped for y.

Any general linear integer optimization problem can be transformed into a problem of type (P). Nearly all practical integer optimization problems are naturally of type (P). Moreover, many non-convex or nonlinear optimization problems can be translated into (P) by the introduction of artificial 0-1 and continuous variables. Typical examples are problems with: a separable nonlinear objective function, disconnected or non-convex regions, fixed charges, logical interrelations of decisions, sequencing or scheduling of activities. For an overview of typical models and a variety of modeling techniques see /18/.

While many practical decision problems can be formulated as integer optimization problems, the solution of such problems is difficult; i.e., the problems are NP-hard. As a consequence, large integer optimization problems with thousands of variables and constraints might require huge amounts of computer time.

2. KEY ASPECTS OF SOLVING INTEGER OPTIMIZATION PROBLEMS

Experience shows that the successful solution of large and difficult integer programs requires:

- a state-of-the-art optimizer
- proper model formulation
- preprocessing of the model.

We describe these aspects in more detail. The main part of the paper concentrates on algorithms for optimizing (P).

2.1 STATE-OF-THE-ART OPTIMIZER

Despite a lot of theoretical and applied research (see /7/ for an introduction) the most successful algorithms for solving general large integer optimization problems are still based on enumerative algorithms, either in form of LIFO implicit enumeration schemes (usually for pure 0-1 problems; but see /14/ and this paper for extensions to mixed 0-1 problems), or of LP-based branch-and-bound algorithms. These algorithms require a tree search, with a node of the tree corresponding to a subproblem in which some of the integer variables are assigned integer values. The task is to resolve a subproblem. This can be done by various techniques, the most inefficient being the enumeration of the values of the remaining integer variables. When a subproblem is solved, another unresolved subproblem is chosen and the process is repeated. In the original form the first type of algorithm (see /1/) is characterized by logical tests for resolving subproblems, whereas the second type of algorithm solves the LP-relaxations of the subproblems,i.e., the subproblems with the integrality requirements on the unassigned integer variables dropped. For a survey of integer programming codes see /13/.

In our opinion a state-of-the-art optimizer must include at least the following features:

- An enumerative master algorithm with the capability to solve LP subproblems.
- Powerful algorithmic features to explore the pure 0-1 part of (P), primarily of logical nature.
- A fast and stable LP-system for solving LP-subproblems of (P).
- Heuristics for rounding LP-solutions of subproblems to generate quickly good integer solutions.
- An efficient implementation of the algorithms. This means appropriate data structures and detailed program design to reduce the average time complexity of the programs.

2.2 PROPER MODEL FORMULATION.

To get the most out of the solution of the LP subproblems we have to stress how important it is to formulate the model so that it has a strong LP-relaxation. In general, a strong LP-formulation is characterized (relative to the original formulation) by more constraints and a smaller LP-polytope. On the solution side this implies a small gap between the objective function value of an initial LP-solution and an optimal integer solution. Furthermore, many of the integer variables will be integer valued in the initial LP-solution. Although a strong LP-formulation might increase the LP solution time at a given node due to a larger number of constraints, this is in general more than offset by a smaller number of required LP's.

An aggregated representation of constraints is a common cause of long running times for a model. For example, the constraint $y1 + y2 \leqslant 2y$, with all variables 0-1, should be replaced by the two constraints $y1 \leqslant y$ and $y2 \leqslant y$. In the second formulation, the corresponding LP-polytope is locally the convex hull of all feasible 0-1 solutions of the constraints, i.e., is as small as possible. A case study in /15/ shows how important a model formulation can be: A mixed 0-1 integer problem which ran for over twenty hours on a fast machine, without any integer solution, was solved in about one minute after disaggregation of the constraints!

2.3 PREPROCESSING THE MODEL.

During this phase several algorithms are applied to the model to obtain a tighter LP-formulation and to fix integer variables. These algorithms are based on recognition of special structures in the 0-1 part of the model (such as special ordered sets, cliques, implications), dynamic inactivation of pure 0-1 constraints, coefficient reduction, generation of sparse cutting planes. Since some of these algorithms are based on repetitive LP optimizations of the model it would be virtually impossible for a user to do this type of preprocessing on his own. On the other hand the automatic preprocessing of a model, in general, does not make a good model formulation obsolete. It can be compared to an optimizing compiler which improves the machine representation of a program in the framework provided by algorithm and program design. Recent research on preprocessing of integer programs can be found in /3, 5, 9, 16/.

3. OPTIMIZATION OF (P)

In this chapter we discuss an algorithm for solving (P) and its implementation, a software system to be called MIPEX (mixed-integer programming experimental), which has the following characteristics:

- MIPEX is an experimental system for solving (P) which consists of a preprocessor and an optimizer.
- The preprocessor is used to improve the strength of the LP relaxation for certain classes of pure and mixed-integer programs.
- The master algorithm is based on enumeration, with optional resolution of the LP subproblem at every node of the tree.
- MPSX/370 (IBM's Mathematical Programming system product) is used for solving the LP subproblems.
- Logical tests of degree 0, 1, 2 are performed at every node of the tree on original and generated (cuts) pure 0-1 constraints.

- Benders cuts with integer coefficients are derived from feasible LP subproblems and added to the coefficient matrix.
- A heuristic based on a certain rounding of the LP-solutions often finds good integer solutions in the initial phase.
- Sophisticated data structures are used to produce an efficient implementation of the system.

In the following sections we describe the optimization part of MIPEX in more details.

3.1 THE SEARCH

Enumerative algorithms for solving (P) proceed by assigning, sequentially, values of 0 or 1 to the zero-one variables. The assigned variables are recorded in an ordered set S. Assigned variables partition naturally into two sets, those variables which are fixed and those which are set. A variable is fixed to a value whenever one ascertains that the opposite value would result in an infeasible subproblem. A variable is set to a value when a branch is taken in the enumeration, and the opposite value will then have to be explored later. The search can be visualized in form of a tree for which each node represents an index set S (corresponding to the assigned, or to the set variables, as one prefers). At a successor of a node exactly one more variable is set. The level of the tree at a node S is defined as the number of variables set in S. Let F(S) be the index set of the 0-1 variables which are free, i.e. are not assigned a value. Let S1 be the index set of the variables in S which are assigned a value of 1. To each node S corresponds a problem (P(S)) which is a subproblem of (P). The following example shows a search tree with a level of 3.

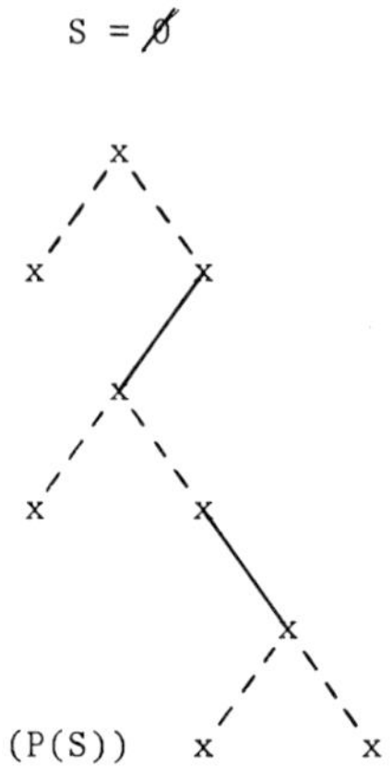

$$\text{Min } c'x + \sum_{j \in F(S)} d(j)\, y(j) + z(S)$$

$$(P(S)) \qquad C\,x + \sum_{j \in F(S)} d(.,j)\, y(j) \begin{Bmatrix} \le \\ \ge \end{Bmatrix} b(S)$$

$$x \ge 0, \quad y(j) \in \{0,1\}, \quad j \in F(S)$$

$$z(S) = \sum_{j \in S1} d(j)\, y(j),$$

$$b(S) = b - \sum_{j \in S1} d(.,j)$$

—— fixed 0-1 variable
--- set 0-1 variable

The task at node S is to fathom (P(S)), i.e. to solve this subproblem. As a first step in this direction one solves the corresponding LP-problem, say (LP(S)). If the LP is infeasible (P(S)) is infeasible. The term infeasible is meant to include the case in which the objective function is no less than a known upper bound. If the LP-solution is integer one has an optimal solution of (P(S)). In any other case one continues to explore (P(S)), using the LP-solution and performing logical tests.

3.2 LOGICAL TESTS

Nearly all practical problems have pure 0-1 side constraints, i.e. constraints of the form $l \leqslant A\ y \leqslant u$ where l,u are real vectors and A is a real matrix of appropriate dimensions. Enumeration has always used logical tests to:

- recognize infeasibility of (P(S))
- fix free 0-1 variables
- Derive either implicitly or explicitly, in the form of logical inequalities, a set of either-or conditions for a subset of the free 0-1 variables at (P(S)).

More precisely, at a given node S, define the vectors a(S), am(S), ap(S) by:

$$a(S) = \sum_{j \in S1} a(.,j), \quad am(S) = -u + a(S) + \sum_{j \in F(S)} na(.,j) \text{ and}$$

$$ap(S) = -l + a(S) + \sum_{j \in F(S)} pa(.,j).$$

Let JP, JN be nonempty disjoint subsets of F(S). If

$$\sum_{j \in F(S)-JP} pa(.,j) + \sum_{j \in JN} na(.,j) \not\geqslant l - a(S)$$ then at least one variable

in JP can be fixed to one or one variable with index in JN can be fixed to zero. Similarly, if

$$\sum_{j \in F(S)-JN} na(.,j) + \sum_{j \in JP} pa(.,j) \not\leqslant u - a(S)$$ then at least one variable

in JN can be fixed to one or one variable with index in JP can be fixed to zero. Algebraically this can be expressed as:

$$\sum_{j \in JP} y(j) + \sum_{j \in JN} (1 - y(j)) \geqslant 1.$$ This constraint can be added to

(P(S)) and act as a cutting plane. It can also be exploited in a logical sense, see /8,14/. The cardinality of the set JP $\cup$ JN is called the degree of the logical tests. Most important are logical tests of degree 0,1,2. Tests of a higher degree are difficult to exploit due to their combinatoric nature. We have:

- Degree 0: If $am(S) \not\leqslant 0$ or $ap(S) \not\geqslant 0$, then (P(S)) is infeasible.
- Degree 1: We assume the degree is not 0. If $a(i,j) > 0$ then $ap(S) - a(i,j) < 0$ implies $y(j) = 1$ and $am(S) + a(i,j) > 0$ implies $y(j) = 0$. If $a(i,j) < 0$ then $ap(S) + a(i,j) < 0$ implies $y(j) = 0$ and $am(S) - a(i,j) > 0$ implies $y(j) = 1$. These tests are performed most efficiently when the a(i,j) are stored in descending order of absolute value for a given row i. Then the non-verification of any of the above tests for some j implies that the test can also not be verified by subsequent variables in that order.
- Degree 2: We use in MIPEX a special form of degree 2 tests, called probing /8/, to be described below.

3.3 PROBING

We assume that the degree at the considered node is larger than 1. A free 0-1 variable is tentatively set to either 0 or 1 (the values are considered in turn). One then pursues the logical consequences by executing logical tests of degree 1 applied to all pure 0-1 side constraints. This test might determine that free variables have to be fixed. If setting y(j) to 0 (or 1) results in the fixing of other variables we call this an implication (we keep track of both 0-implications and 1-implications). We also maintain (0- and 1-) implication counts for all free variables. These tools are used in MIPEX to:

- determine whether a subproblem is infeasible if the implications are followed. In that instance the variable which was tentatively set can be fixed to its opposite value. Experiments show that many variables can actually be fixed and that in many cases the resulting subproblems are proved to be infeasible (see section 5, Table 5).

- branch on variables with large implication counts because it is likely that the resulting subproblems can be more easily fathomed.

Since we have an efficient implementation, probing is used in MIPEX at all nodes over all free 0-1 variables.

The following small example demonstrates the use of probing. We consider a pure 0-1 minimization problem. The degree is higher than 1, because no individual variable can be fixed. The LP-solution is: y1 = y2 = 0, y3 = y4 = y5 = 0.5, z(LP) = 17.5

y1	y2	y3	y4	y5		
200.	300.	5.	10.	20.	Min.	RHS
1.		-20.			≤	0.0
1.		1.		1.	=	1.
	1.	1.	1.		=	1.
	1.		1.	1.	=	1.

Probing determines that this problem is infeasible: Setting y3 = 1 implies that y1, y2, y4 and y5 have to be fixed to 0 resulting in an infeasible problem. Therefore, we have to fix y3 to 0. This implies that y1 = 0 and therefore y5 = 1, so that y2 = y4 = 0. The remaining problem is infeasible. Therefore, neither y3 = 1 nor y3 = 0 is possible.

3.4 ROUNDING OF LP-SOLUTIONS

During a phase I the main emphasis is to find good integer solutions. We use the information of an optimal LP solution at a node for rounding purpose. This is done in several steps:

- The free nonbasic 0-1 variables of the LP solution of (LP(S)) are ordered according to decreasing absolute reduced costs. Experience shows (see /12/) that variables with high absolute reduced costs will have values which frequently result in good integer solution if the remaining free 0-1 variables are properly set.

- The free nonbasic 0-1 variables are set, in the sequence determined by the ordering in step 1, to their values in the LP solution. Then the quasi-integer variables are set to the closest integer. Any such setting is followed by logical tests of degree 0,1. These tests may fix free variables and also lead to infeasible subproblems or backtracks in the search.

3.5 DERIVING BENDERS CUTS

Let z' be the value of the best integer solution found, or some other imposed upper bound on the integer objective function. For a fixed 0-1 vector y which is substituted in (P) we obtain an LP optimization problem (LP(y)). Solving this LP we obtain a Benders cut of the first type $z' > d'y + u'(b - Dy)$ if the LP is optimal or a cut of the second type $0 \geqslant v'(b - Dy)$ if the LP is infeasible. u (resp. v) is an extreme point (resp. extreme ray) of the polytope defined by the dual LP problem of (LP(y)), see /2/. Benders original algorithm requires the solution of one integer optimization problem at each iteration of the algorithm. The direct enumeration scheme in /14/ uses the same cuts but solves only one overall integer optimization problem. For mixed 0-1 problems both methods suffer from the fact that at given node S, the LP problem solved is not a relaxation, i.e. if the LP is integer or infeasible a backtrack is not possible. This explains also why both methods have never been competitive to branch and bound methods for mixed integer problems.

In MIPEX we generate Benders cuts from the optimal solution of the LP relaxation (LP(S)). It is easy to see that we can generate a Benders cut of the first type whenever this LP has an optimal solution. The coefficients of the cuts are the reduced costs of the 0-1 variables in the primal LP solution. By rounding the coefficients to integer values we weaken the cuts somewhat but reduce the storage requirements, due to our supersparse storage scheme and the fact that many of the rounded reduced costs have the same values, even in different LP's. We always allocate enough space to store at least the Benders cut from the initial LP solution. Depending on the amount of storage available we allow at most fifty cuts to be stored. If the space is exhausted the last generated cut replaces the last one stored. If we cannot store at least two cuts we use the reduced costs for a well known ceiling test: If r(j) is the reduced cost in an optimal LP solution of (LP(S)) and z(LP(S)) its LP value we can fix a 0-1 variable j if $z(LP(S)) + |r(j)| > z'$.

3.6 BRANCHING STRATEGY

In the following we shall distinguish two phases of the enumeration. In Phase I, we search for integer solutions; in Phase II we assume that the best, or an adequate, integer solutions has been found, and we attempt to guide the search such that it terminates soon (that is, the strategy is to take branches which lead to the fixing of variables and to quick backtracks. As the default option we consider Phase I to consist of the first 50 nodes of the enumeration. This can be increased for difficult problems.

After the solution of the initial LP problem we create a branching stack to be used later on during the search for the selection of a branching variable if the other branching criteria to be described below fail. The nonbasic 0-1 variables are ordered according to decreasing absolute values of their reduced costs. We compute the rounded value $p = nb * h$ where nb is the number of nonbasic 0-1 variables and h is an input parameter between .4 and .8 with a default value of .7. As we mentioned under rounding of LP-solutions the first p 0-1 variables are likely to have the correct values in an optimal integer solution. The remaining free nonbasic 0-1 variables (with the smallest reduced costs) might have wrong values. Their signed indices are placed as the first nb - p entries in the branching stack (the signs denoting branches of 0 or 1, respectively). We complement their values; i.e. if a variable j is nonbasic at its upper bound we store -j, if it is at the lower bound we store j in the stack. If j is selected later on as branching variable from the stack it is set to zero if $j < 0$ and to one otherwise.

In a second step we add to the branching step the indices of rounded basic variables which are not quasi integer. If a variable j is larger than .5 j is stored in the stack otherwise -j is stored. We store the number of entries in the branching stack and maintain a stack pointer for the next branching variable. This stack pointer is decreased after each selection. An entry is not erased, however. If the stack is empty we start again from the original top of the stack. This strategy tries to use all entries in the stack as branching variable.

In general a branching variable is selected as a byproduct of probing:

- If the maximum implication count of a free 0-1 variable exceeds 0.25 * n where n is the number of 0-1 variables in (P) we take that branch immediately without further probing.

- If full probing is terminated without proving infeasibility of the current subproblem and the maximum implication count is at least 0.2 * n we take a branch for which this maximum is realized, breaking ties arbitrarily.

- In all other cases we select a variable from the branching stack. In phase I (and in phase II as long as no integer solution has been found) we branch opposite to the value which the given nonbasic or quasi-integer variable had or was close too. In phase II we branch on a variable in the candidate stack but opposite to its value in the best found integer solution so far.

3.7 BACKTRACK STRATEGY

When a subproblem (P(S)) at level k + 1 has been resolved the search backtracks to the subproblem at level k. Two cases are distinguished: 1. the backtrack is the result of logical tests (probing). 2. The backtrack is the result of an infeasible LP subproblem. In the first case we proceed via logical tests followed by a branching step. In the second case, however, we always resolve the corresponding LP. This has been found to lead often to further infeasible LP subproblems and, therefore, to further backtracks. The LIFO strategy is used in the current version of MIPEX for selecting a node, i.e. we continue search at the node where the last variable was set by fixing it to its complement.

3.8 SKELETON OF THE MIPEX ALGORITHM

We give now a simplified outline of the underlying algorithm of MIPEX. Details of obvious actions such as update operations, backtracks due to infeasibilities and storing of better integer solutions etc. are omitted.

1. Initialization

   - Initialize parameters and establish all data structures.

   - Execute logical tests of degree 0,1 and full probing on the pure 0-1 side constraints.

   - Solve the initial LP, taking fixed variables into account

   - If the LP is integer or infeasible terminate, since (P) has been resolved.

- Derive and store the initial Benders cut (3.5).
- Execute logical tests of degree 0,1. If further variables are fixed, execute full probing.
- Derive the branching stack and round the LP solution (3.4).

2. Branching step
   - Select a branching variable among the free 0-1 variables according to the strategy described in 3.6.
   - Update S, F(S) a(S), am(S), ap(S), node and level numbers.
3. Logical Tests
   - Execute logical tests of degree 0 and 1.
   - Execute full probing.
4. Linear Program solution, testing and rounding
   - Solve the LP.
   - Test for integrality and infeasibility.
   - Generate a Benders cut. This cut replaces the last one generated if this cut was not generated at level 0. If the cut could not be stored, use the reduced costs for the ceiling test (3.5).
   - In phase I, round the LP solution (3.4).
   - Go to 2.
5. Backtrack
   - Fix the last branching variable to its complementary value.
   - Update S, F(S) a(S), am(S), ap(S), node and level numbers.
   - Execute logical tests.
   - If the backtrack was due to an infeasible LP problem goto 3.
   - Perform full probing.
   - Go to 2.

4. IMPLEMENTATION

Our goal was to design a reliable software system for the testing of algorithmic features which would yield information as to how a production code could eventually be implemented. The implementation is designed for a conventional von Neuman scalar machine architecture. We tried to design the system in an efficient manner, but the actual coding was done in a high level language. PL/I was chosen as programming language because it has interfaces to MPSX/370 allowing to use its algorithmic tools. Furthermore our implementation benefitted from the wide variety of storage classes, file handling and data types available in PL/I. Significant performance improvements would be possible if PL/I is replaced by a systems programming language (such as PLAS). We describe in the following implementation aspects which are important for the efficient execution of logical tests of degree 0, 1 and probing.

The data structures used in the most recent version of MIPEX were chosen as the result of extensive experimentation. When a partial solution S changes a(S), ap(S), am(S) may have to be updated, so that the basic requirement is the high speed addition of a sparse column vector or its positive or negative part to a full vector. Such operations are carried out best in a column oriented sparse matrix scheme in which the nonzeros in each column are grouped by sign and pointer vectors define starting addresses for a column resp. its negative part.

Unfortunately, the column oriented test scheme is not well suited for the execution of logical tests, in particular probing. Adding a rowwise stored copy of the pure 0-1 side constraints, in which the nonzeros in each row are sorted into decreasing order of absolute values of their coefficients, permits the elimination of superfluous degree 0 and degree 1 tests of rows for which am(S) or ap(S) have not been altered during the update due to the sparsity of column vectors. Moreover we keep the nonzeros in each row partitioned into those which belong to assigned variables and those which belong to free variables, see Fig. 1 and 2. To reduce the storage requirements column and rowwise stored sparse matrix use a common storage pool of the different nonzeros. This requires 4 bytes for a nonzero in each representation plus 4 bytes (single precision) for each floating point number. Since the different nonzeros are typically less than 10% of the total number of nonzeros significant memory savings are achieved at a slight expense of execution speed.

Whenever a node changes we update a(S), am(S) or ap(S) using the columnwise structure. During an update those rows for which am(S) resp. ap(S) is increased resp. decreased are tested if further variables can be fixed. Such variables - called candidates - are marked and stored in a queue (see Fig. 3). If the update of the current variable has been completed and the queue is nonempty the next variable of the queue is processed. This may result in an increase of the queue if additional variables (candidates) are to be fixed. The whole update algorithm terminates if either the queue becomes empty or the subproblem becomes infeasible i.e. if there is a row i for which am(S)(i) > 0 or ap(S)(i) < 0.

FIG. 1 DATA STRUCTURES FOR THE PURE 0-1 CONSTRAINTS

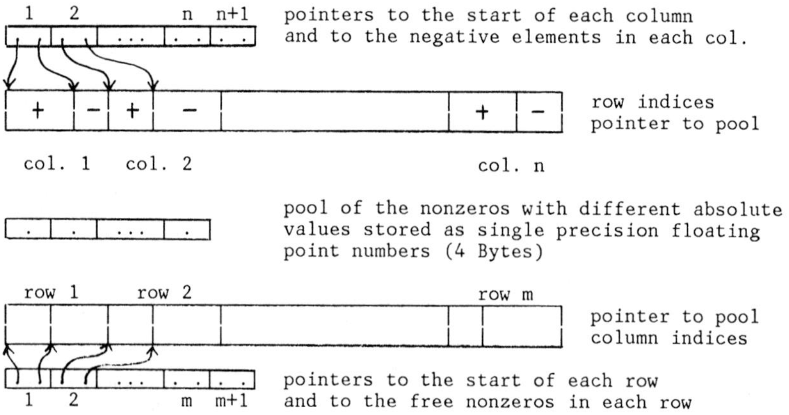

FIG. 2 UPDATE OF THE ROWWISE STRUCTURE

Before update

pointers to pool
column indices

After update

pointers to pool
column indices

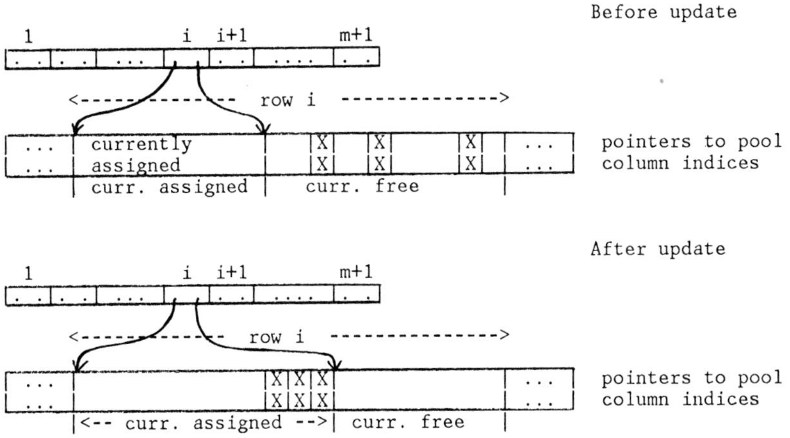

FIG. 3 IMPLEMENTATION ASPECTS OF THE LOGICAL TESTS

```
am                           j                              ap
+-+   +------------------------------------------------+   +-+  +-+
| |   | x            | + | (j1)   x   x(j2)   x        |   |*|  |*|
| |   |              |   |              x               |   | |  |*|
| |   |        x(j3) | + |    x   x           x x       |   |*|  |*|
| |   |   x     x    |   |    x           x             |   | |  | |
| |   |           x  |   |    x       x                 |   | |  | |
| |   |      x       |   |                x             |   | |  |*|
|*|   |   x     x  x x| - |    x  (j4)         x         |   | |  |*|
|*|   |     (j5)      | - |          (j6)         x      |   | |  |*|
+-+   +------------------------------------------------+   +-+  +-+
```

```
        assigned            candidates to       stack for those rows to
        0-1 Variables       be fixed            be reorganized in the
                                                rowwise stored struct.
        ------------------><----------------->
        +-----------------|-------------------|---------------+
        |.................|j|j1|j2|j3|j4|j5|j6|               |
        +-----------------|-------------------|---------------+

                stack     |     queue        |
```

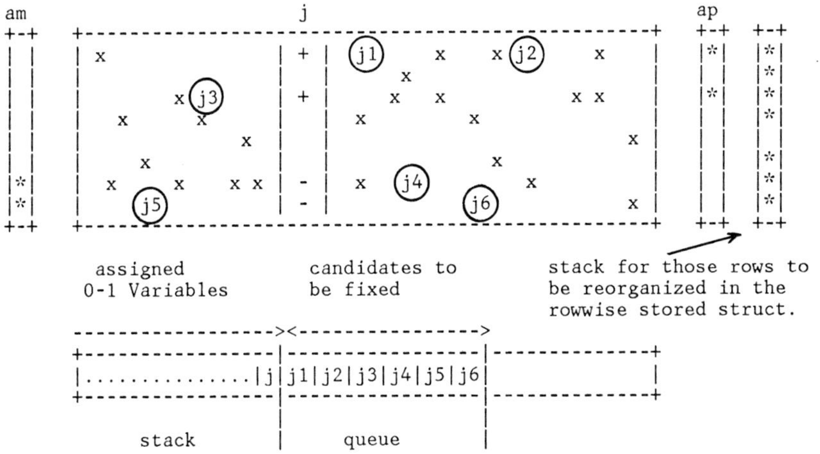

The advantages of this implementation are:

- Only those rows are tested which belong to nonzeros of assigned 0-1 variables.

- Since the free nonzeros are kept sorted in each row a logical test is executed at most once without fixing a variable.

- Probing can be implemented efficiently

5. EXPERIMENTAL RESULTS

We conducted a number of numerical experiments with pure and mixed 0-1 problems as shown in table 1. The problems P02, P04, P06 and P07 correspond to the well known test problems BM23-BM25, L-S-D and L-S-E /14/. They are small but difficult problems and are listed here for comparison of performance with other codes. All other problems come from real life applications from inside and outside IBM and are difficult as well as large.

As performance measure to reach certain stages in the optimization we used: number of nodes in the search tree, number of LP-iterations, and CP-time. This time includes preparation of I/O operations. The first two indicators are relatively machine independent and measure the success with new algorithmic ideas. The last indicator measures the total amount of work and is also strongly influenced by the quality of the implementation. All times and the number of LP-iterations include the solution of the initial LP.

TABLE 1 PROBLEM DATA

Name	n	0-1	m	nz	zlp	zi
P01	33	33	16	131	2,521.57	3,089.00
P02	27	27	21	505	20.57	34.00
P03	28	28	21	523	24.57	38.00
P04	31	31	21	561	29.03	43.00
P05	46	46	42	288	773.75	802.14
P06	74	74	51	567	-559.58	-540.00
P07	89	89	37	530	992.18	1120.00
P08	201	201	21	2,124	6,875.00	7,615.00
P09	313	313	459	4,539	254,862.38	258,411.00
P10	1,994	1,994	114	11,120	3,472.23	3,534.00
P11	2,655	2,655	147	13,858	6,532.10	6,548.00
P12	2,765	2,765	1,068	17,689	3,116.46	3,124.00
P13	1,052	56	397	3,983	2,246,496	2,900,650
P14	1,410	25	485	5,298	81,360,152	85,360,294
P15	1,547	20	1,127	6,678	-82,425,679	-81,439,350
P16	10,910	48	4,475	45,078	-64,995,368	-57,464,368 ?

n : total number of variables in the problem
0-1 : number of 0-1 variables
nz : number of nonzero matrix entries
m : total number of constraints
zlp : optimal objective function value for LP-relaxation
zi : optimal objective function value for integer program if known
? : integer optimum is not known, value of best solution recorded

Of particular interest for us was the historical progress in solving the L-S-E problem (P07 in table 1). This very difficult pure 0-1 problem was unsolved for quite a long time. In table 2 we summarize the results of all codes which to our knowledge obtained improved solution times, starting with Breu and Burdet's code MILPOD /4/. If several strategies were used we reported the best published times. We believe that the L-S-E problem should be included in any numerical investigation about integer programming codes. The best result was obtained by using the preprocessor in /5/, i.e. without using MIP/370 for solving the preprocessed problem.

TABLE 2 HISTORICAL PROGRESS IN SOLVING THE L-S-E PROBLEM

CODE/ (AUTHORS)	Year		NODES	LP ITERAT.	CP-TIME (MINS.)	MACHINE MODEL
MILPOD (BREU/ BURDET)	1973	F	200	?	?	UNIVAC 1108
		P	6481	45411	28.70	
(PIPER)	1974	F	?	?	?	UNIVAC 1108
		P	3458	?	26.13	
HYBRID (JOHNSON/ SUHL)	1978	F	241	836	0.68	IBM 370/ 168-3
		P	2463	7601	7.72	
ENUM (SPIELBERG/ SUHL)	1980	F	61	748	1.07	IBM 370/ 168-3
		P	2101	7041	36.71	
MIPEX (SUHL/ SPIELBERG	1982	F	126	446	0.50	IBM 3081/ D
		P	546	1774	2.98	
MIPEX based on prep. problem	1982	F	29	137	0.02	IBM 3081/ D
		P	137	700	0.69	

F: Finding an optimal integer solution P: Proving optimality

In assessing the tables, one has to keep in mind that MIPEX was written in PL/I, whereas MPSX/MIP/370 is written in assembly language. Furthermore some time is spent in the interface to MPSX/370 whenever an LP is solved which could be avoided in a production code. Nevertheless the tables 2, 3 and 4 speak for themselves. In nearly all cases MIPEX found an optimal solution and proved optimality more effectively than MPSX/MIP/370. More important, the first integer solution is in most instances of very good quality. This is highly important for large problems, for which it may be impossible, or too costly, to prove optimality.

All numerical results reported in tables 2, 3 and 4 are based on executing full probing at each node. Table 5 shows a statistic gathered during proof of optimality for pure 0-1 problems. As can be seen from the table, between 35 to 85% of all probings prove infeasibility of the current subproblem. Also, the maximum number of variables fixed during a single probing, over all trials, was always more than 30% of the total number of 0-1 variables. Although we do note present numerical results our experience shows that for pure 0-1 problems probing is a very important technique for solving difficult problems. We need to point out that Benders cuts play a crucial role in the success of logical tests and probing for pure 0-1 problems.

TABLE 3 FINDING AN OPTIMAL INTEGER SOLUTION

Name	nodes		LP-Iterations		CP-time (min.)	
	MPSX	MIPEX	MPSX	MIPEX	MPSX	MIPEX
P01	2136	0	7467	36	0.46	0.02
P02	427	0	2137	54	0.22	0.02
P03	93	0	478	71	0.06	0.02
P04	168	0	2454	77	0.25	0.01
P05	5000*	0	16201*	121	1.01*	0.03
P06	4112	0	12317	157	1.64	0.14
P07	5312	0	13412	137	2.12	0.02
P08	243	0	2417	379	1.06	0.10
P09	3151	4	18116	744	6.89	0.74
P10	337	21	13979	3776	11.06	3.16
P11	261	4	6432	4001	4.06	2.98
P12	231	314	9978	10865	5.06	11.05
P13	145	93	1056	1300	0.57	0.56
P14	1347	226	6481	2141	3.39	1.63
P15	271	120	8596	4236	7.79	6.27
P16	214!	39?	12063!	8997?	49.48!	31.27?

?: integer optimum is not known, value of best solution used
*: optimal solution was not found in specified time
!: no integer solution was found
All times in minutes on an IBM 3081/D

TABLE 4 PROVING OPTIMALITY

Name	nodes		LP-Iterations		CP-time (min.)	
	MPSX	MIPEX	MPSX	MIPEX	MPSX	MIPEX
P01	5002	78	14267	237	0.96	0.10
P02	463	18	2356	167	0.24	0.07
P03	295	23	1537	215	0.18	0.07
P04	782	49	4209	442	0.45	0.14
P05	5000*	128	16201*	595	1.01*	0.26
P06	8136	27	24366	416	2.41	0.15
P07	11316	137	28131	700	4.81	0.69
P08	894	473	17266	10029	3.67	10.09
P09	6011	29	36144	1294	15.06	1.72
P10	594	251	17037	12021	12.26	8.15
P11	449	311	14316	11701	11.16	15.48
P12	1781	317	19932	11086	12.06	11.40
P13	253	101	2149	1433	0.91	0.64
P14	1567	521	8142	4536	4.57	2.67
P15	272	177	9742	7151	9.08	6.27
P16	214!	39?	12063!	8997?	49.48!	31.27?

?: integer optimum is not known, value of best solution used
*: optimal solution was not found
!: no integer solution found in specified time
All times in minutes on an IBM 3081/D

TABLE 5 PROBING STATISTIC DURING PROOF OF OPTIMALITY

NAME	NODES	LP ITER.	CP-time (MINS)	NUMB. OF PROBINGS	OF THEM INFEAS.	MAX. FIX. VARIABL.	0-1 VAR.
P01	78	237	0.10	210	84	21	33
P02	18	167	0.07	61	27	17	27
P03	23	215	0.07	76	35	19	28
P04	49	442	0.14	149	64	20	30
P05	128	595	0.26	284	141	37	46
P06	27	416	0.15	98	40	32	74
P07	137	700	0.69	351	149	61	89
P08	473	16029	19.09	1239	434	145	201
P09	29	1294	1.72	248	212	103	313
P11	311	11701	20.48	309	198	2192	2655
P12	317	11086	11.40	470	196	1343	2765

All times in minutes on an IBM 3081/D

6. CONCLUSIONS

The enumerative code, though of experimental nature proved to have the potential of solving large 0-1 problems or providing solutions with a small percentage gap. On pure 0-1 problems it proved to be much better than such a polished product code as the latest version of MPSX/MIP/370. This demonstrates that the logical techniques are essential for the solution of large and difficult pure 0-1 problems. To use these techniques effectively at every node of the tree an efficient implementation is necessary. We believe that we found proper data structures to accomplish this task.

For mixed 0-1 problems our improvements compared to a branch and bound code such as MPSX/MIP/370 are not as significant as for pure 0-1 problems. We believe, that the logical techniques are less important for this problem class, except on problems where there are a large number of pure 0-1 side constraints.

It became also clear that preprocessing techniques used to tighten the LP-relaxation are essential for the solution of large and difficult integer programs. Unfortunately, it will be very difficult to use these techniques during the search, since the problem data at a subproblem might have to be changed. This implies that instead of storing only a bit map for the optimal LP-basis one has to store additional data to reconstruct a subproblem during the search.

We believe that the impact of pipelined vector machines for integer programming will be small due to the great sparsity of typical models. However, substantial speedups will be possible by using highly parallel machines (MIMD) with many processing elements due to the high granularity of our algorithm. This is so because subproblems can be optimized independently with very little communication and synchronization overhead. To our knowledge this has not been actually done due to the lack of commercial machines available.

7. REFERENCES

1. E. Balas, An additive algorithm for solving linear programs with zero-one variables, Opns. Res. 13(1965), 517-546

2. J.F. Benders, Partitioning procedures for solving mixed variables programming problems, Numerische Mathematik 4(1961), 238-252

3. A. Brearley, G. Mitra, and H.P Williams, Analysis of mathematical programming problems prior to applying the simplex method, Math. Programming 8 (1975), 54-83

4. R. Breu and C.A. Burdet, Branch and Bound experiments in 0-1 programming, Mathematical Programming Study, 2(1974), 1-50

5. H.P. Crowder, E.L Johnson and M.W. Padberg, Solving Large-Scale Zero-One Linear Programming Problems, Opns. Res. 31(5)(1983), 803-834

6. M. Guignard and K. Spielberg, Logical Reduction Methods in Zero-One Programming, Opns. Res. 29(1981), 49-74

7. R. S. Garfinkel and G. L. Nemhauser, Integer Programming, Wiley, N.Y., 1972

8. M. Guignard, K. Spielberg and U. Suhl, Survey of enumerative methods for integer programming, in Proc. Share 51, (ACM 1978) 2161-2170

9. E.L. Johnson, M.M Kostreva and U.H. Suhl, Solving 0-1 Integer Programming Problems arising from Large Scale Planning Models, Opns. Res. 33(4), 803-819

10. E.Kalan, Aspects of large-scale in-core linear programming ACM Proc. of annual conference (1971), 304-313

11. E.L. Johnson and M.W. Padberg, Degree-Two Inequalities, Clique Facets and Biperfect Graphs, Ann. Discrete Math. 16(1982), 169-187

12. E.L. Johnson and U.H. Suhl, Experiments in integer programming, Discrete Applied Mathematics 2(1980), 39-55

13. A.H.Land and S. Powell, Computer codes for problems of Integer Programming, Annals of Discrete Mathematics 5(1979), 221-269

14. C.E. Lemke and K. Spielberg, Direct search algorithms for zero-one and mixed-integer programming, Opns. Res. 15(1967), 892-914

15. T.G. Mairs, G.W. Wakefield, E.L. Johnson and K. Spielberg, On a production allocation and distribution problem, Man. Science 24(1978), 1622-1630

16. L.A. Oley and R.S. Sjoquist, Automatic Reformulation Of Mixed and Pure Integer Models To Reduce Solution Time In APEX IV, Paper presented at the ORSA/TIMS Meeting in San Diego, California, October 1982

17. C.J. Piper, Implicit enumeration: a computational study, Rep. 115, School of Bus. Adm., Univ. of West. Ont., 1974

18. H.P. Williams, Model Building in Mathematical Programming, John Wiley, 1978

NUMERICAL TREATMENT OF STATE AND CONTROL CONSTRAINTS IN THE COMPUTATION OF OPTIMAL FEEDBACK LAWS FOR NONLINEAR CONTROL PROBLEMS

Peter Krämer-Eis and Hans Georg Bock

0. Introduction

Mathematical modeling of dynamical processes in engineering, economy or science often yields nonlinear control problems in which *control functions* entering a model ODE system for the *state variables* are chosen to *optimize a cost function* subject to initial and end conditions.

Practically any industrial application also demands *state and control constraints* to be met. A particularly general and powerful solution method is the indirect approach: The optimal control is expressed in terms of adjoint and state variables by a maximum principle leading to a multipoint boundary value problem with jump and switching conditions (MPBVP), cf. [3], which - if solved by a suitable BVP algorithm - yields an *open loop* or *nominal control law* $\hat{u}$ as a function of time.

In realization, however, the *real process* state x will deviate from the nominal *model* state $\hat{x}$, and the *real process* parameters p will deviate from their nominal *model* values $\hat{p}$, since reality can at best be a perturbed version of a model problem.

For the practical implementation of an optimal control law it is therefore necessary to adapt it to model perturbations, i.e. to obtain it as a function of state and process parameters. And adaptive law u* which is optimal w.r.t. the original cost function is called *optimal feedback law*.

The present paper develops an algorithm for the computation of optimal feedback control laws, which allows as a unique feature the treatment of state control constraints including even non-connected control sets as the appear e.g. in dynamic processes with "gears". It is a generalization of the multiple shooting algorithm previously developed by the authors [7 ,20], fully retaining its advantages of generality, superior stability properties and minimal amount of on-line computation and storage. Its realization in the code FAUST2 has proven very effective and powerful in extensive numerical tests [21].

1. Numerical computation of constrained open - loop controls

The following constrained Optimal Control Problem is considered

$$L(T,x(T),\hat{p}) = \min \tag{1.1}$$

$$\dot{x}(t) = f(t,x(t),u(t),p) \tag{1.2}$$

$$x(t_0) = x_0 \tag{1.3}$$

$$r(T,x(T),\hat{p}) = 0 \tag{1.4}$$

including

$$u(t) \in K \subset \mathbb{R}^m \qquad \text{control constraints} \tag{1.5}$$

$$g(t,x(t),u(t),\hat{p}) \leq 0 \qquad \text{state constraints} \tag{1.6}$$

Associated MPBVP

An optimal open loop control $\hat{u}$ can be determined in terms of the state variables $\hat{x}$ and adjoint variables $\hat{\lambda}$, which must satisfy the canonical ode - system

$$\begin{aligned} \dot{x} &= f(t,x,u,p) \\ \dot{\lambda}^T &= -\lambda^T f_x(t,x,u,p) + \mu g_x(t,x,u,p) \ . \end{aligned} \tag{1.7}$$

Defining the *Hamiltonian* H

$$H := \lambda^T f - \mu g \ , \tag{1.8}$$

the well-known maximum principle (cf. [11], [19])

$$\begin{aligned} &H(t,\hat{u}(t),\hat{x}(t),\hat{\lambda}(t),\hat{\mu}(t),\hat{p}) = \\ &= \max_v H(t,v,\hat{x}(t),\hat{\lambda}(t),\hat{\mu}(t),\hat{p}) \quad \text{subject to } (1.5,1.6) \end{aligned} \tag{1.9}$$

yields

$$\begin{aligned} \hat{u}(t) &= \arg\max_{v\in K} H = u^{free}(t,x,\lambda,p) \\ \mu(t) &= 0 \end{aligned} \tag{1.10}$$

as long as $g(t,\hat{x},\hat{\lambda},\hat{p}) < 0$ (*free arc*) , and

$$\hat{u}(t) = u^{bound}(t,x,p) \quad (\text{from } g(t,x,u,p) \equiv 0)$$

$$\hat{\mu}(t) = (\lambda^T f_u / g_u)|_{(t,\hat{x},\hat{\lambda},\hat{p})} \tag{1.11}$$

on *boundary arcs* with $g(t,\hat{x},\hat{\lambda},\hat{p}) \equiv 0$.

The transition points t_i from boundary arcs to free arcs and vice versa are determined by *switching functions* (cf. [3])

$$Q_i(t,x(t),\lambda(t),p) = 0 \ . \tag{1.12}$$

The differential equations (1.7) together with the switching conditions (1.10 - 1.12) for the right hand sides, initial and end conditions (1.3,4) for the state variables and, additionally, transversality conditions for the adjoint variables at the end point yield a Multipoint-Boundary Value-Problem (m p b v p) with jumps and switching conditions.

There is a variety of possible formulations of such a m p b v p, which differ in their numerical properties, but are theoretically essentially equivalent (cf. [3]).

Numerical Solution of MPBVP's by OPCON

For the solution of m p b v p's of the above kind hardly any stable and efficient algorithms can be found. The authors used the programm package OPCON [4,5,6] which is based on the multiple shooting technique as described by Bulirsch [13], Stoer, Bulirsch [24] and Deuflhard [14]. OPCON allows to treat m p b v p's of the class

$$\dot{z}(t) = F(t,z(t),p,\text{sign}\, Q_i(t,z(t),p))$$
$$R(z(t_0),z(t_1), \ \dots \ ,z(t_1),z(T),p) = 0 \tag{1.13}$$

where $F := (\partial/\partial\lambda, -\partial/\partial x)\, H$, $z = (x,\lambda)$ denotes state and adjoint variables, p certain system parameters and R involves boundary, transversality and interior point conditions for the parameters and possibly the free endtime.

The system parameters are fixed ($p = \hat{p}$) during the computation of the nominal solution, therefore they may be neglected in this case. Subsequently however, in the feedback case perturbations of these parameters are considered, which requires their explicit treatment.

At the state-dependent zeroes of the switching functions Q_i the right hand side and the trajectory itself may be discontinuous. These points are called *switching points* of z .

$$z(t_k^+) = z(t_k^-) + \phi_k(t_k, z(t_k^-), p)$$
$$t_k \text{ defined by } Q_k(t_k, z(t_k^-), p) = 0 \tag{1.14}$$

The multiple shooting technique

A sufficiently fine grid is chosen on the interval $[t_0, T]$

$$t_0 = \tau_0 < \tau_1 < \ \dots\ < \tau_{m-1} < \tau_m = T \tag{1.15}$$

and on each subinterval the initial value problems (i v p)

$$\dot{z}(t) = F(t, z(t), p, \text{sign}\, Q_i) \quad t \in [\tau_i, \tau_{i+1}]$$
$$z(\tau_i) = s_i \qquad\qquad i = 0, \dots, m-1 \tag{1.16}$$

are solved. $s_i = (s_i^1, s_i^2)$ represents the trajectory $(x(\tau_i), \lambda(\tau_i))$ $(S := (s_0, \dots, s_m))$. The numerical integration of (1.16) is performed in accordance with the jumps and switching condition (1.12,14). This requires the explicit determination of switching points by an iterative scheme.

Starting from initial estimates, S is iterated in such a way that boundary, transversality and interior point conditions are fulfilled on the one hand and the trajectory is continuous at the nodes τ_i on the other hand. This leads to the typical multiple shooting equation

$$M(S) := \begin{bmatrix} R(s_0, z(t_1;S), \dots, z(t_l;S), z(T;S), p) \\ \\ z(\tau_i;S) - s_i \quad (i = 1, \dots, m) \end{bmatrix} := \begin{bmatrix} h_0 \\ \\ h_i \end{bmatrix} = 0 \tag{1.17}$$

The operator M is highly nonlinear, has complicated differentiability properties and is represented only numerically. (1.17) is efficiently solved by a modified Newton method, where the stepsize parameter is chosen by means of the so-called natural level functions (cf. Deuflhard [15], Bock [5]).

In each iteration step the following structured linear subproblem has to be solved

$$\begin{bmatrix} D_0 & D_1 & \cdots & D_{m-1} & D_m \\ G_0 & -I & & & \\ & G_1 & -I & 0 & \\ & & \ddots & \ddots & \\ 0 & & & G_{m-1} & -I \end{bmatrix} \begin{bmatrix} \Delta s_0 \\ \cdot \\ \cdot \\ \cdot \\ \Delta s_m \end{bmatrix} = - \begin{bmatrix} h_0 \\ \cdot \\ \cdot \\ \cdot \\ h_m \end{bmatrix} \tag{1.18}$$

where

$$D_i = \partial R / \partial s_i \ , \tag{1.19}$$

$$G_i = \partial z(\tau_{i+1}^{+}; S)/\partial s_i \quad (\textit{transition matrix} \text{ from } \tau_i \text{ to } \tau_{i+1})$$

By means of a block - Gauß - elimination (1.18) can be reduced to the *condensed system*

$$E_0 \Delta s_0 =: \begin{bmatrix} I & 0 \\ E_x^0 & E_\lambda^0 \end{bmatrix} \Delta s_0 = -u_0 \ ,$$

$$\Delta s_{i+1} = G_i \Delta s_i + h_{i+1} \qquad (i = 0, \dots, m-1) \tag{1.20}$$

where E_0, u_0 are recursively defined by

$$E_m := D_m \ , \ E_i = \begin{bmatrix} 0 & 0 \\ E_x^i & E_\lambda^i \end{bmatrix} = D_i + E_{i+1} G_i$$

$$u_m := h_0 \ , \ u_i = u_{i+1} + E_{i+1} h_{i+1} \qquad i = m-1, \dots, 0 \ . \tag{1.21}$$

As a result of the iteration, a solution of (1.17) and thus a nominal trajectory $\hat{x}(t)$, $\hat{\lambda}(t)$ and a nominal control $\hat{u}(t) = \tilde{u}(t, \hat{x}(t), \hat{\lambda}(t), \hat{p})$ is obtained.

2. Numerical computation of constrained closed loop controls

Some remarks about the theoretical solution approach of the computation of the feedback control appear to be appropriate (for details see [21]):

(i) As shown above, the nominal control $\hat{u}$ in the constrained case is characterized by switching functions. Under certain regularity assumptions $\hat{u}$ can be embedded into a piecewise C^1 feedback control $u^*(t,x,p)$ which exists in the neighbourhood of the nominal solution

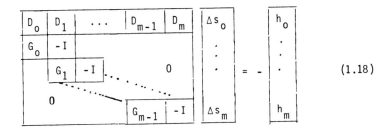

$$\hat{u}(t) = u^*(t,\hat{x}(t),\hat{p}) \tag{2.1}$$

(ii) By means of an embedding theorem the adjoint variables can be shown to be piecewise differentiable functions of state and system parameters ([20], Theorem 1.3)

$$\lambda(t,x,p) \in C^1 \tag{2.2}$$

(iii) Expanding the Hamiltonian (1.7) in terms of controls, states, parameters and adjoint variables and maximizing this expansion yields an approximating feedback control of the following kind (assuming K is open *)

$$\begin{aligned} u^* &= u^*_{free}(t,x,\lambda,p) = \\ &= \hat{u}(t) - H_{uu}^{-1} \, [(H_{ux} + f_u^T\lambda_x)(x(t)-\hat{x}(t)) + (H_{up}+f_u^T\lambda_p)(p-\hat{p})] \, , \end{aligned} \tag{2.3}$$

as long as it satisfies the constraints, and

$$u^* = u^*_{bound}(t,x,p) \quad (\text{such that} \;\; g(t,x,u,p) \equiv 0 \,) \, , \tag{2.4}$$

as long as u^*_{free} is outside the admissible region, where x,p denote actual state and parameters. As in the nominal case, the transition surfaces from free to boundary arcs are determined by switching functions according to (1.12) depending on actual state and expanded adjoints instead of nominal state and adjoints

$$\overline{Q}_i = Q_i(t,x(t),\hat{\lambda}(t) + \lambda_x(x(t) - \hat{x}(t)) + \lambda_p(p - \hat{p}),p) = 0 \, . \tag{2.5}$$

Note, that the treatment of switching conditions in the feedback case is not unique, under certain circumstances an expanded or partially expanded version of (2.5) may be preferable (Example 3.1).

In (2.3,4,5) *all terms* are *explicitely computable* along with the nominal solution except for the partial derivatives of the adjoint variables with respect to states and parameters. These partial derivatives λ_x, λ_p will be called *feedback matrices* in the following.

Computation of feedback matrices at nodes τ_i

Under the same regularity conditions as mentioned above a perturbation

---

* if K is closed, a more complex form is obtained, as in example 3.1.

theorem for nondifferentiable m p bv p's can be proven ensuring that in an appropriate neighbourhood of the nominal solution, the discrete solution of the m p bv p is a piecewise differentiable function of initial states and system parameters (Bock [2], Th.3.2.12; Krämer-Eis [21], Th. 2.1,2.2). Replacing $x(\tau_0) - x_0 = 0$ by $x(\tau_i) - x_i = 0$, one obtains

$$S^i(x_i,p) \in C^1 \quad \text{at} \quad (x_i,p) = (\hat{x}(\tau_i),\hat{p}) \;, \; S^i := (s_i, \ldots, s_m) \;. \tag{2.6}$$

In particular, the derivatives of interest are

$$ds_i^1 / d(x_i,p) \quad , \quad ds_i^2 / d(x_i,p) \tag{2.7}$$

which are obtained from

$$(\partial M^i/\partial S^i)(\partial S^i / \partial(x_i,p)) + \partial M^i / \partial(x_i,p) \equiv 0 \;. \tag{2.8}$$

Here, $\partial M^i / \partial S^i$ can be identified with the submatrix of (1.18) for the nominal solution, dropping all elements corresponding to nodes with index less than i (*reduced* m p bv p).

Application of the condensing algorithm to (2.8) yields the required quantities (2.7) as solution of

$$\begin{bmatrix} I & 0 \\ E_x^i & E_\lambda^i \end{bmatrix} \begin{bmatrix} \frac{ds_i^1}{dx_i} & \frac{ds_i^1}{dp} \\ \frac{ds_i^2}{dx_i} & \frac{ds_i^2}{dp} \end{bmatrix} = - \begin{bmatrix} -I & 0 \\ 0 & K_i \end{bmatrix} \tag{2.9}$$

with E_x^i, E_λ^i from (1.21) and K_i recursively defined by

$$K_i = D^p + \Sigma_{j=i+1}^m (E_x^j \; E_\lambda^j) \; G_{j-1}^p$$
$$D^p = \partial R / \partial p \;, \; G_i^p = \partial z(\tau_{i+1}^-; S) / \partial p \;. \tag{2.10}$$

The derivation holds strictly under the assumption that the linear systems (2.9) are nonsingular. In case of control constraints and end conditions, this is not necessarily the case and the solution of (2.9) is replaced by pseudo - inversion (see [21] for details).

Solving (2.8) leads to the required feedback matrices at the nodes of the multiple shooting discretization scheme:

$$\lambda_x(\tau_i) = -(E^i_\lambda)^{-1} E^i_x$$
$$\lambda_p(\tau_i) = -(E^i_\lambda)^{-1} K_i \qquad (2.11)$$

The feedback matrices are piecewise differentiable, and along the nominal trajectory, they are discontinuous at the switching points only. Therefore, on each subintervall from switching point to switching point the feedback matrices can be suitably represented e.g. by splines or polygons.

Computation of feedback matrices at switching points

The transition matrix $G = (G^z, G^p)$ of any intervall $[t_i, t_j]$ along the nominal trajectory $\hat{z}$ is

$$G^z(t_j, t_i) = \partial\hat{z}(t_j^-) / \partial\hat{z}(t_i^+)$$
$$G^p(t_j, t_i) = \partial\hat{z}(t_j^-) / \partial p \qquad (t_i, t_j \in [t_0, T]; t_i < t_j) \qquad (2.12)$$

Under the assumption that there is exactly one switching point t_s within the i-th subinterval of the nominal trajectory $\hat{z}$ causing a discontinuity described by the jump function

$$\gamma(t, \hat{z}(t), p) = \begin{pmatrix} \hat{z}(t_s^-) \\ p \end{pmatrix} + \begin{pmatrix} \phi(t, \hat{z}(t_s), p) \\ 0 \end{pmatrix} \qquad (2.13)$$

the transition matrix over the whole interval $[\tau_i, \tau_{i+1}]$ is easily seen to be (cf. [21])

$$G(\tau_{i+1}, \tau_i) = G(\tau_{i+1}, t_s) \cdot P_G \cdot G(t_s, \tau_i) \qquad (2.14)$$

where

$$P_G := (P^z_G, P^p_G) = \partial\gamma / \partial(z_s, p) - \{(\partial\gamma / \partial(z_s, p))\, F^- - F^+ +$$
$$+ \partial\gamma / \partial t) \cdot \partial Q / \partial(z_s, p)\} / \dot{Q} \qquad (2.15)$$

$$F^{\pm} := (F(t_s^{\pm}, z(t_s^{\pm}), Q(t_s^{\pm}, z(t_s^{\pm}), \hat{p}), \hat{p}), 0)^T, \; z_s = \hat{z}(t_s^-), \; Q(t_s^-, z_s, \hat{p}) = 0 .$$

Correspondingly, the derivative of the interior point condition at a switching point is obtained

$$D_s = \partial R / \partial(z_s, p) \cdot P_D$$
$$P_D = (P^z_D, P^p_D) = I - \{F^- \cdot \partial Q / \partial(z_s, p)\} / \dot{Q} \qquad (2.16)$$

P_G takes the influence of the change of the switching point and the jump function on the trajectory into account, P_D the influence of the change of the switching point on the interior point condition.

Using (2.14,15,16) one obtains the following matrices E_s, K_s at a switching point

$$\begin{aligned} E_s^+ &= E_{i+1}\, G^z(\tau_{i+1}, t_s) \qquad\qquad \left(E_s^\pm =: \begin{bmatrix} 0 & 0 \\ E_x^{s\pm} & E_\lambda^{s\pm} \end{bmatrix} \right) \\ E_s^- &= D_s + E_{i+1}\, G^z(\tau_{i+1}, t_s)\, P_G^z \\ K_s^+ &= D^p + \Sigma_{j=i+2}^m (E_x^i E_\lambda^i) G^p(\tau_j, \tau_{j-1}) + (E_x^{i+1} E_\lambda^{i+1}) G^p(\tau_{i+1}, t_s) \\ K_s^- &= K_s^+\, P_G^p \end{aligned} \tag{2.17}$$

with E_{i+1}, E_x^i, E_λ^i as defined in (1.21). Formally, (2.16) leads to the same recursion as for the nominal condensing algorithm (1.21 and 2.10, resp.). In (2.17) the recursion is split up explicitly at the switching point, yielding the right and left hand limits of the feedback matrices if the right hand side of the o.d.e. system or the trajectory itself is discontinuous:

$$\begin{aligned} \lambda_x(t_s^\pm) &= -\,(E_\lambda^{s\pm})^{-1}\, E_x^{s\pm} \\ \lambda_p(t_s^\pm) &= -\,(E_\lambda^{s\pm})^{-1}\, K_s^\pm \end{aligned} \tag{2.18}$$

With (2.11,18) the feedback control is completely off-line computable at arbitrary nodes of the multiple shooting discretization scheme and at the nominal switching points.

In practice one proceeds in the following way: Firstly, the nominal solution is usually computed on a rather coarse grid to restrict storage consumption. The grid size only depends on the stability of the i v p's (1.16). After this, the subdivision (1.15) is refined and the feedback matrices are computed on a finer grid to get a sufficiently accurate global representation. This requires essentially just one recursion and m' (fine grid) matrix decompositions.

Remark: Under extreme stability conditions the condensing algorithm is replaced by a modified version based on multistage least squares decomposition as described in [5].

3. Numerical applications

3.1 A control problem with nonconnected control region: Energy minimization of a subway car

The dynamics of a Westinghouse R - 42, which is a quite typical New York subway car, are described by the following ode - system (Bock, Longman [8]; Viswanathan, Longman, Domoto [25]) :

$$\dot{s} = v$$

$$\dot{v} = \begin{cases} g/w \cdot (eT(v,k,w) - R(v,w,c)) & \text{if } k = 1,2 \\ -g/w \cdot R(v,w,c) - C & \text{if } k = 3 \\ -d \ , \ d \in [d_0, d_1] & \text{if } k = 4 \end{cases} \tag{3.1.1}$$

For an implementation of energy optimal control strategies, a feedback adaptation with respect to perturbations of the system parameters w (weight of the car), c (number of cars) and e (engines that work) is of interest.

There are *discontinuities in the dynamics* because external resistors are in the circuit below some critical velocities which are taken out in steps. This is shown in fig. 1 and 2 which give the characteristic lines of tractive effort and input power

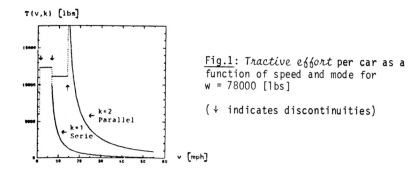

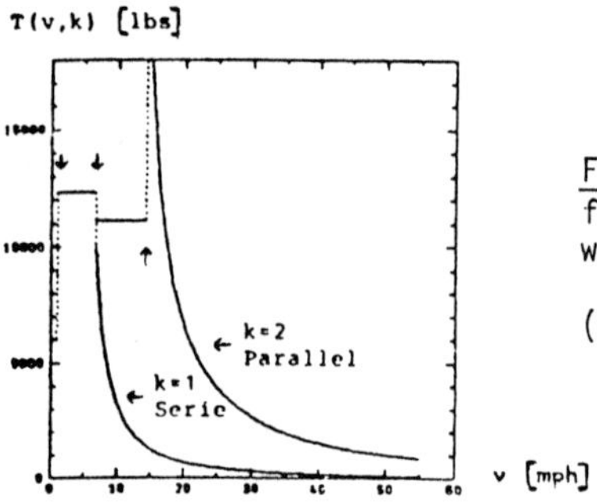

Fig.1: *Tractive effort* per car as a function of speed and mode for w = 78000 [lbs]

(↓ indicates discontinuities)

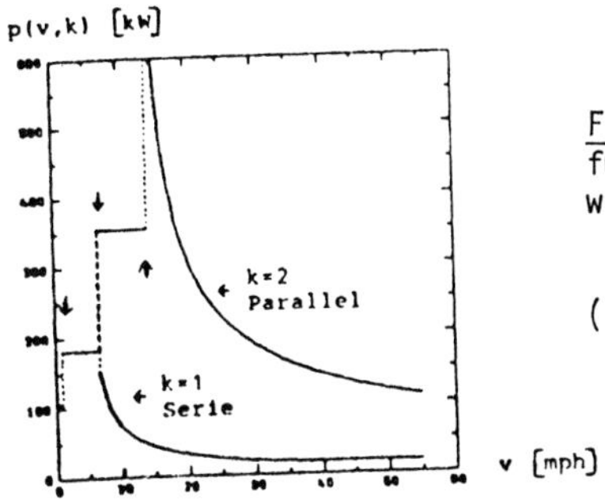

Fig.2: *Input power* per car as a function of speed and mode for w = 78000 [lbs]

(↓ indicates discontinuities)

The system (3.1.1) is governed by 4 different modes

$$k = \begin{cases} 1 & \text{series} \\ 2 & \text{series - parallel - shunt} \\ 3 & \text{coasting} \\ 4 & \text{braking within} \quad d \in [0,3] \end{cases} \qquad (3.1.2)$$

The subway train has to be driven from one station to the next within a prescribed transit time. The example given here is a typical local run with .4 miles (= 2112 ft) distance between the stations and 65 seconds transit time.

$$\begin{matrix} v(0) = 0 & v(T) = 0 \quad [\text{ mph }] \\ s(0) = 0 & s(T) = 2112 \; [\text{ ft }] \end{matrix} \qquad T = 65 \; [\text{sec}] \qquad (3.1.3)$$

The operating mode is to be chosen in order to minimize the total energy expended going from one station to the next

$$\int_0^T ep(v,k,w)\,dt = \min \qquad (3.1.4)$$

Because problem (3.1.1 - 4) has a nonconnected control region the nominal solution has been calculated by means of the *Competing Hamiltonians Algorithm* (Bock, Longman [8],[9]) .

The energy optimal method of operating a subway train results in *energy savings of* 26 - 38% over normal operation, depending on weight and distance ([8],[9]) .

The following results are part of a detailed investigation in co-operation with R.W. Longman, which is presently in preparation. For the computation of the feedback matrices (corresponding to the nominal values w = 78000 , e = 1 , c = 10 of the system parameters) an equidistant mesh of stepsize 0.05 sec. was chosen.

In order to simulate the real life performance of the feedback control law, the maximum perturbation from nominal trajectory and nominal system parameters were computed for which the feedback control still brings the subway car exactly into the final state (3.1.3). Only a deviation of two seconds in the final time is permitted. For the state variables these results are shown in fig. 3,4 .

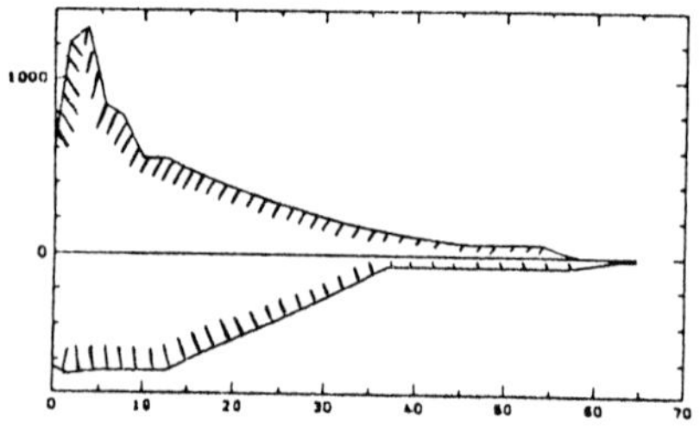

Fig.3: Maximum perturbations of distance

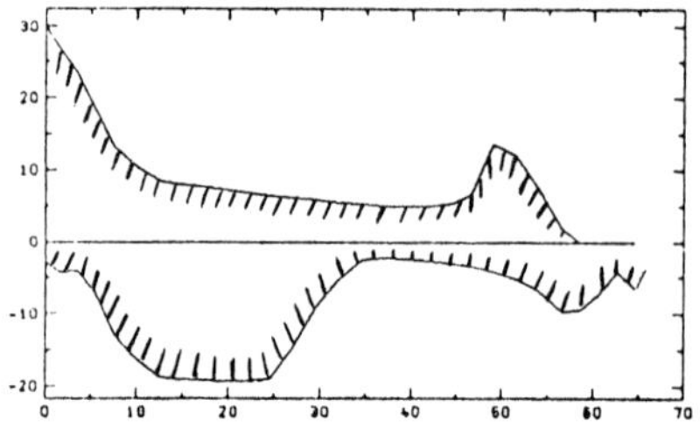

Fig.4: Maximum perturbations of velocity

and the system parameters are permitted to vary within

w:	72000 - 110000	(empty to crush loaded car)	
c:	5 - 10	(cars)	(3.1.5)
e:	70 - 100	(% of engines that work)	

This means, that *the complete range of variations to be expected in practice is covered by the feedback control!*

Fig. 5a, 5b show an example where only 70% of the motors work, fig. 6a,6b show a case in which a drastic drop of velocity occurs. In both cases the train comes to a stop in the tunnel if the nominal control is used, whereas the next station is reached exactly with the feedback control.

Fig. 5a: distance s [ft]

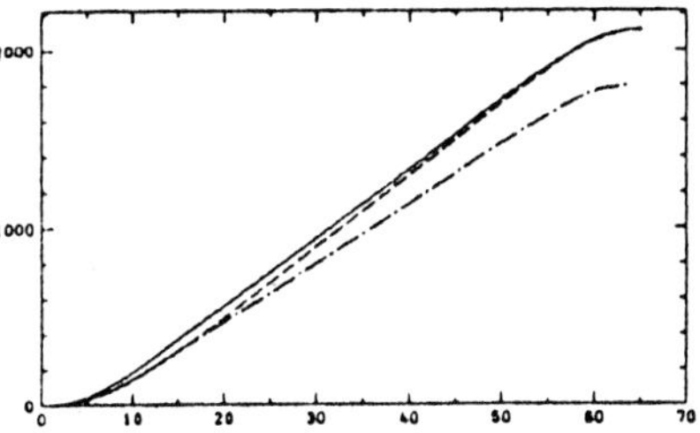

Fig 5b: velocity v [mph]

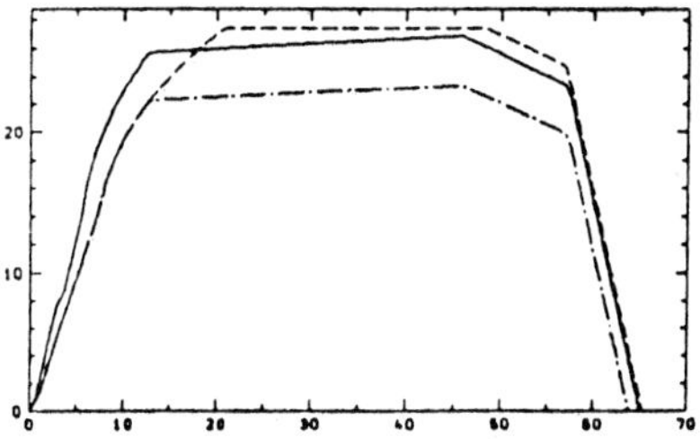

Comparison feedback vs. nominal for $\Delta e = -.3$

(—— undisturbed, ---- feedback disturbed, -·-· nominal disturbed)

Fig. 6a: distance s [ft]

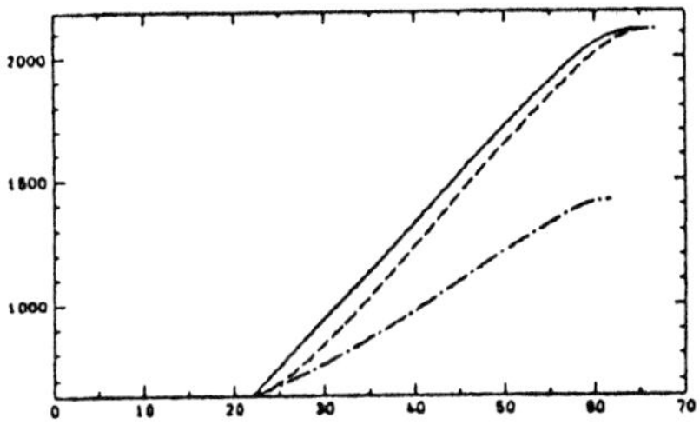

Fig. 6b: velocity v [mph]

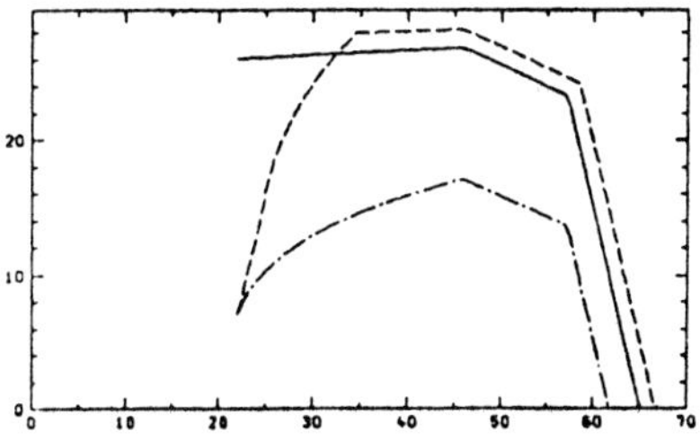

Comparison feedback vs. nominal for $\Delta v(22) = -19$ [mph]

(—— undisturbed, ---- feedback disturbed, -·-· nominal disturbed)

Fig 7a,7b show the situation when the train is 10 seconds late. In order to get back to the schedule the transit time is decreased to 55 seconds (Note, that this is almost time optimal). With the feedback control the prescribed final time is missed by only 2 seconds which is a gain of 8 seconds by a near - energy - optimal solution.

Fig. 7a: distance s [ft]

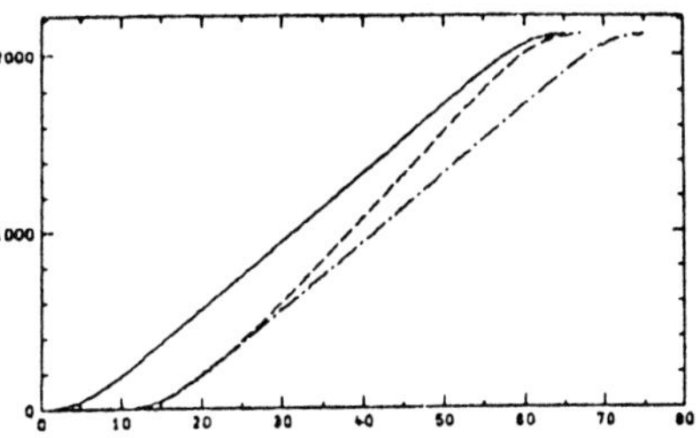

Fig. 7b: velocity v [mph]

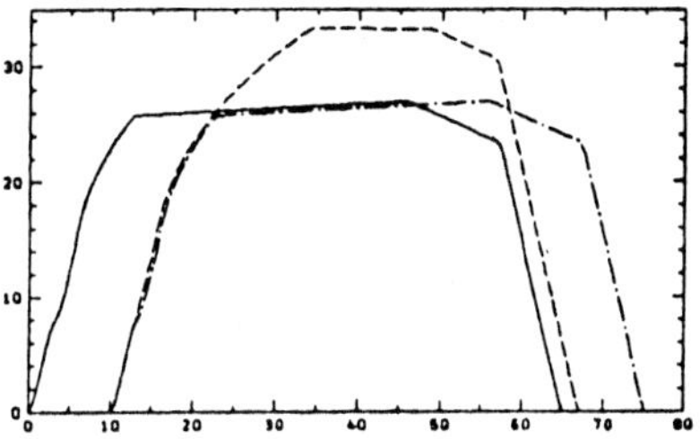

Comparison feedback vs. nominal for $\Delta T = -10$ [sec]

(—— undisturbed, ---- feedback disturbed, -·-· nominal disturbed)

It is important to point out, that the feedback control is no classic LQ-control, which brings the perturbed trajectory back to the nominal trajectory. Instead, it follows a trajectory which satisfies the boundary conditions and is second order energy optimal under the perturbed conditions.

The quality of this feedback approximation is shown in Table 1 which compares energy consumption for the feedback control with the optimum value possible under the perturbed condition in seven selected cases.

Table 1: Comparison feedback vs. optimal cost functional

perturb absolut	opt. cost function $T=T^{nom}$	opt. cost function $T=T^{fb}$	feedback cost funct	perturb in %	deviation feedback-opt (T^{fb})	seconds behind schedule
$\Delta W=-6000$	1.075	1.074	1.075	8 %	0.1 %	0.04
$\Delta W=32000$	1.540	1.504	1.513	41 %	0.6 %	0.66
$\Delta n=-5$	1.154	1.151	1.151	50 %	0.0 %	0.08
$\Delta e=-.3$	1.228	1.214	1.216	30 %	0.2 %	0.29
$\Delta T=-10$	2.010	1.672	1.674	15 %	0.1 %	1.82
$\Delta s(0)=-640$	2.613	2.116	2.143	30 %	1.3 %	1.69
$\Delta v(22)=-19.3$	2.225	2.068	2.097	74 %	1.4 %	1.65

Although there are perturbations up to 74% in state and parameters, feedback cost and minimum cost (computed for the feedback transit time too) differ by less than 1.5% .

3.2 A control problem with state constraints: Optimal descent of a Space Shuttle subject to reradiative heating constraints

A Space Shuttle is a glider, the flight of which can be approximately described by the following model [16,17,22]

$$\dot{v} = -c_w(c_A)\frac{F}{2m}\rho(h)\,v^2 - g_o\left(\frac{R}{R+h}\right)^2 \sin\gamma \qquad \text{(tangential velocity)}$$

$$\dot{\chi} = \frac{F}{2m}c_A\,\rho(h)\,v\frac{\sin\alpha}{\cos\gamma} - \frac{v}{R+h}\cos\gamma\cos\chi\tan\Lambda \qquad \text{(heading angle)}$$

$$\dot{\gamma} = \frac{F}{2m}c_A\,\rho(h)\,v\cos\alpha - \left(\frac{g_o}{v}\left(\frac{R}{R+h}\right)^2 - \frac{v}{R+h}\right)\cos\gamma \qquad \text{(flight path angle)} \qquad (3.2.1)$$

$$\dot{\Lambda} = \frac{v}{R+h} \cos\gamma \sin\chi \qquad \text{(cross-range angle)}$$

$$\dot{h} = v \sin\gamma \qquad \text{(altitude)}$$

with $\rho(h) = \rho_0 \exp(-\beta h)$ (air density) and $c_w(c_A) = c_{w_0} + c_A^n$ (drag coefficient).

The system is controlled by the lift coefficient c_A and the aerodynamic bank angle α. The critical part of the re - entry manoevre brings the Shuttle from an altitude of 95 down to 35 km, while reducing the speed from 7.85 to 1.116 km/sec. Thus the boundary conditions are

$$v(0) = 7.85\ ,\ h(0) = 95.\ ,\ \gamma(0) = -1.25 \cdot \frac{\pi}{180}\ ,\ \chi(0) = 0\ ,\ \Lambda(0) = 0$$
$$v(T) = 1.116\ ,\ h(T) = 30.,\ \gamma(T) = -2.7 \cdot \pi / 180 \tag{3.2.2}$$

The final time T is free, the cross range angle Λ at the final time is to be maximized

$$-\Lambda(T) = \min . \tag{3.2.3}$$

The lift coefficient c_A is restricted, and there is a state constraint which describes different levels of maximum permitted skin temperatures.

$$c_A \in [.0,.6] \qquad \text{(control constraint)}$$
$$c_A - G(v,h) \le \Delta c_A \qquad \text{(state constraint)} \tag{3.2.4}$$

As the nominal situation, $\Delta c_A = 0$ was chosen, which corresponds to a reradiative heating of 1093^0 C. As shown in [3],[21], the minimum possible reradiative heating is 890^0 C $\hat{=}$ $\Delta c_A = -.0546$ (G is a highly nonlinear function of altitude and velocity).

The feedback matrices were computed on an equidistant mesh of size $0.01 \cdot T$. As before, the maximum perturbations from the nominal trajectory are computed, for which the feedback control still brings the shuttle into the desired final state within the strict tolerances

$$\Delta v(T) = 5 \quad [m/sec]$$
$$\Delta\gamma(T) = 0.03 \quad [\ ^0\ ] \tag{3.2.5}$$
$$\Delta h(T) = 80 \quad [\ m\ ]$$

The corrective aereas shown in fig. 8 - 12 were computed separately for perturbations of each state variable, assuming that at time t all other components are undisturbed.

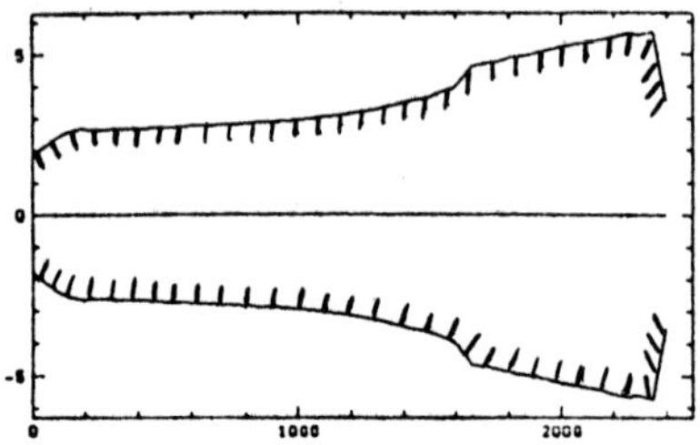

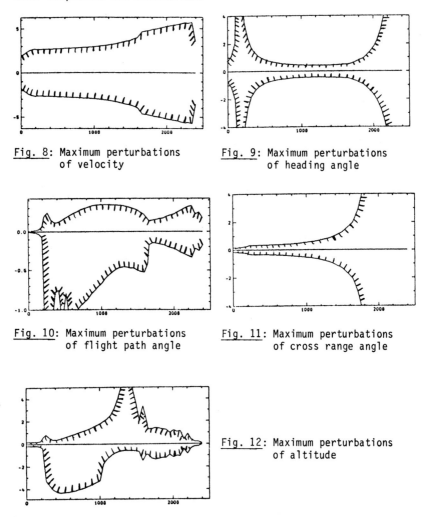

Fig. 8: Maximum perturbations of velocity

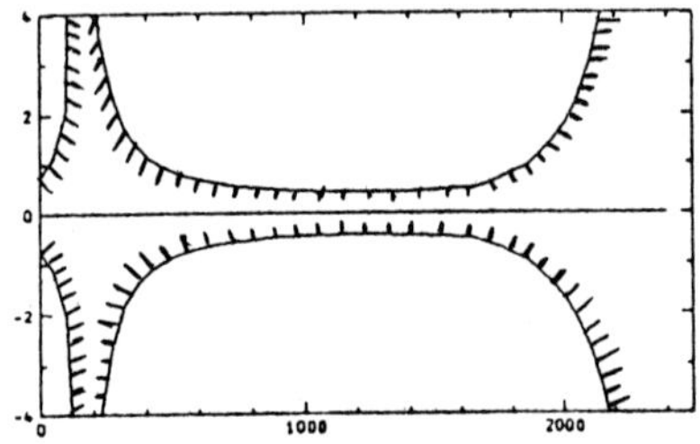

Fig. 9: Maximum perturbations of heading angle

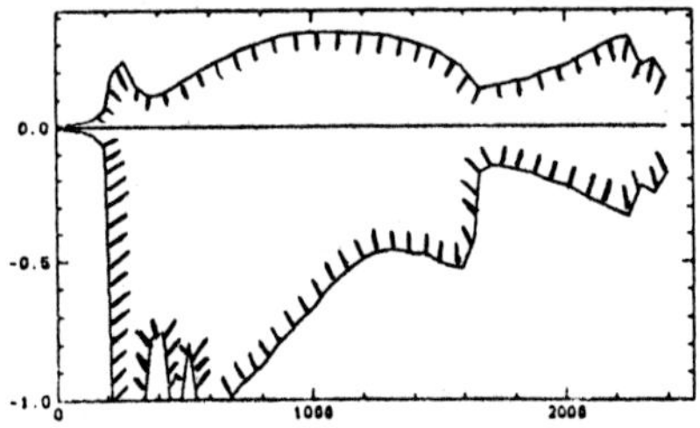

Fig. 10: Maximum perturbations of flight path angle

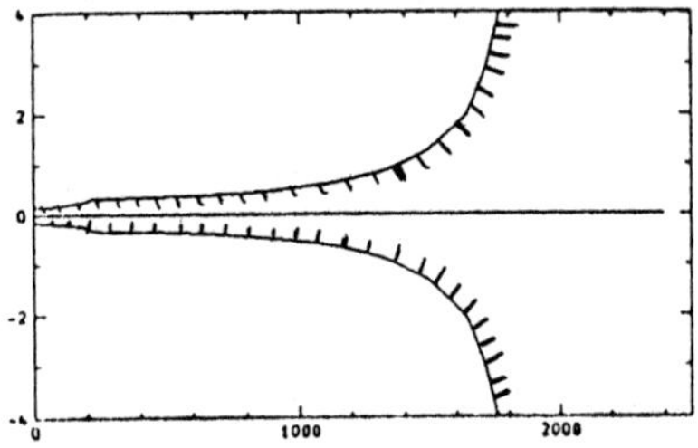

Fig. 11: Maximum perturbations of cross range angle

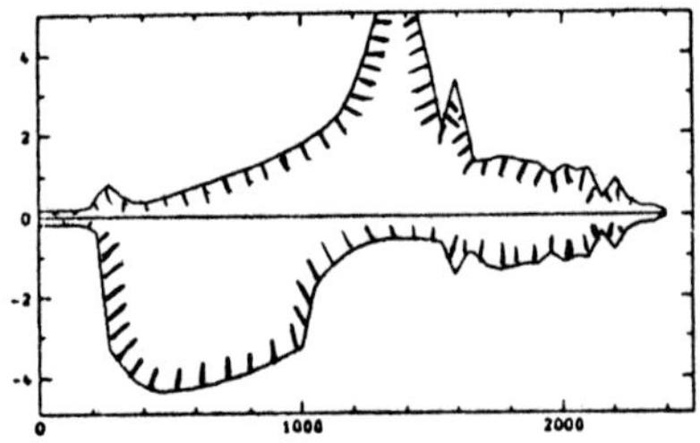

Fig. 12: Maximum perturbations of altitude

Figures 13 - 15 show a comparison of feedback and nominal situation (flight path angle and altitude) for three cases of perturbations of state variables: (i) $\Delta h(464) = -4.34$ [km] , (ii) $\Delta\gamma(758) = -.85$ [^{o}] , (iii) $\Delta\chi(146) = 24.9$ [^{o}] . Each case results in a crash trajectory if the nominal control is used, whereas the perturbed trajectory with feedback meets the boundary conditions within the prescribed accuracy.

The examples clearly show, that the use of an open - loop control (without feedback) for the Shuttle-Reentry is insufficient in practice, whereas the use of the feedback approximation allows a safe and stable descent *which strictly satisfies the heating constraint!*

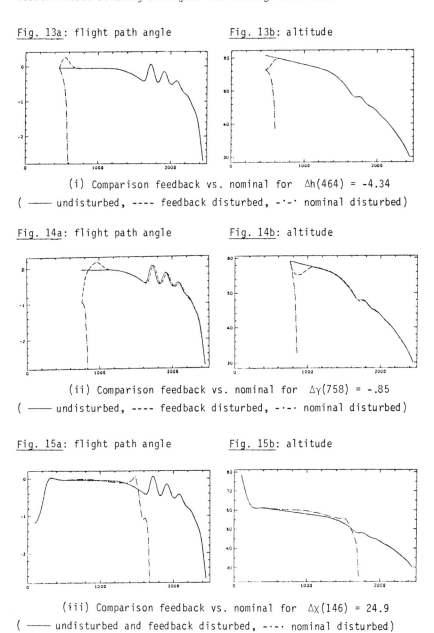

Fig. 13a: flight path angle

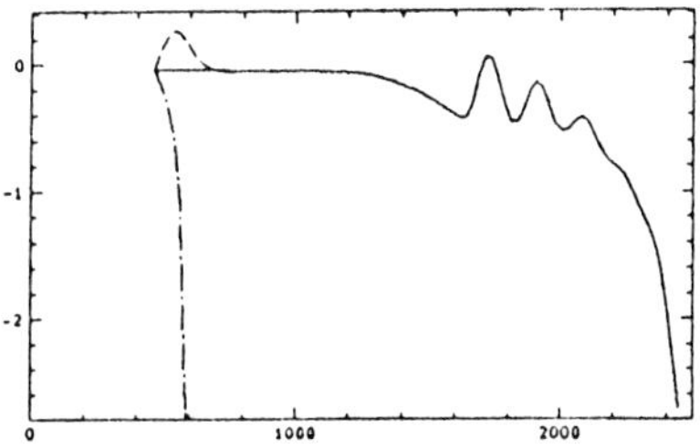

Fig. 13b: altitude

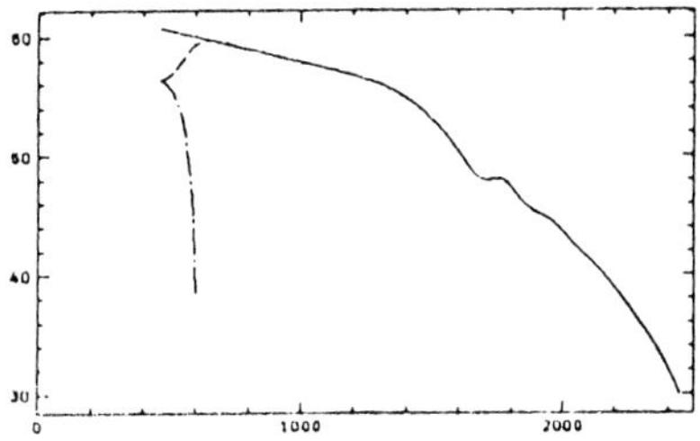

(i) Comparison feedback vs. nominal for $\Delta h(464) = -4.34$
(—— undisturbed, ---- feedback disturbed, -·-· nominal disturbed)

Fig. 14a: flight path angle

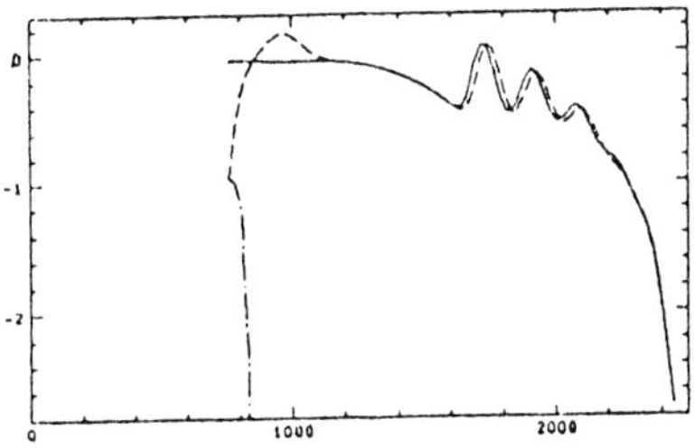

Fig. 14b: altitude

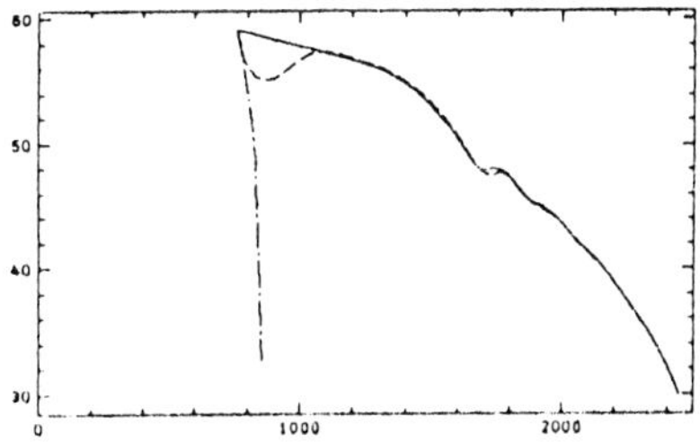

(ii) Comparison feedback vs. nominal for $\Delta\gamma(758) = -.85$
(—— undisturbed, ---- feedback disturbed, -··· nominal disturbed)

Fig. 15a: flight path angle

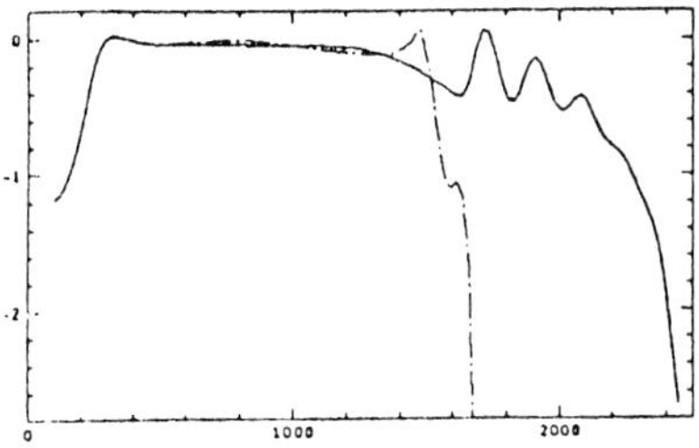

Fig. 15b: altitude

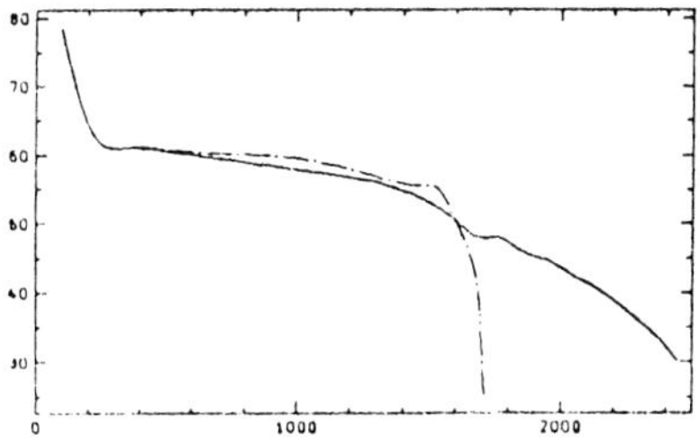

(iii) Comparison feedback vs. nominal for $\Delta\chi(146) = 24.9$
(—— undisturbed and feedback disturbed, -··· nominal disturbed)

Table 2 again shows the comparison of feedback and optimal cost functional for selected cases. As in the subway problem the feedback control not only fulfills the boundary conditions but does this nearly optimal with respect to the cost functional. Perturbations up to 30% cause a deviation of less than 0.1% in the cost functional.

Table 2: Comparison feedback vs. optimal cost functional

perturb. absolut	opt. cost functional	feedback cost. funct.	perturbat in %	deviation feedb.-opt
$\Delta h(464.) = -4.34$	42.30	42.24	5 %	0.14 %
$\Delta\gamma(758.) = -.85$	41.73	41.69	31 %	0.10 %
$\Delta\chi(146.) = +24.9$	68.13	68.12	31 %	0.01 %

4. Conclusions

The new algorithm described in this paper yields first order feedback approximations in addition to a nominal control with minimum computational and programming effort.

In contrast to "classical" methods based on a Riccati - approach to solve Bellman's equation (cf. Bryson, Ho [12], Dyer, Mc Reynolds [18]) it is stable, because the stable multiple shooting method is used (or, in general, any other stable bvp - method suitable for mpbvp's with jumps and switching conditions! - such as the collocation method implemented by Bär [1]).

The necessary computations can all be performed off - line (a priori), time and storage consuming on - line computations are avoided. Thus, it needs considerably less computational effort than other stable methods (cf. Branca [10], Pesch [23]) .

No second order derivatives are needed, which makes it very convenient for the user and eliminates a source of programming errors.

Moreover, the algorithm has a very broad field of applicability: The *free - endtime case* (omitted above for simplicity) can be immediately included by a transformation technique (see [21]), and the *inclusion of system parameters* in the formalism allows an efficient treatment of specific perturbations (of the differential equations, the boundary conditions etc.).

Above all, the new algorithm is the only one known to the authors capable of the treatment of control problems with control and state constraints and problems with nonconnected control regions.

Acknowledgement: This work has been supported by the Deutsche Forschungsgemeinschaft (Sonderforschungsbereich 72).

References

[1] V. Bär: Ein Kollokationsverfahren zur numerischen Lösung allgemeiner Mehrpunktrandwertaufgaben mit Schalt- und Sprungbedingungen mit Anwendungen in der optimalen Steuerung und der Parameteridentifizierung, diploma thesis, Bonn,1984 .

[2] H.G. Bock: Numerische Optimierung zustandsbeschränkter parameterabhängiger Prozesse mit linear auftretender Steuerung unter Anwendung der Mehrzielmethode, diploma thesis, Köln, 1974.

[3] H.G. Bock: Numerische Behandlung von zustandsbeschränkten und Tschebyscheff-Steuerungsproblemen, Carl - Cranz-Gesellschaft,1981.

[4] H.G. Bock: Recent advances in parameteridentification techniques for o.d.e., Progress in scientific computing 2, Deuflhard, Hairer (eds.), Birkhäuser, Boston, 1982.

[5] H.G. Bock: Randwertproblemmethoden zur Parameteridentifizierung in Systemen nichtlinearer Differentialgleichungen, to appear in Bonner Mathematische Schriften, 1985.

[6] H.G. Bock: Numerical Solution of Nonlinear Multipoint Boundary Value Problems with Application to Optimal Control, ZAMM 58, 1978.

[7] H.G. Bock, P. Krämer-Eis: A multiple shooting method for numerical computation of open and closed loop controls in nonlinear systems, Proceedings of the 9th IFAC-World Congress, Budapest, 1984.

[8] H.G. Bock, R.W. Longman: Optimal control of velocity profiles for minimization of energy consumption in the New York subway system, Proceedings of the 2nd IFAC-Workshop on control applications of nonlinear programming and optimization, Oberpfaffenh., 1980.

[9] H.G. Bock, R.W. Longman: Computation of optimal controls on disjoint control sets for minimum energy subway operation, Proc. Am. Astr. Soc. Symposium on engineering science and mechanics, Taiwan, 1982.

[10] H.W. Branca: Konstruktion optimaler Steuerungen für Flugbahnkorrekturen bei Echtzeitrechnung, diploma thesis, Köln, 1973.

[11] A.E. Bryson, W.F. Denham, S.E. Dreyfuss: Optimal programming problems with inequality constraints I, AIAA Journal 1, 1963.

[12] A.E. Bryson, Y.C. Ho: Applied optimal control, Ginn and Company, Waltham, Mass., 1969.

[13] R. Bulirsch: Die Mehrzielmethode zur numerischen Lösung von nichtlinearen Randwertproblemen und Aufgaben der optimalen Steuerung, Carl-Cranz-Gesellschaft, 1971.

[14] P. Deuflhard: Recent Advances in Multiple Shooting Techniques, Computational Techniques for ODE's (edited by Gladwell, Sayers), Academic Press (1980).

[15] P. Deuflhard: A modified Newton method for the solution of ill-conditioned systems of nonlinear equations with application to multiple shooting, Num. Math. 22, 1974.

[16] P. Deuflhard, H.J. Pesch, P. Rentrop: A modified continuation method for the numerical solution of two-point boundary value problems by shooting techniques, Num. Math. 26 , 1976.

[17] E. D. Dickmanns: Optimale Steuerung für Gleitflugbahnen maximaler Reichweite beim Eintritt in Planetenatmosphäre, in Bulirsch et al. (eds.), Optimization and optimal control, LN Math. 477, 1975.

[18] P. Dyer, S.R. Mc Reynolds: The computation and theory of optimal control, Academic Press, New York, 1970.

[19] D.H. Jacobson, M.M. Lele, J.L. Speyer: New necessary conditions of optimality for control problems with state-variable inequality constraints, J. Math. Anal. Appl. 35, 1971.

[2o] P. Krämer-Eis: Numerische Berechnung optimaler Feedback-Steuerungen bei nichtlinearen Prozessen, diploma thesis, 1980.

[21] P. Krämer-Eis: Ein Mehrzielverfahren zur numerischen Berechnung optimaler Feedback-Steuerungen bei beschränkten nichtlinearen Steuerungsproblemen, Bonner Mathematische Schriften Nr. 164, 1985.

[22] H.J. Pesch: Numerische Berechnung optimaler Steuerungen mit Hilfe der Mehrzielmethode, dokumentiert am Problem der optimalen Rückführung eines Raumgleiters unter Berücksichtigung von Aufheizungsbegrenzungen, diploma thesis, 1973.

[23] H.J. Pesch: Numerische Berechnung optimaler Flugbahnkorrekturen in Echtzeitrechnung, Dissertation, TU München, 1978.

[24] J. Stoer, R. Bulirsch: Einführung in die Numerische Mathematik II. Springer, Berlin, Heidelberg, New York, 1973

[25] C.N. Viswanathan, R.W. Longman, G.A. Domoto: Energy conservation in subway systems by controlled acceleration and deceleration, Intern. J. Energy Research, Vol. 2, 1978.

OPTIMAL PRODUCTION SCHEME FOR THE GOSAU HYDRO POWER PLANT SYSTEM

(W.Bauer/H.Gfrerer/E.Lindner/A.Schwarz/Hj.Wacker)

1. Introduction

In Austria an important tool for controlling energy production is given by a suitable use of hydro energy storage plants. About 22 % of the yearly electric energy production is done by plants of this type which serve for two main purposes:

(i) peak power production

(ii) shifting energy production into those periods of time where the energy consumption is high, for instance, in winter.

In our paper we are primarily concerned with (ii). We want to maximize the income the power company OKA (Oberösterreichische Kraftwerke AG) gets for selling the energy produced within a certain interval of time, say one year.

Here we confine ourselves to solve this problem for a small system, the Gosau power plant system. In this special case an isolated solution - i.e. a solution which does not interfere with activities concerning other power plants - can be accepted. Peak power demands exist and may be included by additional constraints.

In literature a large number of papers is concerned with the optimal control of storage power plant systems. General solution approaches are proposed, e.g. by Theilsiefje [19], Kalliauer [12], and Harhammer/Muschik/Schadler [11] discuss large systems using both nonlinear and linear models. Special systems are treated by Gaillard/Brüngger [4] who are working with a combination of linear and dynamic optimization (see also Yeh/Trott/Asce [24]). Roefs/Bodin [15] compare different solution methods including implicit stochastic processes and decomposition techniques. The multipurpose storage plant system Bhumipol/Sirikit in Thailand was analysed by Rhode/Naparaxawong [16]

who applied implicit stochastic optimization methods. Many authors use linear programming techniques, e.g. Habibollazadeh [11] for the Krängede storage plant system.
In our paper special attention is paid to modelling and solution techniques. Mathematically one has to solve a problem of optimal control with a nonlinear nonconvex objective and a large number of (mostly linear) constraints. After discretization we have a large scaled nonlinear optimization problem of dimension N (up to 1ooo) and about 4N constraints. Beside classical dynamic programming some new solution methods are applied, e.g. a decomposition/convexification technique proposed by Gfrerer [6] and a technique based on imbedding combined with an active index set strategy [9]. A simplified version of the control problem can be analysed theoretically leading to an equivalent low dimensional optimization problem [7].

2. Description of the storage plant system "Gosau"

The system consists of three serially connected storage power plants each with a reservoir of different capacity.
Figure 1 gives a schematical description (for details see Barwig/Peßl [1]).

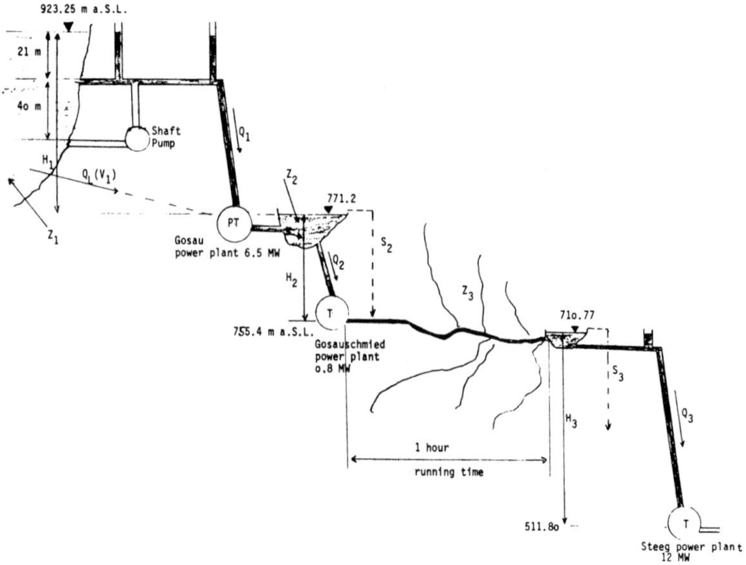

System Gosau (i = 1): Gosau power plant (G) - reservoir: Lake Gosau with a capacity of one year.

System Gosauschmied (i = 2): Gosauschmied power plant (GS) - reservoir Gosauschmied with a capacity of one day.
System Steeg (i = 3): Steeg power plant (S) - reservoir Klaushof with a capacity of some hours.

Notations:

$H_i(t)$	actual head of reservoir i: height difference between surface of the reservoir and below water surface [m]
$h_i(t)$	level of reservoir i above sea level (= level above Adria) [m a.A.]
$H_{iM}(t)$	reduced (manometric) head: H_i minus losses in the tunnel conduction system i [m]
$Q_i(t)$	discharges through the turbine(s) of plant i [m^3/s]
$\eta_{iT}(H_{iM},Q_i)$	efficiency function of the turbine(s) of plant i
$P_{iT}(t)$	output power of the turbine(s) of plant i [KW]
$\eta_{iG}(P_{iT})$	efficiency function of the generator(s) of plant i in dependence of P_{iT}
$P_i(t)$	output power of the generator(s) of plant i [KW]
$Z_i(t)$	influx to the reservoir i [m^3/s]
$V_i(t)$	volume of reservoir i [m^3]
$S_i(t)$	spilling water at reservoir i [m^3/s]
J_i	reservoir characteristics of reservoir i [m^3]
$a(t)$	tariff function [Austrian Shilling = Ö.S./KWh]
g:	gravity constant: 9.81 [m/s^2]
ρ:	density of water: 1ooo [kg/m^3]
$Q_L(V_1)$:	seepage losses of the Lake Gosau in dependence on the volume V_1 resp. the height h_1. (We use the same symbol Q_L in both cases). [m^3/s]

Remark: The reservoir characteristics describe the volume of the reservoir in dependence on the head H:

$$V = J(H)$$

To eliminate H we use the inverse function:

$$H = f(V)$$

3. The Problem

Our aim is to maximize the monetary value of the yearly energy production (= rated energy production) respecting all restrictions, technical and other ones.
The Gosau Lake is situated in a karst region of the Austrian Alps and from this there result heavy seepage losses Q_L (up to 1.5 m^3/s). The influx to Lake Gosau and the seepage losses are subterraneous and cannot be measured directly. Q_L is lost for production at the Gosau plant but adds to the influx of Gosauschmied. A special shaft pumping system has to be activated for exploiting Lake Gosau below a level of 9o2 m a.A. Spilling water is possible for Gosauschmied and Steeg.
For touristic reasons the level of Lake Gosau during summer must be at least 915 m a.A. and 92o m a.A. (July, August), respectively. In Steeg a power of 1.5 MW has to be produced from 6 a.m. until 1o p.m. because of a supply contract with the ÖBB (Austrian Railway Company).

3.1 Splitting of the Problem - Phase I

Figure 2 shows the principal production scheme for Gosau

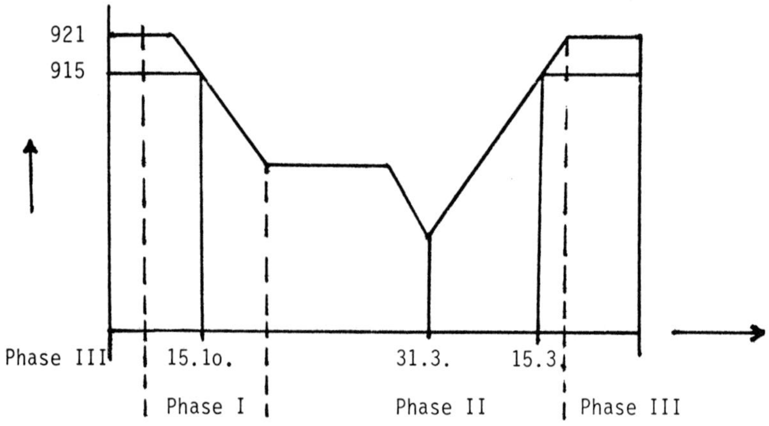

Phase I:

The earliest starting point for sinking the Gosau reservoir in autumn is fixed at September 1st because of tourism. The results of the optimization show that this can be done without loss of generality. The sinking process ends also w.l.o.g at December 31st at the level of 9o2.5 m a.A. Below that level the seepage losses remain constant. Additionally, the shaft pump system must be activated below that level.

Phase II:

First the level of 9o2.5 m is kept fixed because of purposes of reserve - only the natural influx is used for production. Then the sinking process is continued using the shaft pump system where both the starting point and the deepest level are determined by the optimization. Phase II ends at June 15th when a level of 915 m a.A. must be reached both by the natural influx coming from the Dachstein glacier and by pumping activities (the water is pumped back from the Gosauschmied reservoir to Lake Gosau).

Phase III:

Lake Gosau is stored up to a level of 92o m a.A. and kept at this height using only the natural influx. Additionally in case of heavy rainfall sinking activities are prescribed for flood protecting purposes.

Therefore, our problem may be split into three independent subproblems defined by the phases I, II, III. While Phase III is trivial the two other problems lead to nonlinear problems of optimal control. To preserve clearness we confine ourselves to describe the solution of Phase I only.

4. The Mathematical Model: Phase I

4.1 The Model (M)

(M1) The rated energy

$$E(T) := \int_0^T a(t)\,\{P_1(t) + P_2(t) + P_3(t)\}\,dt = \text{Max!}$$

where $P_i(t)$ describes the power produced by plant i

$$P_1(t) := g.\rho\, Q_1(t) f_1(V_1(t),V_2(t)) \eta_1(V_1(t),V_2(t),Q_1(t))$$

$$P_i(t) := g.\rho\, Q_i(t) f_i(V_i(t)) \eta_i(V_i(t),Q_i(t)) \quad i = 2,3$$

Here f_i describes the actual head for plant i; for i = 1 the head depends both on the volume of Lake Gosau and of the reservoir Gosauschmied. Usually one describes the efficiency in dependence on the head H (and Q), here we have changed this into a dependence on V (and Q) using the same symbol. t = 0, and t = T, correspond to September 1st, and December 31st, respectively.

(M2) The water balance can be described by the so-called continuity equation:

$$\dot{V}_1(t) = Z_1(t) - Q_1(t) - Q_L(V_1(t))$$

$$\dot{V}_2(t) = Z_2(t) + Q_1(t) + Q_L(V_1(t)) - Q_2(t) - S_2(V_2(t))$$

$$\dot{V}_3(t) = Z_3(t) + Q_2(t-\tau_0) + S_2(V_2(t-\tau_0)) - Q_3(t) - S_3(V_3(t))$$

$$V_1(o) = 18.5 \,.10^6 \,[m^3] \qquad V_1(T) = 9.5 \,.\, 10^6 \,[m^3]$$

$$V_2(o) = V_2(T) = V_{2max} := 24 \,.\, 10^4 \,[m^3]$$

$$V_3(o) = V_3(T) = V_{3max} := 2 \,.\, 10^4 \,[m^3]$$

The seepage losses Q_L depend monotonically on V_1 and add to the influx of Gosauschmied. Spilling water S_2, and S_3, may occur at Gosauschmied, and Klaushof, respectively. The discharges of plant i add to the influx of the following plant. The running time from Gosauschmied to Klaushof is given by τ_0 (about one hour).

$V_1(o)$, and $V_1(T)$, correspond to a height of 92o m a.A., and 9o2.5 m a.A., respectively, for Lake Gosau. Because of the small capacities both of the reservoir Gosauschmied and Klaushof we may assume periodicity.

(M3) We have to observe the following constraints

$$V_1(T) \le V_1(t) \le V_1(o)$$

$$V_1(t) \ge V_{T,T} := 1.5 \,.\, 10^6 \,[m^3] \text{ for } t \in [1.9.,15.10.]$$

$V_{2min}: = 6{,}6 \,.\, 10^4\ [m^3] \leq V_2(t) \leq V_{2max}$

$V_{3min}: = 1{,}6 \,.\, 10^4\ [m^3] \leq V_3(t) \leq V_{3max}$

$0 \leq Q_1(t) \leq 6 \quad [m^3/s]$

$0 \leq Q_2(t) \leq 6{,}6\ [m^3/s] \quad 0 \leq S_2(t)$

$0 \leq Q_3(t) \leq 7{,}4\ [m^3/s] \quad 0 \leq S_3(t)$

$V_{T,T}$ corresponds to a level of 915 m a.A. As in phase I we do not consider pumping activities we have $Q_i \geq 0$. Spilling water is nonnegative by definition.

(M4) $P_3(t) = g.\rho\ Q_3(t)\ f_3(V_3(t))\ \eta_3(V_3(t),Q_3(t)) \geq 1{,}5 \,.\, 10^6 \quad [W]$

$t \in [6 \text{ a.m.}, 10 \text{ p.m.}]$

(M4) describes the demand given by the supply contract with the ÖBB.

4.2 Principal Solution Strategies for Phase I

Model (M) presents a highly nonlinear problem of optimal control where V(t) describes the state and Q(t) the control variable. Even neglecting the losses, the spilling water, the ÖBB contract and assuming constant efficiencies we have a nonconvex objective because of the tariff function.

Solution was achieved by the following steps:

Step 1: Optimization for the Gosauschmied System (1 week).

Though the reservoir of Gosauschmied has only a capacity of one day optimization was done for one whole week to understand the influence of the weekend.

Under some simplifications this model may be solved partly by purely theoretical methods resulting from optimal control. One gets a low dimensional model which then is solved by a homotopy method combined with a suitable active index set strategy.

The result gives information on the optimal strategy for Gosauschmied (resp. Steeg) in principle.

Step 2: Sinking of the Gosau reservoir

Here we determined the optimal sinking strategy for Lake Gosau neglecting both the other plants. Solution was done by dynamic programming.

Step 3: Optimization of the chain Gosau - Gosauschmied - Steeg
First, optimization was performed for the whole system by dynamic programming. We used two time steps a day. To reduce the dimension of the problem the small capacity of the Klaushof reservoir was neglected.
Based on these results we used refined models for planning periods ranging from a) some weeks down to b) one day.
For a) we used both dynamic programming and a special decomposition/convexification technique proposed by Gfrerer.
For b) a special nonlinear optimization model was built which allows even to include peak power demands called for by the load dispatcher of the power company.

Step 4: Input data: seepage losses, determination of the influx, efficiencies.
In our model (M) we assume those input parameters to be known. Considerable amount of work, including careful measurements by the engineers of the power company, was necessary to determine the input data. To determine the seepage losses, e.g. some 3o years of production had to be analysed.

In the following we describe shortly Step 1, Step 3 and parts of Step 4.

5. Numerical Models

5.1 Weekly Optimization of Gosauschmied

5.1.1 Reduction to an equivalent finite dimensional model

We meet the following simplifications:

(i) $\eta_2 = \text{const.}$ (ii) $0 < Z(t) < Q_{2max}$

While (i) is essential for our theory, (ii) is met only for ease of presentation.

The following model arises (index 2 omitted)

$$\int_0^T a(t)\, f(V(t))Q(t)\, dt = \text{Max!}$$

$$\dot{V}(t) = Z(t)-Q(t),\ V(o) = V_0,\ V(T) = V_T$$

$$V_{min} \leqq V(t) \leqq V_{max};\quad 0 \leqq Q(t) \leqq Q_{max}$$

We search for an optimal $(V^*(t), Q^*(t)) \in C[0,T] \times L^\infty[0,T]$

Notations:

$\bar{V}: = \{t | V^*(t) = V_{max}\}$, $\quad \underline{V}: = \{t | V^*(t) = V_{min}\}$

$\bar{Q}: = \{t | Q^*(t) = Q_{max}\}$, $\quad \underline{Q}: = \{t | Q^*(t) = 0\}$

$\tau_k = 6^h$ (Monday) $+ k \,.\, 24^h \quad k = 0,\dots,5$

$\xi_k = 22^h + (k-1).24^h$, $k = 1,2,\dots,5$, $\xi_6 = 13^h$ (Saturday)

$$a(t) = \begin{cases} a_H & t \in (\tau_{k-1}.\xi_k] \quad k = 1,\dots,6 \\ a_L & t \in (\xi_k,\tau_k] \quad k = 1,\dots,6 \end{cases}$$

Using the results from the theory of optimal control one can show the following result, where $s(t)$ is the switching function for $Q^*(t)$. (Details are given in [7]).

Theorem:

(i) $t \in \bar{V} \cup \underline{V} \cup \bar{Q} \cup \underline{Q}$ a.e. for $t \in [0,T]$

(ii) $\underline{V} \subset \{\xi_i | i = 1,\dots,6\}$

(iii) $s(t_0) < 0 \Rightarrow s(t) < 0$ in $(t_0,\bar{t}]$

$s(t_0) > 0 \Rightarrow s(t) > 0$ in $(\underline{t},t_0]$

$s(t_0) = 0 \wedge t_0 \notin \bar{V} \Rightarrow s(t) < 0$ in $(t_0,\bar{t}]$, $s(t) > 0$ in $(\underline{t},t_0)$

where

$\bar{t}: = \min\{\xi_i | \xi_i \geqq t_0\}$ resp. $\bar{t} = T$ for $t_0 > \xi_6$

$\underline{t}: = \max\{\xi_i | \xi_i < t_0\}$ resp. $\bar{t} = 0$ for $t_0 \leqq \xi_1$

(iv) $s(t_0) < 0$ for $t_0 \in (\xi_i,\tau_i] \Rightarrow s(t) < 0$ in $[t_0,T]$

$s(t_0) > 0$ for $t_0 \in (\tau_{i-1},\xi_i] \Rightarrow s(t) > 0$ in $[0,t_0]$

This theorem gives already the structure of the optimal solution $(V^*(t), Q^*(t))$:

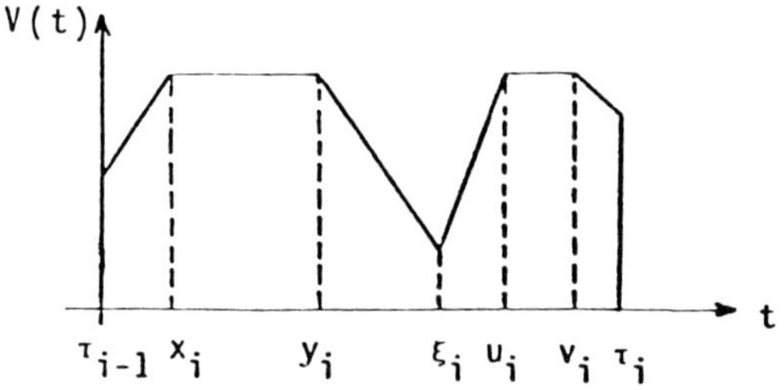

$$Q^*(t) = \begin{cases} 0 & t \in (\tau_{i-1}, x_i] \cup (\xi_i, u_i] \\ Z(t) & t \in (x_i, y_i] \cup (u_i, v_i] \\ Q_{max} & t \in (y_i, \xi_i] \cup (v_i, \tau_i] \end{cases}$$

Mathematically one gets an equivalent finite dimensional model:

$$\phi(x,y,u,v) = \text{Max!}$$

with 24 variables (x,y,u,v) and 55 constraints. In our special case with $V_0 = V_T = V_{max}$ the model reduces to 12 variables (we have: $x_i := \tau_{i-1}$ and $v_i := \tau_i$) and 36 constraints.

5.1.2 Solution of the finite dimensional model

Our model now received is of the following type:

$$P: \min \{f_0(x) | f_i(x) \leq 0, \quad j = 1,\dots,m\} \quad x \in \mathbb{R}^n$$

To dispose of the difficulty to get a starting value for a local iteration procedure we use imbedding:

$$P(s): \min \{h_0(x,s) | \quad h_j(x,s) \leq 0, \quad j = 1,\dots,m\} \; s \in [0,1] := S.$$

for instance: $h_0(x,s) := s\, f_0(x) + (1-s)\, \|x-x_0\|^2$

$$h_j(x,s) := f_j(x) + (s-1)|f_j(x_0)|$$

We assume that there exists a locally unique solution of P(s): $x(s) \in C(S)$.

To reduce the numerical effort we use an active index set strategy.

$I(s) := \{j \in \{1,\dots,m\} \mid h_j(x,s) = 0\}$

Thenonly equality constrained problems have to be solved:

$P^I(s)$: $\min \{h_0(x,s) \mid h_j(x,s) = 0, \ \ j \in I(s)\}$

Assumptions:

V1: $h_i(x,s)$, $i = 0,1,\dots,m$ sufficiently smooth

V2: $\{\nabla h_j(x,s)\}_{j \in I}$ linearly independent

V3: there holds the strong second order condition for $P^I(s)$

V4: the number of critical points (i.e. points where the I(s) changes) is finite.

Then the following three stages algorithms for the nonlinear system of the first order Kuhn-Tucker condition is feasible in the sense of Avila and converges quadratically:

(i) $z(s) = (x(s),u(s))$ is determined by classical continuation inside a stability set ($I(s)$ = const.) using a Newton like iteration procedure: Euler predictor - Newton corrector

(ii) Computation of a critical point

(iii) Determination of the new active index set after passing a critical point

For details see Gfrerer/Guddat/Wacker [5] and Gfrerer/Guddat/Wacker/Zulehner [9] .

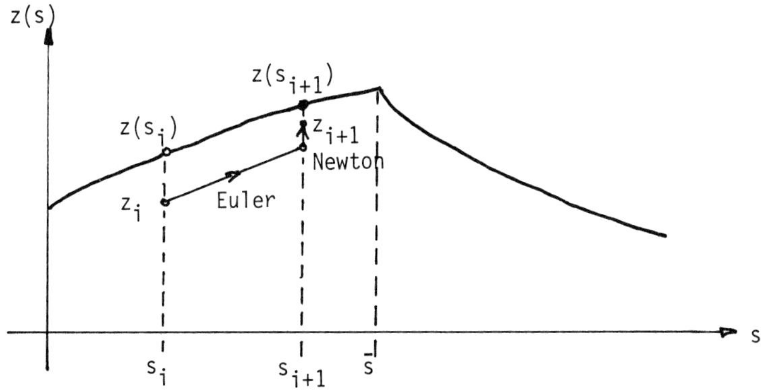

Numerical Example

i(day)	1	2	3	4	5	6
y_i(h)	9.51	33.62	57.4	81.1	98.3	12o
u_i(h)	24	48	72	96	12o	141.7

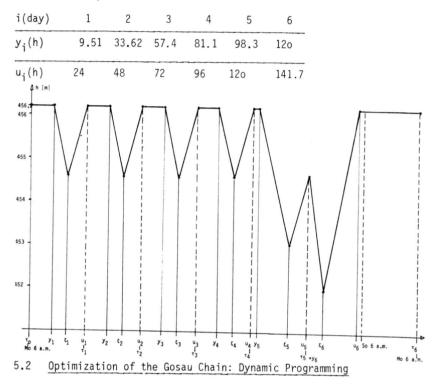

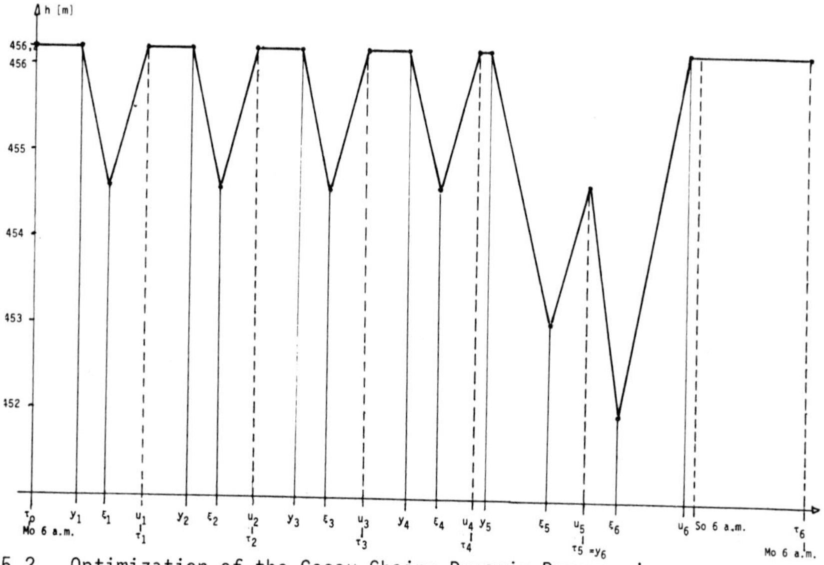

5.2 Optimization of the Gosau Chain: Dynamic Programming

To get the optimal control for the sinking process for the Gosau chain during phase I (1.9. - 31.12.) dynamic programming was used. The time steps defining the levels were chosen according to the two tariff periods a day. We get 24o steps. The discharges, the influxes, the seepage losses and the spilling water were assumed to be constant during one time period. We neglect the running time τ_0 and the small capacity of the reservoir Klaushof. For the state variables we use the heights of the reservoirs Gosau and Gosauschmied, for the control we take the discharges.

Figure 3 results from the 7th, 8th and 1oth week (containing 15.1o.) (d: day, n: night, all influxes were simulated according to reality. E_i: energy produced by plant i [KWh])

h_1	Q_1	E_1	h_2	Q_2	E_2	Q_3	E_3	day/ night
915.24	-	-	77o.31	3.o	5741	2.65	67871	d
915.18	-	-	769.3o	-	-	o.63	8266	n
915.14	-	-	77o.76	4.6	82o3	4.o6	1o3941	d
915.o5	-	-	766.95	o.6	488	1.o5	13767	n
915.oo	5.4	84329	768.38	6.6	1o936	6.o1	155486	d
914.39	4.4	35252	769.67	5.4	5195	5.82	76576	n
914.14	4.8	75743	77o.39	6.6	12124	6.o8	157385	d
913.59	3.4	26965	77o.58	5.6	5522	6.o9	8o23o	n
913.38	4.8	753o2	77o.43	6.6	121o9	6.21	16o63o	d
912.86	3.6	28385	77o.5o	5.6	55o5	6.22	81915	n
9o8.96	4.4	66883	769.77	5.6	1o966	6.36	16437o	d
9o8.42	-	-	771.o2	4.8	4692	6.58	86371	n
9o4.56	5.o	71885	77o.39	5.8	11259	6.38	164733	d
9o3.96	-	-	77o.41	o.1	1oo	1.69	223o9	

Figure 3 describes the situation when the demand of the tourist traffic ($h_1 \geqq 915$ m a.A.) becomes invalid.

Before October 1st, Gosau doesn't work and the level of Lake Gosau is sinking only because of the seepage losses. Gosauschmied acts exactly as in the theoretical model discussed in 5.1. and consequently so does Steeg. Exactly at October 1st (winter tariff!) Gosau is activated at day. Shortly before October 15th production is stopped, the exact time depending on the influx. After that date, Gosau first works day and night to avoid losses.

This implies that both Gosauschmied and Steeg do the same even accepting the lower night tariff. Then the strategy changes more and more to a clear day/night activity rythm.

Figure 4 shows the seepage losses.

Figure 4

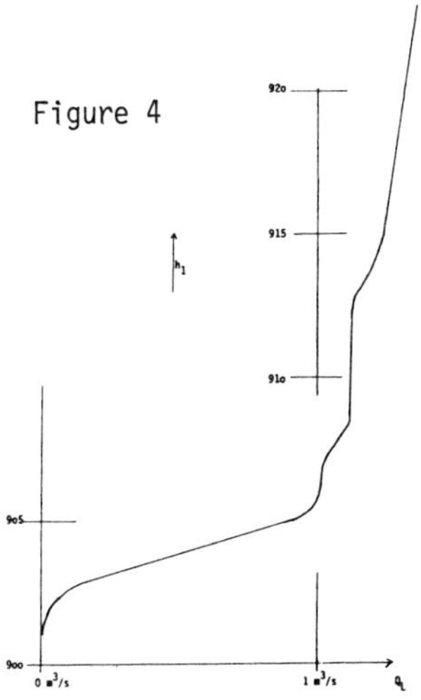

5.3 Short Time Optimization by Decomposition/Convexification

The results from 5.2. were refined by a short time model (e.g. one fortnight using a time discretization of one or two hours). From the medium term model we took only the boundary values V_{1B}, V_{1E}. The results gave not only a refinement but are an additional check for the cruder optimization model from Section 5.2. For convenience we confine ourselves to describe the decomposition technique for one plant only, say Gosau. For details and generalizations compare Gfrerer [6], [8]. Of course, the whole problem, i.e. Phase I for all three plants, may be solved by this method, too.

5.3.1 A separable version of (M): (MS)

As in 5.2 we use a time discretization ($\Delta t_k := t_k - t_{k-1}$ down to one hour). We assume Z_k, Q_k constant in $(t_{k-1}, t_k]$, η constant in $[t_B, t_E]$.

$$\bar{H}_k := f\,\frac{(V_{k-1}+V_k)}{2}\ ;\ Q_{Lk} := Q_L(\bar{H}_k);$$

$$Q_k := \frac{V_{k-1}-V_k}{\Delta t_k} + Z_k - Q_{Lk}$$

By $E_k(V_{k-1}, V_k)$ we denote the value of the energy produced in $(t_{k-1}, t_k]$. We get (omitting the constants g, ρ, η):

$$E = \sum_{i=1}^{N} E_i(V_{i-1}, V_i) := \sum_{i=1}^{N} a_i Q_i \int_{t_{i-1}}^{t_i} f\left(V_{i-1} + \frac{(V_i - V_{i-1})}{\Delta t_i}(t - t_{i-1})\right)dt = \text{Max!}$$

By splitting up V_i: V_i^+, V_i^- we get a separable model: (MS)

$$\text{(MS1)}\qquad -\sum_{i=1}^{N} E_i(V_{i-1}^+, V_i^-) = \text{Min!}$$

$$\text{(MS2)}\qquad V_i^+ = V_i^- \qquad i = 1,\ldots,N-1$$

$$\text{(MS3)}\qquad V_0^+ = V_B,\ V_N^- = V_E$$

$$V_{min} \le V_i^-,\ \ V_i^+ \le V_{max} \qquad i = 1,\ldots,N-1$$

$$0 \leq \frac{V_{i-1}^{+}-V_{i}^{-}}{\Delta t_i} + Z_i - Q_L(f\ \frac{(V_{i-1}^{+}+V_{i}^{-})}{2}) \leq Q_{max} \qquad i = 1,\ldots,N$$

5.3.2 Convexification of (MS)

To apply decomposition techniques one needs convexity for (MS) at least in a neighbourhood of the optimal solution. In our case convexity is hurt in the objective (at least by the tariff function even for a constant efficiency) and in the Q-constraint because of Q_L.

For our case of short term optimization convexification for Q_L is achieved by linearization with respect to V. The objective is convexified by adding a sufficiently convex term controlled by the positive constants c_i^{+}, c_i^{-}:

$$\bar{E}_k :\, = -\, E_k(V_{i-1}^{+},V_i^{-}) + c_{i-1}^{+}(Y_{i-1}^{+}-V_{i-1}^{+})^2 + c_i^{+}(Y_i^{-} - V_i^{-})^2$$

5.3.3 Decomposition

Now we may solve the resulting convex problem by decomposition. The equality constraints (MS2) are included in the objective by duality:

$$\phi(Y,\lambda) :\, = \min\ \{ \sum_{i=1}^{N} \bar{E}_i + \sum_{i=1}^{N-1} \lambda_i(V_i^{+}-V_i^{-}) \mid V \text{ subject to (MS3)} \}$$

We have to solve the unconstrained Minimax problem:

$$\min_{Y\in \mathbb{R}^{2N-2}} \ \max_{\lambda\in \mathbb{R}^{N-1}} \ \phi(Y,\lambda)$$

To determine $\phi(Y,\lambda)$ N linearly constrained optimization problems of dimension two only are to be solved which can be done quite efficiently:

$$\min\{\varphi_i := \bar{E}_i(V_{i-1}^{+},V_i^{-},Y_{i-1}^{+},Y_i^{-})+\lambda_{i-1}V_{i-1}^{+}-\lambda_i V_i^{-} \mid V \text{ subject to (MS3)}\} \quad (*)$$

The following iteration procedure is globally and superlinearly convergent:

Start: $k: = 0, Y^k \in \mathbb{R}^{2N-2}$

Step 1: Maximization of $\phi(Y^k,\lambda)$ with respect to λ: λ^k

V^k: corresponding solution of the problem (∗) when evaluating $\phi(Y^k,\lambda^k)$

Step 2: Set $Y^{k+1}: = V^k$, $k: = k+1$, Goto Step 1

The efficiency of this technique is demonstrated by the following results:

N	Iterations (It)	CPU(IBM 36o-155)	Constraints
42	1o3	14"	166
168	172	1'27"	67o
1oo8	739	35'17"	4o3o

5.3.4 Application to the Gosau Power Plant Chain

Assumptions: constant efficiencies, $\Delta t_k = 1$ hour, Z_{ik}= const.= 1 m^3/s, $i = 1,2,3$

The problem can be split up both with respect to the System Gosau/Gosauschmied and Steeg and to the time intervals as described above. $\phi(Y,\lambda)$ is determined by solving N five dimensional problems and N four dimensional problems. For details see Gfrerer [8].

Results from an one week optimization

t [h]	$Q_1[m^3/s]$	$Q_2[m^3/s]$	$Q_3[m^3/s]$
6	4.3	6.6	6.1
7	4.3	6.6	7.4
⋮	⋮	⋮	⋮
15	4.3	6.6	7.4
16	o.3	5.6	7.4
⋮	⋮	⋮	⋮
2o	-	6.1	7.4
21	-	-	7.4
22	-	-	-

To get an accuracy of $\varepsilon_{rel}(Q) \leq 1$ % we need 14' (IBM 3o31)

6. Determination of the Efficiency

6.1 The Problem

Special attention was paid to determine the input data. We describe a black box model by which the efficiency η can be determined. We have:

$$\eta = \frac{P}{P_{Th}} = \frac{\text{output power of the generator}}{\text{theoretical power}}$$

The theoretical power P_{Th} is reduced by losses in the pipe system, by losses in the turbine and by losses in the generator (including those of the transformer). The losses in the pipe system are assumed by the engineers to be proportional to the squares of the discharges in each of the sections of the pipe system:

$$H \to H_M := H - c\,Q^2 \qquad (H_M\text{: manometric head})$$

The efficiency η_T of the turbine is given by the producer as shell like curves:

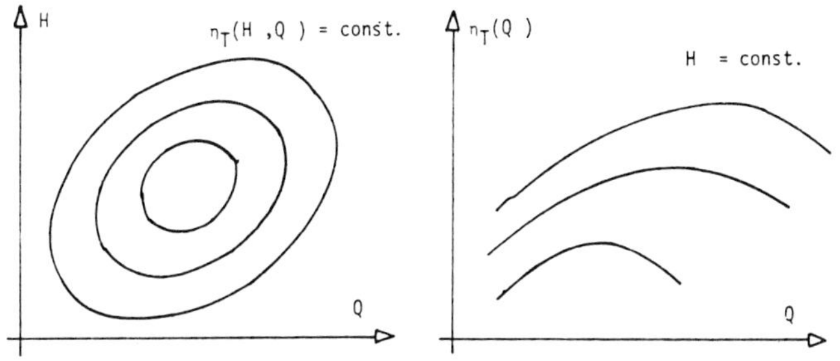

Therefore, we may describe η_T in the following way:

$$\eta_T = \eta_T(H_M,Q) := [a_0 + a_1 \;\; Q + a_2 H_M + a_3 H_M Q + a_4 Q^2 + a_5 H_M{}^2]$$

The efficiency of the generator is given in dependence on the output power P_T of the turbine and can be described by

$$\eta_G = \eta_G(P_T) = b_0 + b_1 P_T + b_2 P_T{}^2$$

In practice the efficiency of the generator given by the producer is acceptable, though the information available both for η_T and the losses in the pipe system seem doubtful, at least by our experiencies. For the case of Gosau we accepted η_G and determined the unknown parameters (c,a) by measurements taken by the engineers of the power company.

6.2 Parameter identification by Tiller/Schwetlick and Deuflhard

Schwetlick/Tiller [18] considered the following equation in $\mathbb{R}^m$, depending on a parameter $\theta \in \mathbb{R}^p$

$$y = r(x,\theta), \quad x \in \mathbb{R}^n$$

Based on measurements (X^i, Y^i), $i = 1,\dots,l$ one searches not only for the unknown parameters θ (resulting in a standard nonlinear least squares problem) but also for "exact" points x where the measurements were taken. This leads to the minimization of the following objective:

$$S(z) := S(x,\theta) := \frac{1}{2} \sum_{i=1}^{l} \begin{pmatrix} r(x^i,\theta)-Y^i \\ x^i-X^i \end{pmatrix}^T W_i \begin{pmatrix} r(x^i,\theta)-Y^i \\ x^i-X^i \end{pmatrix}$$

W_i are positive definite weight-matrices.

Using Cholesky factorization $W_i = R_i^T R_i$ we get a nonlinear minimum norm problem:

$$S(z) = \frac{1}{2} \|f(z)\|_2^2 \text{ with } f(z) := [f_1(z), f_2(z)]^T$$

where $f_1(z) := \begin{bmatrix} R_{11}(x^1-X^1) \\ R_{11}(x^l-X^l) \end{bmatrix}$ and $f_2(z) := \begin{bmatrix} R_{12}(x^1-X^1)+R_{13}[r(x^1,\theta)-Y^1] \\ R_{12}(x^l-X^l)+R_{13}[r(x^l,\theta)-Y^l] \end{bmatrix}$

and $R_i = \begin{pmatrix} R_{i3} & R_{i2} \\ 0 & R_{i1} \end{pmatrix}$

Using Gauß-Newton technique we get by linearization

$$\|J\Delta z-f\|^2 = \left\| \begin{bmatrix} J_{11} & 0 \\ J_{21} & J_{22} \end{bmatrix} \begin{bmatrix} \Delta x \\ \Delta\theta \end{bmatrix} - \begin{bmatrix} f_1 \\ f_2 \end{bmatrix} \right\|^2 =$$

$$= \|Q(J\Delta z-f)\|^2 = \|K_{11}\Delta x+K_{12}\Delta\theta-g_1\|^2 + \|K_{22}\Delta\theta-g_2\|^2 = \underset{(\Delta x,\Delta\theta)}{\text{Min!}}$$

where the orthogonal matrix Q is suitable chosen. Assuming full rank p for J_{22} we have full rank for J. Then the problem can be split:

(i) Solve $\|K_{22}\Delta\theta - g_2\|^2 = \underset{\Delta\theta}{\text{Min!}} \rightarrow \overline{\Delta\theta}$

(ii) Solve $K_{11}\Delta x = (g_1 - K_{12}\overline{\Delta\theta})$ in the following way:

$$K_{11,i}\Delta x^i = g_{1,i} - (K_{12,i}\Delta\theta) \qquad i = 1,\dots,l$$

(iii) $z^{k+1} := z^k - \gamma_k \Delta z^k$

For (iii) the damping technique proposed by Deuflhard, using the natural level function $T(z^k|J^k)$, see [2], proved successful. In case of rank $(J_{22}) < p$ the following regularization may be used:

$$\left\| \begin{bmatrix} K_{11}(\lambda) & K_{12}(\lambda) \\ 0 & K_{22} \\ 0 & \lambda \end{bmatrix} \Delta z - \begin{bmatrix} g_1(\lambda) \\ g_2 \\ 0 \end{bmatrix} \right\|^2 = \underset{\Delta z}{\text{Min!}} \quad \text{with } K_{1i}(\lambda) := \sqrt{1+\lambda^2}\, K_{1i}, \quad g_1(\lambda) := g_1/\sqrt{1+\lambda^2}$$

6.3 Application: efficiency of Gosau

To determine the efficiency of Gosau we used the output power P of the turbine:

$$\eta(H,Q) = \left(1-\frac{cQ^2}{H}\right)\left[a_0 + a_1 Q + a_2 H_M + a_3 Q^2 + a_4 H_M^2 + a_5 H_M Q\right]$$

with $H_M := H - c\,Q^2$

We have: $x = (H,Q)$, $\theta = (c; a_0,\dots,a_5)$, $y = P$

$$y = P(H,Q) = g.\rho.H.Q.\eta(H,Q)$$

$$W_i = \mathrm{diag}\left(\left[\frac{1}{\sigma_H H^i}\right]^2, \left[\frac{1}{\sigma_Q Q^i}\right]^2, \left[\frac{1}{\sigma_P P^i}\right]^2 \right)$$

$n = 2$, $m = 1$, $p = 7$ and $l = 43$ measurements

$\sigma_H = 0{,}00025$, $\sigma_Q = 0{,}00007$, $\sigma_P = 0{,}5$

Using a (very concise but complicated) direct method [23] we get for our"black box efficiency" the following result:

η	H [m]	Q [m^3/s]	η_{BB}(black box)
	155	6.1o	o.66
o.66	144	5.87	o.66
	132	5.19	o.67
	132	3.64	o.66
	155	5.68	o.71
o.68	146	5.51	o.68
	133	4.3o	o.68
	155	5.5o	o.72
o.71	146	4.8o	o.7o
	141.4	4.23	o.68

For details see [13].

7. Realization in Practice

The results for Phase I were put into practice in 1983 and 1984. For comparison we give some information on production, start of the sinking process and sinking time.

year	sinking time (days)	start	production (GWh)
1968	75	5.11.	1.75
1969	12o	2o.o9.	1.55
1971	164	1o.o8.	o.65
1974	1o9	13.11.	1.2
1979	65	15.1o.	2.o9
1982	78	22.o9.	1.93
1983	51	1.1o.	2.65
1984	45	1.1o.	2.6o

Acknowledgements:

The data were made available to us by the OKA.
Thanks are given to the management of the OKA, Dir.Dipl.-Ing.W.Barwig, and both to the engineers, Ing. Feichtinger and Ing. Peßl, for their substantialhelp which also included laborious experiments.
Our practical work has been supported both by the Austrian Ministry of Science and Research and the OKA.
Our theoretical work has been supported partly by the Austrian Fonds of Research. (P3472, P4786, P5729).

References:

[1] W.Barwig/R.Peßl, Optimization of the Gosau System, in "Applied Optimization Techniques in Energy Problems (Proceedings), Teubner pp. 94-118, 1985

[2] P.Deuflhard/V.Apostolescu, A Study of the Gauss-Newton Method for the Solution of Nonlinear Least Squares Problems, in: Frehse/ Pallaschke/Trottenberg (ed.): Special Topics of Applied Mathematics, Amsterdam, North-Holland Publ. pp.129-15o (198o)

[3] H.Engl/Hj.Wacker/E.Zarzer: Ausgewählte Kapitel der Kontrolltheorie, in: Computergestützte Planungssysteme (ed.Noltemeier), Physica Verlag, Würzburg, pp. 155-2o2 (1976)

[4] A.Gaillard/H.Brüngger, Speicherbewirtschaftung, Studienberichte Heft 17, Institut für Operations Research IFOR, ETH-Zürich, pp. 1-15, 1984

[5] H.Gfrerer/J.Guddat/Hj.Wacker, A globally convergent algorithm based on imbedding and parametric optimization, Computing 3o, pp. 225-252 (1983)

[6] H.Gfrerer, Globally convergent decomposition methods for nonconvex optimization problems, Computing 32, 199-227 (1984)

[7] H.Gfrerer, Optimization of hydro energy storage plant problems by variational methods, ZOR/B, Bd. 28, 87-1o1 (1984)

[8] H.Gfrerer, Optimization of storage plant systems by decomposition, in: "Applied Optimization Techniques in Energy Problems (Proceedings), Teubner, 215-226, 1985

[9] H.Gfrerer/J.Guddat/Hj.Wacker/W.Zulehner, Path following methods for Kuhn-Tucker-Curves by an active index set strategy, in: Proc. Parametric Optimization, Springer Lecture Notes (ed.Jongen), 1985

[1o] H.Habibollazadeh, Optimal Short-Time Operation Planning of Hydro-electric Power Systems, Royal Institute of Technology, Department of Electric Power System Engineering, Stockholm, Sweden, 1983

[11] P.G.Harhammer/M.A.Muschik/A.Schadler, Optimization of Large Scale MIP Models, Operation Planning of Energy Systems, in: "Applied Optimization Techniques in Energy Problems, Teubner, 286-322,1985

[12] A.Kalliauer, Compactification and decomposition methods for NLP problems with nested structures in hydro power system planning, in: "Survey of Mathematical Programming", Proc.of the 9th Intern. Mathematical Programming Symposium, Budapest, August 23-27, 1976 (ed.Prêkopa, Publ.House of the Hungarian Academy of Sciences, pp. 243-258.

[13] E.Lindner/Hj.Wacker, Input Parameter for the Optimization of the System Partenstein: Forecasting of the Influx, Determination of the Efficiency Function, in:"Applied Optimization Techniques in Energy Problems (Proceedings), Teubner, 286-322, 1985

[14] A.Prêkopa et al., Studies in Applied Stochastic Programming (ed. A.Prêkopa), Tanulmânyok 8o, 1-2o9 (1978)

[15] T.G.Roefs/L.D.Bodin, Multireservoir Operation Studies, Water Resources Research, Vol. 6, Nr. 2, pp. 41o-42o, 197o

[16] F.G.Rhode/K.Naparaxawong, Modified Standard Operation Rules for Reservoirs, Journal of Hydrology, 51, pp. 169-177, 1981

[17] G.Schmitz/J.Edenhofer, Flood Routing in the Danube River by the New Implicit Method of Characteristics (IMOC) 3rd Int.Appl.Math. Modelling Conference, Hamburg, pp. 1-3, 198o

[18] H.Schwetlick/Tiller, Numerical Methods for Estimating Parameters in Nonlinear Models with Errors in the Variables, Preprint Nr.69, Univ.Halle, Sektion Mathematik, 1-25 (1982)

[19] K.Theilsiefje, Ein Beitrag zur Theorie der wirtschaftlichen Ausnutzung großer Speicherseen zur Energieerzeugung, ETZ-A, Bd.82 H17, pp.538-545, 1981

[2o] Hj.Wacker, Mathematical Techniques for the Optimization of Storage Plants, in "Applied Optimization Techniques in Energy Problems (Proceedings), Teubner, 4o5-448, 1985

[21] Hj.Wacker, Applied Optimization Techniques in Energy Problems (Proceedings) Teubner, 1-485, 1985

[22] Hj.Wacker/W.Bauer/S.Buchinger, Einsatz mathematischer Methoden bei der Hydroenergiegewinnung, ZAMM 63, pp.227-243 (1983)

[23] Hj.Wacker/W.Bauer/H.Gfrerer/E.Lindner/A.Schwarz, Erzeugungsoptimierung in Wasserkraftwerken, Arbeitsbericht für das BMfWF/OKA, pp. 1-22o (1985)

[24] W.G.Yeh/W.Z.Trott/A.M.Asce, Optimization of Multiple Reservoir System, Journal of the Hydraulics Division, pp.1865-1884, 1973

PART VI

ALGORITHM ADAPTATION ON SUPERCOMPUTERS

THE USE OF VECTOR AND PARALLEL COMPUTERS IN THE SOLUTION OF LARGE SPARSE LINEAR EQUATIONS

Iain S. Duff

Abstract

We discuss three main approaches that are used in the direct solution of sparse unsymmetric linear equations and indicate how they perform on computers with vector or parallel architecture. The principal methods which we consider are general solution schemes, frontal methods, and multifrontal techniques. In each case, we illustrate the approach by reference to a package in the Harwell Subroutine Library. We consider the implementation of the various approaches on machines with vector architecture (like the CRAY-1) and on parallel architectures, both with shared memory and with local memory and message passing.

1 Introduction

We discuss the direct solution of large sparse linear equations on vector and parallel computers. To be specific, we examine algorithms based on Gaussian elimination for solving the equation

$$\mathbf{Ax} = \mathbf{b} \quad , \tag{1.1}$$

when the coefficient matrix **A** is unsymmetric, large, and sparse. We will look at three classes of algorithms and comment on how effective they are at exploiting vector and parallel architectures.

The classes into which we divide our algorithms are general methods, band and frontal techniques, and multifrontal approaches. We summarize the features of each approach in Section 2 and discuss each in more detail in the succeeding three sections. Finally, we draw some conclusions in Section 6.

All of the methods considered are direct methods based on Gaussian elimination. That is to say, they all compute an **LU** factorization of a permutation of the coefficient matrix **A**, so that

$$\mathbf{PAQ} = \mathbf{LU} \ , \tag{1.2}$$

where **P** and **Q** are permutation matrices, and **L** and **U** are lower and upper triangular matrices respectively. These factors are then used to solve the system (1.1) through the forward substitution

$$\mathbf{Ly} = \mathbf{P}^T\mathbf{b} \tag{1.3}$$

followed by the back substitution

$$\mathbf{U}(\mathbf{Q}^T\mathbf{x}) = \mathbf{y} \ . \tag{1.4}$$

The study of algorithms for effecting such solution schemes when the matrix **A** is large and sparse is important, not only for the problem in its own right, but also because the type of computation required makes this an ideal paradigm for large-scale scientific computing in general. In other words, we believe that a study of direct methods for sparse systems encapsulates many issues which appear widely in computational science and which are not so tractable in the context of really large scientific codes. The principal issues can be summarized as follows :

(i) floating-point calculations themselves form only a small proportion of the total code,

(ii) there is a significant data-handling problem,

(iii) storage is often a limiting factor and auxiliary storage is frequently used,

(iv) although the innermost loops are often well defined, there is usually a significant amount of time spent in computations in other parts of the code, and

(v) the innermost loops can sometimes be very complicated.

Issues (i)–(iii) are related to the manipulation of sparse data structures. The efficient implementation of techniques for handling these are of crucial importance in the solution of sparse matrices. Similar issues arise when handling large amounts of data in other large-scale scientific computing problems. Issues (ii) and (iv) serve to indicate the sharp contrast between sparse and non-sparse linear algebra. In code for large full systems, well over

90 per cent of the time (on a serial machine) is typically spent in the innermost loop whereas, as we shall see in Section 3, a substantially lower fraction is spent in the innermost loops of sparse codes. The lack of dominance of a single loop is also characteristic of a wide range of large-scale applications.

2 The principal features of the methods considered

The methods that we consider can be grouped into three main categories: general techniques, frontal methods, and multifrontal approaches. In all cases, we assume that the coefficient matrix is unsymmetric. Hence, for example, we do not consider methods specially designed for symmetric positive-definite systems. However, in such special cases, algorithms can also be divided into three similar groups so that most of our remarks are still pertinent in such environments. It is not our intention to describe the methods in any detail. We merely wish to draw attention to the features which are important when considering the influence of vector and parallel architectures. For readers requiring further background on these techniques, the book by Duff, Erisman, and Reid (1986) is recommended. Further references on each approach are given in the appropriate section of this paper. In each case, we illustrate the approach by reference to a package in the Harwell Subroutine Library. We also use runs of these packages as evidence for some of our conclusions.

We first consider a completely general approach typified by the Harwell Subroutine MA28 (Duff 1977). The principal features of this general approach are that numerical and sparsity pivoting are performed at the same time so that dynamic data structures are used in the initial factorization, and that sparse data structures are used throughout — even in the innermost loops. As we will see when we discuss such a technique in Section 3, these features must be considered drawbacks with respect to vectorization and parallelism. Although special purpose methods can be employed to take advantage of a particular structure (for example, band solvers), the strength of the general approach is that it will give a very satisfactory performance over a wide range of structures.

Frontal schemes can be regarded as an extension of band or variable-band schemes and will perform well on systems whose bandwidth or profile is small. From the point of view of using such methods to solve grid-based problems (for example, discretizations of partial differential equations), the efficiency of frontal schemes will depend crucially on the underlying geometry of the problem. It is, however, possible to write frontal codes so that any system can be solved; sparsity preservation is obtained from an initial ordering and numerical pivoting can be performed within this ordering. A characteristic of frontal methods is that no indirect addressing is required in the innermost loop. We use the code MA32 from the Harwell Subroutine Library (Duff 1981, 1983) to illustrate this approach when we discuss it further in Section 4.

The final class of techniques, which we study in Section 5, is an extension of the frontal methods termed multifrontal. The extension permits efficiency for any matrix whose nonzero pattern is symmetric or nearly symmetric and allows any sparsity ordering techniques for symmetric systems to be used. The restriction to nearly symmetric patterns arises because the initial ordering is performed on the sparsity pattern of the Boolean sum of $\mathbf{A}$ and $\mathbf{A}^T$. The approach can, however, be used on any system (for example, Harwell Subroutine Library code MA37 — Duff and Reid 1984). As in the frontal method, full matrices are used in the innermost loop so that indirect addressing is avoided. There is, however, more data movement than in the frontal scheme and the innermost loop is not so dominant. In addition to the use of direct addressing, multifrontal methods also differ from the first class of methods because the sparsity pivoting is entirely separated from the numerical pivoting.

There have been some recent efforts to design algorithms which require only static data structures during the matrix factorization (for example, George and Ng 1984, 1985). We do not discuss these but note that they use indirect addressing in their innermost loops and so many of the considerations of Section 3 will apply.

3 General sparse codes

In general sparse codes, we use sparse data structures throughout. We thus begin this section by describing the most common sparse data structure, which is the one used in our archetypal code MA28 (Duff 1977). The structure for a row of the sparse matrix is illustrated in Figure 3.1. All rows are stored in the same way and a pointer is used to identify the location of the beginning of the data structure for each row. If the pointer in Figure 3.1 was that for row i, then entry a_{ij_1} would have value ξ. Clearly access to the entries in a row is very straightforward although indirect addressing is required to identify the column of an entry.

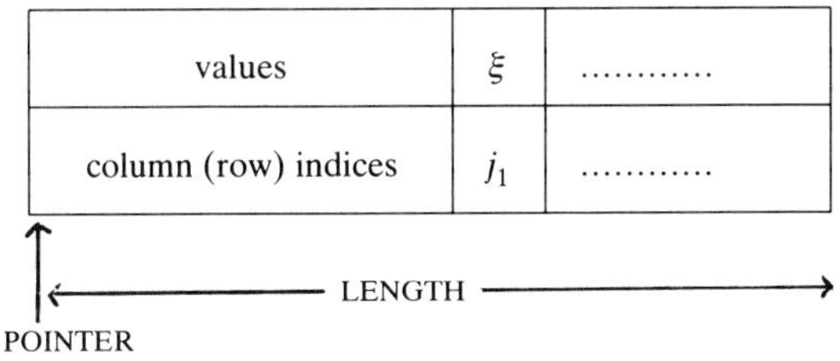

Figure 3.1. General sparse matrix storage scheme.

The innermost loop of the general sparse code consists of adding a multiple of the pivot row to a non-pivot row with a nonzero in the pivot column. Since, in general, there could be fill-in in the non-pivot row, the data structures and coding must allow for the length of the non-pivot row to increase. This means that the innermost loop can become quite complicated. We illustrate this with the Fortran code in Figure 3.2, which is reproduced from the current version of the Harwell code MA28. Although it would be possible to split the loop to reduce the number of IF statements, much of the complication would remain, particularly the double indirect addressing for the entry A(PIVROW).

On early vector machines, like the CRAY–1, loops like that shown in Figure 3.2 would only run at scalar speeds and this was very much borne out by the performance of MA28 on the CRAY–1 at Harwell (Duff 1984a). Indeed, even on much less complicated loops with only one level of indirect addressing, the first generation of vector supercomputers would exhibit essentially scalar behaviour.

```
          DO 590 JJ = J1,J2
C Access columns of non-pivot row
            J = ICN(JJ)
C Jump if corresponding entry in pivot row is zero
            IF (IQ(J).GT.0) GO TO 590
C Increment counter to number of pivot row entries used
            IOP = IOP + 1
C Find location in array A of pivot row entry in column J
            PIVROW = IJPOS - IQ(J)
C Perform Gaussian elimination operation
            A(JJ) = A(JJ) + AU * A(PIVROW)
C Keep a count of the largest entry in reduced matrices.
C  Used in stability check
            IF (LBIG) BIG = DMAX1(DABS(A(JJ)),BIG)
C Count entries less than drop tolerance
            IF (DABS(A(JJ)).LT.TOL) IDROP = IDROP + 1
C Flag ICN entry to show that pivot row entry has been used
            ICN(PIVROW) = - ICN(PIVROW)
  590     CONTINUE
```

Figure 3.2. Fortran 77 code for an innermost loop of MA28

Most recent vector processors, however, have a facility for hardware indirect addressing and one might be tempted to believe that all problems associated with indirect addressing have been overcome. Dongarra and Duff (1986) have done an extensive study of several typical loops from sparse codes on a number of machines with hardware indirect addressing including the CRAY X-MP, the CRAY–2, the Convex C–1, the NEX SX–2, and the Fujitsu FACOM VP 100. Their findings show that the asymptotic rate for most loops is indeed very high, usually around half the peak performance of the machine.

```
          DO 100 I=1,M
            A(ICN(I)) = A(ICN(I)) + AMULT * W(I)
      100 CONTINUE
```

Figure 3.3. Sparse SAXPY loop.

For example, on the CRAY X-MP, the loop shown in Figure 3.3 (a sparse SAXPY) ran asymptotically at only 5.5 Megaflops when hardware indirect addressing was inhibited but ran asymptotically at over 80 Megaflops when it was not. On the surface, the manufacturers' claim to have conquered the indirect addressing problem would seem to be vindicated, and we might be led to believe that our sparse general codes would now perform at about half the rate of highly-tuned full matrix code.

There are two flaws in this reasoning which we will now discuss. The first lies in the $n_{1/2}$ value (Hockney and Jesshope 1981) for the sparse loops. That is to say, the length of the loop required to attain half the maximum performance. This measure is directly related to the start-up time for the loop. For the loop shown in Figure 3.3, the $n_{1/2}$ value on the CRAY X-MP is about 50 which, relative to the typical order of sparse matrices being solved by direct methods (greater than 10000), is insignificant. However, the loop length for sparse calculations depends not on the order of the system but rather on the length of the pivot row. We have done an extensive empirical examination of this length using the MA28 code on a wide range of applications. We show a representative sample of our results in Table 3.1. Except in the examples from the analysis of structures (the last three in the table), this length is very low and, even in the small structures examples, is much less than the $n_{1/2}$ value mentioned above. Thus the typical performance rate for the sparse inner loop is far from the asymptotic performance.

Order of matrix	Number of entries	Discipline	Av length of pivot row	% time in inner loops
655	2854	Chemical engineering	3	37
199	701	Stress analysis	3	43
1176	9864	Electrical circuit analysis	8	58
147	2449	Atomic spectra	17	55
363	3157	Linear programming	5	35
541	4285	Atmospheric pollution	10	52
1454	3377	Power system networks	2	39
1107	5664	Computer simulation	19	42
420	7252	Structural analysis	29	49
838	5424	Aerospace	27	49
1005	4813	Ship design	24	50

Table 3.1. Statistics from MA28 on the IBM 3084 for a selection of matrices from different disciplines.

The second problem with the use of hardware indirect addressing in general sparse codes is that the amount of data manipulation in such a code means that a much lower proportion of the time is spent in the innermost loops than in code for dense matrices. Again we have performed an empirical study on MA28 and show these results in Table 3.1 also. The percentage given in that table is for the total time of three loops in MA28, all at the same innermost level of nesting as that in Figure 3.2. In most cases, the loop in that figure accounts for half the total time in these three loops. We see that typically around 50 per cent of the overall time on an IBM 3084 is spent in the innermost loop. Thus, even if that loop were made to run infinitely fast, a speed-up of only about a factor of two would be obtained. Of course, if we can avoid the necessity for numerical pivoting, then the code and the inner-loop structure can be made much simpler. For example, the SPARSPAK code for positive-definite systems has been modified by Boeing Computer Services so that it uses hardware indirect addressing on the CRAY X-MP. On typical problems, they have managed to speed up the execution time by a factor of nearly four. A significant effect of this improvement is that the balance between using general codes and variable-band techniques on vector machines has now tilted towards the former (Lewis and Simon 1986). However, while the speed-up due to hardware indirect addressing is very creditable, it falls a long way short of the factor of over thirty between scalar and vector performance on that machine.

Our conclusion is therefore that vector indirect addressing is of limited assistance for present-generation general sparse codes. Even for general systems, however, easy advantage can be taken of vectorization by using a hybrid approach, where a full matrix routine is used when fill-in has caused the reduced matrix to become sufficiently dense. Duff (1984a) reports on some experiments with such an approach on the CRAY–1. He finds that the switch-over density for overall time minimization can often be very low (typically 20 per cent dense) and that gains of a factor of over four can be obtained even using unsophisticated full matrix code.

When we turn to parallelism in general sparse codes, we encounter the same problem as with full Gaussian elimination, namely that the amount of parallelism is limited by the pivoting (for example, Vavasis 1986). Although some overlap is possible between operations with different pivots (for example, the wavefront method of Kung *et al.* 1981a), the selection of pivots requires n sequential steps for a system of order n. In Section 5, we discuss how the sparsity of the system removes this bottleneck and can lead to exploitation of parallelism in the simultaneous selection of pivots. However, the essence of the general sparse methods that we are discussing here, is that the reduced matrix is updated before the next pivot is chosen so that the gains of Section 5 are not possible. Alaghband and Jordan (1983) have implemented the Harwell MA28 code on the Denelcor HEP. They concluded that the best possible speed-up on any parallel architecture was about 6, largely because of the sequential bottleneck just mentioned.

More minor problems with the parallelism of this general approach relate to some of the problems that we encountered with vectorization, namely that the innermost loops are complicated, involve indirect addressing, and do not dominate the time for computation. A final comment relates particularly to shared memory machines. Most implementations of the general purpose approach, including MA28, involve some form of garbage collection since dynamic storage schemes are in use. The most common garbage collection schemes involve moving all the "active" data to one end of the working array and adjusting pointers appropriately. To do this would, however, involve a join operation and the cessation of parallel processing until completion of the garbage collection. We are currently investigating other garbage collection schemes to avoid this bottleneck, including a modification of the "buddy system" of Knuth (1973) and localized garbage collection.

4 Frontal methods

Frontal methods have their origins in the solution of finite-element problems from structural analysis. One of the earliest computer programs implementing the frontal method is that of Irons (1970), although he only considered the case of symmetric positive-definite systems. The method can, however, be extended to unsymmetric systems (Hood 1976) and need not be restricted to finite-element applications (Duff 1981). Indeed, we prefer to view frontal methods as an extension of band or variable-band schemes.

A common method of organizing the factorization of a band matrix of order n with semibandwidth b is to allocate storage for a full $b \times 2b-1$ matrix, which we call the frontal matrix, and to use this as a window that runs down the band as the elimination progresses. Thus, at the beginning, the frontal matrix holds rows 1 to b of the band system. This enables the first pivotal step to be performed (including pivoting if this is required) and, if the pivot row is then moved out of the frontal matrix, row $b+1$ of the band matrix can then be accommodated in the frontal matrix. The second pivotal step can then be performed within the frontal matrix. It is very common to use a larger frontal matrix since greater efficiency may be possible through being able to move blocks of rows at a time. Additionally, it is then usually possible to perform several pivot steps within the one frontal matrix. The traditional reason for this implementation of a banded solver is for the solution of band systems by out-of-core methods since only the frontal matrix need be held in main storage. This use of auxiliary storage is also one of the principal features of our general frontal method (Duff 1984b).

It is very easy to extend this "windowing" method to variable-band matrices. In this case, the frontal matrix must have order at least $\max_{a_{ij} \neq 0}\{|i-j|\}$. Further extension to general matrices is possible by observing that any matrix can be viewed as a variable-band matrix. This points to the main problem with this technique: for any arbitrary matrix with an arbitrary ordering, the required size for the frontal matrix may be very large. However, for discretizations of partial differential equations (whether by finite elements or finite differences), good orderings can usually be found.

It is not necessary to discuss all the details of frontal codes in this paper; the important feature concerns the implementation of Gaussian elimination on the frontal matrix. The frontal matrix is of the form:

$$\begin{pmatrix} \mathbf{A} & \mathbf{B} \\ \mathbf{C} & \mathbf{D} \end{pmatrix} , \tag{4.1}$$

where **A** and **D** are square matrices of order k and r respectively, where usually $k \ll r$, $(k+r=m)$. The object at this stage is to perform k steps of Gaussian elimination on the frontal matrix (choosing pivots from **A**), storing the factors $\mathbf{L}_A\mathbf{U}_A$ of **A**, $\mathbf{CA}^{-1}$, and **B** on auxiliary storage devices, and generating the Schur complement $\mathbf{D}-\mathbf{CA}^{-1}\mathbf{B}$ for use at the next stage of the algorithm. In the parlance of frontal methods, the rows and columns of **A** are "fully summed", that is there will be no further entries in these rows and columns later in the computation. Typically, **A** might have order 10 to 20 while **D** is of order 200 to 500.

When numerical pivoting is required, the effect is to constrain the choice of the pivots from **A**. In the unsymmetric case, the pivots can be chosen from anywhere within **A** and, in our approach (Duff 1981), we use the standard sparse matrix technique of threshold pivoting, where $a_{ij} \in \mathbf{A}$ is suitable as pivot only if

$$|a_{ij}| \geq u \,.\, \max(\max_s |a_{sj}|, \max_s |c_{sj}|) \tag{4.2}$$

where u is a preset parameter in the range $0 < u \leq 1$.

Notice that this means that large entries in **C** can prevent the selection of some pivots from **A**. Should this be the case, $k_1 < k$ steps of Gaussian elimination will be performed and the resulting Schur complement $\mathbf{D}-\mathbf{C}\mathbf{A}_1^{-1}\mathbf{B}$, where $\mathbf{A}_1$ is a square submatrix of **A** of order k_1, will have order $r+k-k_1$. Although this can increase the amount of work and storage required by the algorithm, the extra cost is typically very low and all pivotal steps will eventually be performed since the final frontal matrix has a null **D** block (that is, $r=0$).

The important aspect of frontal schemes is that all the elimination operations are performed within a full matrix, so that techniques for exploitation of vectorization or parallelism on full systems can be used. It is also important that k is usually greater than 1, in which case more than one elimination is performed on the frontal matrix.

Dave and Duff (1986) used this fact to design a very efficient assembler-language kernel for the CRAY–2 that combines two pivot steps in order to overlay memory fetches with arithmetic. In Table 4.2 we show their results on a version of the MA32 package that uses this kernel. The problems are artificially generated finite-element problems with rectangular elements having nodes at corners, mid-points of sides and centre and with five variables on each node.

Dimension of element grid	16×16	50×50	100×100
Maximum order of frontal matrix	195	536	1035
Total order of problem	5445	51005	202005
Megaflops	100	302	345

Table 4.2. Performance of CRAY–2 version of MA32 on element problems.

The benefits of vectorization are quite apparent in the results shown in Table 4.2. However, the organization of the frontal method is similar to that for a band matrix solver which is inherently sequential. Thus, we still have the limitation of n sequential pivot steps, as we discussed in Section 4. We address this problem by the multifrontal methods that we discuss in the next section.

5 Multifrontal methods

Multifrontal methods are described in some detail by Duff *et al.* (1986) and their potential for parallelism by Duff (1986a, 1986b). We assume here that the matrix has a nonzero pattern that is symmetric but note that any system can be considered if we are prepared to store explicit zeros. Although we will not describe this class of methods in detail here, we will work through a small example shown in Figure 5.1 to give the reader a flavour for the important points and to introduce the notion of an elimination tree (discussed in detail by Duff 1986a and Liu 1985). In Figures 5.1 and 5.2, $\times$ denotes a nonzero entry and zeros are left blank.

$$\begin{array}{cccc} \times & & \times & \times \\ & \times & \times & \times \\ \times & \times & \times & \\ \times & \times & & \times \end{array}$$

Figure 5.1. Matrix used to illustrate multifrontal scheme.

We assume the matrix in Figure 5.1 is ordered so that pivots will be chosen down the diagonal in order, although we are not immediately concerned with the nature of this ordering. At the first step, we can perform the elimination corresponding to the pivot in position (1,1), first "assembling" row and column 1 to get the submatrix shown in Figure 5.2.

$$\begin{array}{ccc} \times & \times & \times \\ \times & & \\ \times & & \end{array}$$

Figure 5.2. Assembly of first pivot row and column.

Here, by "assembling", we mean placing the nonzeros of row and column 1 into a submatrix of order the number of nonzeros in row and column 1. Thus the zero entries a_{12} and a_{21} have caused row and column 2 to be omitted in Figure 5.2, and so an index vector is required to identify the rows and columns that are in the submatrix. The index vector for the submatrix in Figure 5.2 would have entries (1,3,4) for both the rows and the columns. Column 1 is then eliminated using pivot (1,1) in Figure 5.2 to give a reduced matrix of order two with associated row (and column) indices 3 and 4. In conventional Gaussian elimination, updating operations of the form

$$a_{ij} = a_{ij} - a_{i1}[a_{11}]^{-1}a_{1j} \qquad (5.1)$$

would be performed immediately for all (i,j) such that $a_{i1}a_{1j} \neq 0$. However, in this formulation, the quantities

$$a_{i1}[a_{11}]^{-1}a_{1j},\ a_{i1}a_{1j} \neq 0 \tag{5.2}$$

are held in the reduced submatrix, and the corresponding updating operations (5.1) are not performed immediately. These updates are not necessary until the corresponding entry is needed in a later pivot row or column. The reduced matrix can be stored until that time.

Row (and column) 2 is now assembled, the (2,2) entry is used as pivot to eliminate column 2, and the reduced matrix of order two, with associated row (and column) indices of 3 and 4, is stored. Because of the relationship of these submatrices to that of (4.1), they are called frontal matrices. Since there will generally be more than one frontal matrix stored at any time (currently we have two stored), the method is called "multifrontal". Now, before we can perform the pivot operations using entry (3,3), the updating operations from the first two eliminations (the two stored frontal matrices of order two) must be performed on the original row and column 3. This is effected by summing or "assembling" the reduced matrices with the original row and column 3, using the index lists to control the summation. Note that this gives rise to an assembled submatrix of order 3 with indices (3,4,5) for rows and columns. The pivot operation which eliminates column 3 using pivot (3,3) leaves a reduced matrix of order one with row (and column) index 4. The final step sums this matrix with the (4,4) entry of the original matrix. The sequence of major steps in the elimination can be represented by the tree shown in Figure 5.3.

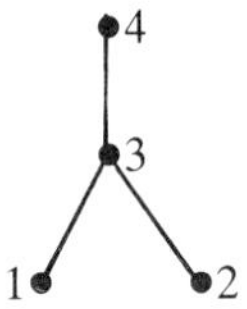

Figure 5.3. Elimination tree for the matrix of Figure 5.1.

The same storage and arithmetic is needed if the (4,4) entry is assembled at the same time as the (3,3) entry, and in this case the two pivotal steps can be performed on the same submatrix. This corresponds to collapsing or

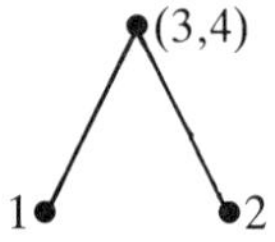

Figure 5.4. Elimination tree for the matrix of Figure 5.1 after node amalgamation.

amalgamating nodes 3 and 4 in the tree of Figure 5.3 to yield the tree of Figure 5.4. On typical problems, node amalgamation produces a tree with about half as many nodes as the order of the matrix. Duff and Reid (1983) employ node amalgamation to enhance the vectorization of a multifrontal approach. In the context of parallelism, node amalgamation creates a larger granularity at each node which can assist in exploiting parallelism with this technique.

The computation at a node of the tree is simply the assembly of information concerning the node together with the assembly of the reduced matrices from its sons followed by some steps of Gaussian elimination. Indeed, each node corresponds to the formation of a frontal matrix of the form (4.1) followed by some elimination steps, after which the Schur complement is passed on for assembly at the father node.

The main feature of general elimination trees is that computation at any leaf node can proceed immediately and simultaneously and computations at nodes not on the same direct path from the root to a leaf node are independent. All that is required for computations to proceed at a node is that the calculations at its sons have been completed. A full discussion of this is given by Duff (1986a).

Clearly the parallelism available through the elimination tree is very dependent on the ordering of the matrix. In general, short bushy trees are preferable to tall thin ones since the number of levels determines the inherent sequentiality of the computation. Two common ordering strategies for general sparse symmetric systems are minimum degree and nested dissection (see, for example, George and Liu 1981, or Duff *et al.* 1986). Although these orderings are very similar in behaviour for the amount of arithmetic and storage, they give fairly different levels of parallelism when used to construct an elimination tree. We illustrate this point with some results from Duff and Johnsson (1986) in Table 5.1, where the maximum speed-up is computed as the ratio of the number of operations in the sequential algorithm to the number of sequential operations in the parallel version, where account has been taken of data movement as well as floating-point calculations.

Ordering	Minimum degree	Nested dissection
Number of levels in tree	52	15
Number of pivots on longest path	232	61
Maximum speed-up	9	47

Table 5.1. Comparison of two orderings for generating an elimination tree for multifrontal solution. The problem is generated by a 5–point discretization of a 10×100 grid.

If the nodes of the elimination tree are regarded as atomic, then the level of parallelism reduces to one at the root and usually increases only slowly as we progress away from the root. If, however, we recognize that parallelism can be exploited within the calculations at each node (corresponding to one or a few steps of Gaussian elimination on a full submatrix), much greater parallelism can be achieved. In Table 5.2, we give some results of Duff and Johnsson (1986) illustrating this effect. Of course, the increase in parallelism comes at the cost of smaller granularity, and the most efficient balance between these opposing effects will depend on the computer architecture.

	30×30 grid	10×100 grid
No parallelism within nodes	3.7	7.5
Parallelism within nodes	30	47

Table 5.2. Illustration of effect of exploiting parallelism within tree nodes. Values given are, in each case, the maximum speed-up.

Duff (1986a) considered the implementation of multifrontal schemes on parallel computers with shared memory. The main difficulty with local memory architectures lies in mapping the elimination tree onto the architecture so that the message passing between son and father nodes can be executed efficiently. A restricted form of nested dissection has been used to generate a binary tree with good load balancing. This is discussed in more detail by Duff (1986b).

6 Conclusions

In conclusion, it is evident from our discussion that frontal methods are highly efficient on vector machines while general purpose codes do not benefit so greatly from vectorization. Multifrontal approaches can achieve some of the benefits of the single front case but, because of the much greater data manipulation, cannot be expected to yield as significant gains.

The advantages that can be obtained from parallelism are less evident, but there is much current research in this area. General purpose codes and frontal methods are not very promising, but we have indicated some of the possibilities presented by multifrontal techniques. Straightforward implementation of sparse factorization techniques on hypercubes is being examined (see for example, George *et al.* 1986), but initial experiments are not very promising.

Multifrontal techniques can be viewed on a range of levels. For example, with automatic subdivision yielding small granularity (Duff 1986a, Liu 1985), or using larger granularity by generating a simple subdivision of an underlying region followed by use of the frontal method on each region (Benner 1986, Geist private communication 1985, Berger *et al.* 1985). This latter approach seems particularly promising for several recent supercomputers that have a few vector processors and so exhibit both a vector and a low-level parallelism capability. In general, partitioning methods can be used to split a matrix into subproblems (for example, Duff *et al.* 1986) that can then be handled in parallel. A good splitting or tearing is one which produces several approximately equal subproblems with only a few variables in the tear set. To our knowledge, the parallel implementation of general tearing techniques has not been investigated in any depth. We feel, however, that schemes of this sort will prove to have good properties for parallelism.

Acknowledgements

I would like to thank John Reid and Nick Gould for their helpful remarks on a draft of this paper. This work was begun while the author was visiting Argonne National Laboratory. The partial support of the Applied Mathematical Sciences subprogram of the Office of Energy Research, U.S. Department of Energy, under Contract W–31–109–Eng–38 is gratefully acknowledged.

References

Alaghband, G. and Jordan, H. F. (1983). Parallelization of the MA28 sparse matrix package for the HEP. Report CSDG–83–3, Department of Electrical and Computer Engineering, University of Colorado, Boulder, Colorado.

Benner, R. E. (1986). Shared memory, cache, and frontwidth considerations in multifrontal algorithm development. Report SAND85–2752, Fluid and Thermal Sciences Department, Sandia National Laboratories, Albuquerque, New Mexico.

Berger, P., Dayde, M., and Fraboul, C. (1985). Experience in parallelizing numerical algorithms for MIMD architectures use of asynchronous methods. *La Recherche Aerospatiale* **5**, 325–340.

Dave, A. K. and Duff, I. S. (1986). Sparse matrix calculations on the CRAY–2. Report CSS 197, Computer Science and Systems Division, AERE Harwell. In Proceedings International Conference on Vector and Parallel Computing, Loen, Norway, June 2–6, 1986. *Parallel Computing* (To appear).

Dongarra, J. J. and Duff, I. S. (1986). Performance of vector computers for direct and indirect addressing in Fortran. Harwell Report. (To appear).

Duff, I. S. (1977). MA28 – a set of Fortran subroutines for sparse unsymmetric linear equations. AERE R8730, HMSO, London.

Duff, I. S. (1981). MA32 – A package for solving sparse unsymmetric systems using the frontal method. AERE R10079, HMSO, London.

Duff, I. S. (1983). Enhancements to the MA32 package for solving sparse unsymmetric equations. AERE R11009, HMSO, London.

Duff, I. S. (1984a). The solution of sparse linear systems on the CRAY–1. In Kowalik (1984), 293–309.

Duff, I. S. (1984b). Design features of a frontal code for solving sparse unsymmetric linear systems out-of-core. *SIAM J. Sci. Stat. Comput.* **5**, 270–280.

Duff, I. S. (1986a). Parallel implementation of multifrontal schemes. *Parallel Computing* **3**, 193–204.

Duff, I. S. (1986b). The parallel solution of sparse linear equations. In Händler, Haupt, Jeltsch, Juling, and Lange (1986), 18–24.

Duff, I. S. and Johnsson, S. L. (1986). Node orderings and concurrency in sparse problems: an experimental investigation. Proceedings International Conference on Vector and Parallel Computing, Loen, Norway, June 2–6, 1986. Harwell Report. (To appear).

Duff, I. S. and Reid, J. K. (1983). The multifrontal solution of indefinite sparse symmetric linear systems. *ACM Trans. Math. Softw.* **9**, 302–325.

Duff, I. S. and Reid, J. K. (1984). The multifrontal solution of unsymmetric sets of linear systems. *SIAM J. Sci. Stat. Comput.* **5**, 633–641.

Duff, I. S., Erisman, A. M., and Reid, J. K. (1986). *Direct methods for sparse matrices.* Oxford University Press, London.

George, A. and Liu, J. W. H. (1981). *Computer solution of large sparse positive-definite systems.* Prentice–Hall, New Jersey.

George, A. and Ng, E. (1984). Symbolic factorization for sparse Gaussian elimination with partial pivoting. CS–84–43, Department of Computer Science, University of Waterloo, Ontario, Canada.

George, A. and Ng, E. (1985). An implementation of Gaussian elimination with partial pivoting for sparse systems. *SIAM J. Sci. Stat. Comput.* **6**, 390–409.

George, A., Heath, M., Liu, J., and Ng, E. (1986). Sparse Cholesky factorization on a local-memory multiprocessor. Report CS–86–01. Department of Computer Science, York University, Ontario, Canada.

Händler, W., Haupt, D., Jeltsch, R., Juling, W., and Lange, O. (Eds.) (1986). *CONPAR 86.* Lecture Notes in Computer Science 237, Springer–Verlag, Berlin, Heidelberg, New York, and Tokyo.

Hockney, R. W. and Jesshope, C. R. (1981). *Parallel computers.* Adam Hilger Ltd., Bristol.

Hood, P. (1976). Frontal solution program for unsymmetric matrices. *Int. J. Numer. Meth. Engng.* **10**, 379–400.

Irons, B. M. (1970). A frontal solution program for finite-element analysis. *Int. J. Numer. Meth. Engng.* **2**, 5–32.

Knuth, D. E. (1973). *The art of computer programming. Second edition. Volume 1. Fundamental algorithms.* Addison–Wesley, Massachusetts, Palo Alto, and London.

Kowalik, J.S. (Ed.) (1984). *High-speed computation. NATO ASI Series. Vol. F.7.* Springer–Verlag, Berlin, Heidelberg, New York, and Tokyo.

Kung, S.–Y., Arun, K., Bhuskerio, D., and Ho, Y. (1981a). A matrix data flow language/architecture for parallel matrix operations based on computational wave concept. In Kung, Sproull, and Steele (1981b).

Kung, H., Sproull, R., and Steele, G. (Eds.) (1981b). *VLSI systems and computations.* Computer Science Press, Rockville, Maryland.

Lewis, J. G. and Simon, H. D. (1986). The impact of hardware gather/scatter on sparse Gaussian elimination. *Supercomputing Forum, Boeing Computer Services* **1** (2), 9–11.

Liu, J. W. H. (1985). Computational models and task scheduling for parallel sparse Cholesky factorization. Report CS–85–01. Department of Computer Science, York University, Ontario, Canada.

Vavasis, S. (1986). Parallel Gaussian elimination. Report CS 367A, Department of Computer Science, Stanford University, Stanford, California.

LOCAL UNIFORM MESH REFINEMENT ON
VECTOR AND PARALLEL PROCESSORS

William D. Gropp

1. Introduction

The numerical solution of two and three dimensional partial differential equations (PDEs) by various discretizations is an important and common problem which requires significant computing resources. In fact, the computational requirements of these problems are so great that the straightforward methods are too expensive in both time and memory for any computer, existing or planned. We will discuss one class of algorithms for this problem with particular emphasis on their suitability for vector and parallel computers.

There are three major ways to organize a discretization: a uniform grid, a completely non-uniform grid, and a locally uniform grid. A uniform grid is a regular mesh of points as in Figure 1a. Such grids are easy to use and efficient for vector and parallel computers. Their problem is in their size: a $256 \times 256 \times 256$ grid, which is a low resolution grid, requires 16 million elements per equation, and over 134 million floating point operations per equation per time step or iteration, even for the simplest difference approximations. The amount of memory required is possibly the most serious constraint; even a computer with a quarter of a billion bytes of memory could only manage two equations on the 3-D grid above.

The most common solution to this problem has been to use an *adaptive* grid. In such a grid, the mesh is either statically or dynamically placed where needed to compute an accurate solution. Such a grid is shown in Figure 1b. These *pointwise adaptive* grids have the advantage of using as few floating point operations and memory as possible [1]. However, there are several problems with this approach. One is that the pointwise nature of the mesh is not as suited to vector and parallel computers. Another is the additional programming difficulty of these methods.

An intermediate position is provided by locally uniform refinements. In this approach, the basic unit of refinement is larger than a single mesh point, as shown in Figure 1c. The advantage of this kind of grid over a pointwise adaptive grid is that the individual refinements are more suitable for vector and parallel computers. Further, though more

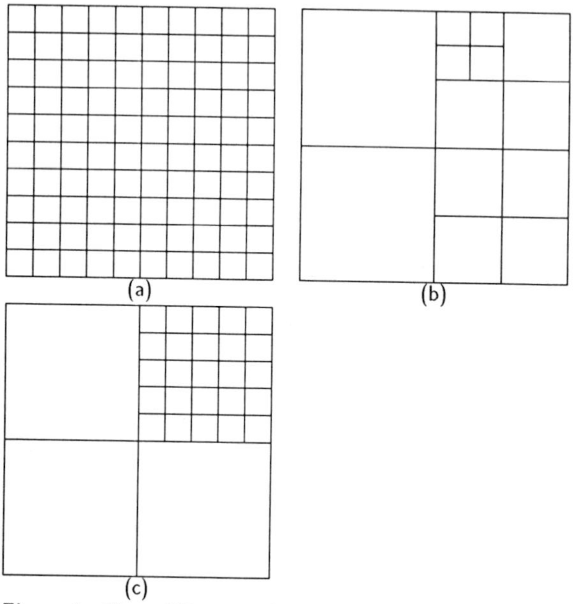

Figure 1: Three different styles of gridding. (a) is a uniform mesh, (b) a pointwise refined mesh, and (c) a LUMR mesh.

mesh points are used than for the pointwise refinement method, the actual amount of memory used may be smaller because of the lower overhead in the method. The problems with this method are the programming difficulty and the slightly less general nature of the grids.

Adaptive methods have been used to solve PDEs for some time. This work for elliptic PDEs includes MLAT of Brandt [7], the adaptive finite element methods developed by Babuska and Rheinboldt [1] and Bank [2]. More recently, many authors have contributed to the solution of time dependent PDEs. For parabolic problems, representative work is Davis and Flaherty [8], Gannon [12], and Bieterman [5]. For hyperbolic problems in one dimension, representative work is Dwyer et al. [10]. LUMR itself began in a paper of Oliger [17] and has been continued by his students in Berger [3, 4], Bolstadt [6], and Gropp [13, 14, 15].

In this paper we will describe in detail LUMR and its application on vector and parallel computers. We will concentrate on the effect a parallel computer has on the development of an algorithm.

2. Serial LUMR algorithm

We briefly describe the algorithm for LUMR in two space dimensions on a single processor; a more complete discussion of some variants of LUMR may be found in [3, 6].

We will also discuss only a single level of refinement, as multiple levels are handled in a natural, recursive fashion.

The coarse grid is made up of a union of rectangles; each rectangle has a discrete, computationally uniform grid. Every R_g (R_g for ReGridding) steps on the coarse grid, a new refinement also consisting of a union of rectangles is chosen. These new grids are refined by a factor R, that is, the space and time step sizes on the refined grids are reduced by a factor of R from those on the coarse grid. Each grid in the refinement is *overlayed* on top of the coarse grid, rather than inserted into the coarse grid, in order to preserve the uniformity of the coarse grid. While this simplifies the computation on both the fine and coarse grids, it does introduce a number of minor but important complications. First, some points in the domain have both a coarse grid point and a fine grid point. This means that we must inject the values computed on the fine grid into the coarse grid wherever a coarse grid point lies inside the refinement. Second, since the fine grid is not patched into the coarse grid, we must do something special at coarse-fine and fine-fine grid boundaries. At coarse-fine grid boundaries, we can use interpolation from values computed on the coarse grid. At fine-fine grid boundaries, we can take the values needed from the appropriate fine grid. Both of these steps are necessary in order to insure an accurate solution. The first is required to prevent the inaccurate solution computed on the coarse grid in the region where refinement is done from affecting the solution away from the refinement. The second is necessary to preserve the accuracy on the fine grid (if the values were taken from the coarse grid, the accuracy would be reduced to that of the original coarse grid computation). Thus, the fine and coarse grids must communicate with each other frequently. The algorithm is shown in Algorithm 1.

The main reason for LUMR is to make all grids computationally uniform. By maintaining computational uniformity, we can make better use of fast vector computers.

As a justification for this choice, consider a simple model of a vector computer. Each operation on n mesh points takes $s + an$ time, where s is the start-up time and a is the time per operation. On a regular $n \times n$ grid, the time required is $s + an^2$. Using pointwise refinements, only one point can be done at a time, and thus the time is $m(s+a)$, where m is the number of mesh points in the refined grid. The amount of work in the two methods is equal if $m \approx n^2a/(s+a)$, and the pointwise refinement is actually more expensive when $m > n^2a/(s+a)$. Since a is typically significantly smaller than s ($s \approx 6a$ for CRAY-1 and $s \approx 100a$ for Cyber 205), m must be significantly smaller than n^2 for the pointwise refinement to be cost effective.

This model is very crude and does not take into consideration the effect of scatter-gather hardware or other methods for cacheing the operands for the pointwise refinement

$i \leftarrow 0.$

Repeat until done:

1. *Integrate the coarse grid*
2. *Regrid if* $i \bmod R_g = 0$
   - 2.1. Decide where to place the refinements.
   - 2.2. Initialize the new fine grids with the data in either the old fine grids or in the coarse (parent) grid. (Must use the old fine grids if possible.)
3. *Integrate*
   - 3.1. For R steps: Integrate each fine grid by integrating
     - 3.1.1. Interior of fine grid
     - 3.1.2. Boundary of fine grid. Fine grids may need to communicate with each other (at overlaps between fine grids) and with the coarse grid (where there is a coarse-fine grid boundary).
   - 3.2. Update the coarse grid with the new solution on the fine grid.

     $i \leftarrow i + 1$

Algorithm 1: Algorithm for serial LUMR.

method. It also ignores the amount of memory traffic, which can have a significant effect on the computational efficiency of the method. Still, it shows that simple operation counts can be misleading.

3. Why are lower order terms important?

Much of the rest of this paper will be concerned with the costs of communication in a parallel processor. Such costs are often asymptotically dominated by the computation costs, even on a parallel processor. In fact, a frequent comment about the analysis of the computational complexity of parallel algorithms is that lower order communication can be neglected simply by making the problem large enough. While this is true asymptotically, the same argument would say that the fastest way to multiply two matrices is with the most sophisticated algorithm ($\mathcal{O}(n^{2.496})$) instead of the obvious algorithm ($\mathcal{O}(n^3)$). The problem is of course one of constants.

As an example, lets consider a simple 3-d PDE problem with no mesh refinement. The mesh is an $n \times n \times n$ cube and there are p processors. If we divide the domain into cubes of size $n/p^{1/3} \times n/p^{1/3} \times n/p^{1/3}$ (to minimize the ratio of communication to computation), the complexity looks like

$$\alpha\frac{n^3}{p} + 6\left(\gamma + \beta\frac{n^2}{p^{2/3}}\right) \tag{3.1}$$

where

$$\alpha = \text{time per node in computation}$$

$$\beta = \text{time per node in communication}$$

$$\gamma = \text{start up time per cube side}$$

$$p = \text{number of nodes.}$$

The $p^{2/3}$ comes from the fact that each cube has side $n/p^{1/3}$. If instead we divide the domain into $n \times n \times n/p$ slabs, the complexity looks like

$$\alpha\frac{n^3}{p} + 2\left(\gamma + \beta n^2\right) \tag{3.2}$$

Note that these have the same asymptotic complexity in n (both are $\mathcal{O}(n^3)$) and the same asymptotic complexity in p (both are $\mathcal{O}(1)$). We will call the term $\alpha n^3/p$ the *computation term* and everything else the *communication term*.

Each of these equations has some interesting things to say. In Equation (3.1), note that if p/n^3 is a constant (constant number of mesh points per processor), the relative size of the computational term $\alpha n^3/p$ and the communication term $6(\gamma + \beta n^2/p^{2/3})$ is also constant. This case could arise in a massively parallel computer with all but the largest problems, since each processor must have at least one mesh point. In Equation (3.2), note that for p/n^3 constant, the communication term eventually dominates computation.

For Equation (3.1), if p is fixed, then as n increases, the communication term will indeed become less important. However, it may become less important only for very large n. Consider the following 10 Gigaflop computer: 1000 processors, each of which is 10 Megaflops. Let the start up time $\gamma = 100\mu secs$, the transfer time $\beta = 10\mu sec$, and the computation time $\alpha = 6\mu secs$ (60 floating point operations/per mesh point). Then the total time for a single step or iteration is

$$.06n^3 + 6(100 + .1n^2)\mu\text{seconds.}$$

For example, the communication time is 10% of the computation time for:

$$.06n^3 = 10 \times 6(100 + .1n^2),$$

or $n = 100$. At an n of 100, there are one million mesh points. This is a large calculation, each step taking about .07 seconds when using 1000 processors.

For Equation (3.2), the situation is far worse. Given the same 10 Gigaflop computer, the total time for a single step or iteration is

$$.06n^3 + 2(100 + 10n^2)\mu\text{seconds}.$$

In this case, the communication time is 10% of computation at $n > 3333$, and the two times are equal for $n \approx 333$ The time for a single step with these n's are 40.7 *minutes* and 4.4 seconds respectively. An adaptive calculation will increase the amount of communication by both increasing the startup time (more complicated messages and routing) and the transfer time (messages have to go farther). Also, this assumes that all processors may be communicating simultaneously. Both of these will increase the amount of communication time relative to the computation time.

In particular, for $n = 256$, which is the example we began this paper with, the slab approach (Equation (3.2)) takes 1.4 seconds per step, while the cube approach (Equation (3.1)) takes only 0.14 seconds, a factor of 10 faster. However, note also that for small p and $\gamma \gg \beta$, slabs may be faster than cubes because of the fewer IO starts.

4. Communication in LUMR

Communication in LUMR takes on three major aspects: within a single grid, between a coarse grid and its fine grid child, and between two grids at the same level. Each of these has an influence on the way LUMR is distributed across a parallel processor.

Communication within a single grid is of the form "get the mesh points adjacent to the current point and use them to advance the solution" (cf. Figure 2a). In terms of memory accesses, there are several memory access per mesh point. Thus, there are roughly as many memory accesses as there are floating point operations. Since both interprocessor communication (on a parallel machine) and random memory accesses (on a vector machine) are expensive, it makes sense to store each grid on a single processor and in a regular storage pattern.

Communication between a coarse grid and its fine child is also on the order of the number of mesh points (cf. Figure 2b). For example, if the amount of refinement is R, then a child with N mesh points must send N/R^d data to the coarse grid, where d is the dimensionality of the problem. This also argues for keeping the coarse grid and its children on the same processor so as to minimize the amount of expensive interprocessor communication. We will compare this to the other obvious choice, that of giving each processor a different grid, in Section 6.

Finally, there is communication between grids at the same level (cf. Figure 2c). Here, if each grid has N^d mesh points, the amount of communication is only $\approx N^{d-1}$. Since

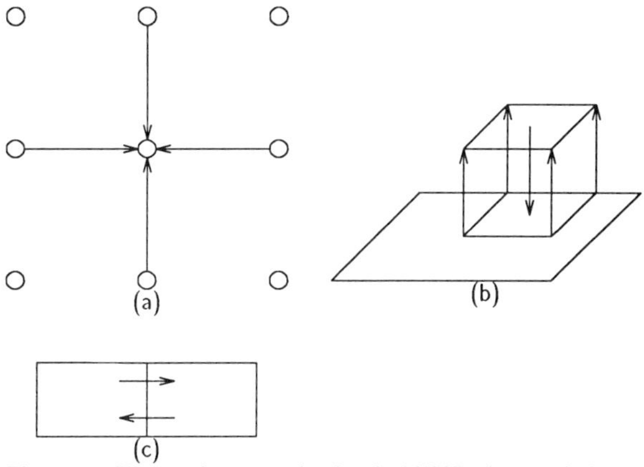

Figure 2: Types of communication in LUMR. Arrows indicate the direction of communication. (a) is within a grid for a 5-point difference scheme, (b) is between fine (smaller) and coarse (larger) grids, and (c) is between grids on the same level.

this is the smallest amount of communication per mesh point, it makes sense to distribute the problem so that only the coarse–coarse grid communication uses the inter-processor communication. In other words, the physical domain of the problem is divided into pieces, and each of those pieces is given to a different processor. Each processor then handles any refinements which may occur on that piece. This approach will reduce the inter-processor communication over using separate processors for each grid.

5. Using parallel computers

Now we discuss some of the issues in using vector and parallel computers for solving PDEs. First, we will set the background by dividing parallel computers into two broad categories. We then discuss some of the problems in using both types of parallel computers, and discuss the impact they have on the structure of efficient algorithms.

5.1. Kinds of parallelism

There are two basic kinds of parallelism—tightly coupled and loosely coupled. A tightly coupled computer has some combination of multiple functional units and shared memory and usually a single program counter. There are many examples of tightly coupled machines; some are Cray-1, CDC Cyper 205, and FPS-264. Such machines are effectively programed even with existing Fortran compilers.

A loosely coupled computer is a collection of individual processors, complete with their own memory and program counter, which are connected by some sort of communication network. An example is the Intel Hypercube.

Naturally, these concepts are fairly vague, and there are machines which fit into neither category. Of more interest to us are machines which fit into *both* categories. Such a machine would be a loosely coupled computer, each of whose processors is a tightly coupled computer. Machines such as the Cray-XMP and Cray-2 are examples.

It is the author's claim that future supercomputers will take this form, a loosely coupled collection of tightly coupled processors. Let's look at the alternatives.

A tightly coupled machine quickly runs into several limitations imposed by physical law. First is the speed of light. A machine with a single clock and a 1 nanosecond cycle time can be no more than roughly $\frac{1}{3}$ meters wide. At such a size, heat dissipation and packing density become severeproblems. In fact, it has been argued that the most important technical contribution of the Cray-1 and Cray-2 is in the cooling systems, not the electronics! A further problem is fault tolerance, as the complexity of the computer increases, there is an increasing likelihood that a critical component will fail, shutting the whole machine down. All of these factors together probably limit the fastest tightly coupled machine to no more than 10–100 Gigaflops (a 100 Gigaflop machine would be less than a centimeter across).

A loosely coupled machine runs into Grosch's law [11]: processor speed within a single technology goes as the square root of the price. For example, by going from a single chip scalar processor to a single chip vector processor will yield significantly greater performance at a smaller increment in price. In today's market, given the choice between N MC68020 processors and $N/2$ MC68020s, each with a 10 Megaflop pipelined floating point chip (and thus a tightly coupled machine with two functional units), the most cost effective per Megaflop is the latter.

5.2. Making use of parallel computers

There are a number of problems in making efficient use of parallel computers. We will discuss them here and indicate how algorithms may be designed to reduce these problems.

In a tightly coupled computer, the major problem is that the amount of parallelism in a general algorithm is limited. For example, in [16], given an infinite number of resources, only roughly 90-fold parallelism could be discovered in some typical Fortran programs from numerical analysis. Offsetting this problem is the relative ease with which tightly coupled computers may be efficiently programmed.

In contrast, a loosely coupled processor is much harder to program, and it is not clear how much parallelism is available. Offseting this problem is the multiple control paths which give a program on a loosely coupled computer more flexibility to exploit what parallelism is present.

To efficiently use a tightly coupled processor, there must be a regular memory access pattern. Specifically, access to memory and/or vector registers must be very regular. This usually means clean "inner loops" (no or only restricted conditionals, no indirect addressing). And the principle bottlenecks in these machines is often the memory bandwidth; e.g. [9]. On many machines, the time to access a single, random element is significantly longer than the time to access consecutive elements in memory. For example, if it takes 5 times as long to access a single element as it does an element in a vector, then a pointwise refinement using m points must have $m < 5 \times 2n^2$ on an $n \times n$ mesh, i.e., less than 1/10 as many points as on a uniform mesh. The 2 comes from first chasing the pointer, then accessing the value. The more recent vector supercomputers come with scatter/gather hardware which is intended to reduce but not eliminate this penalty in random accesses to memory. These requirements are a major reason why adaptive algorithms are not more wide spread; pointwise refinement does not make efficient use of vector processors.

To efficiently use a loosely couple processor, the larger parts of the algorithm must be distributed among the processors. This usually means distributing the "outer loops". In turn, this requires writing the algorithm so that the outer loops are "clean", that is, they don't involve many references to data on another processor. Further, each processor is to some degree isolated from the others. Thus, we must worry about interprocessor communication.

There are physical constraints on what kind of communication is possible. For example, every processor may be connected to every other processor. A direct connection is out of the question for many processors, since the number of connections grows as the square of the number of processors. Thus, a switching network of some kind is often used. However, this imposes a time delay proportional to the log of the number of processors. (This delay can be hidden by making the network sufficiently faster than the processors, or, in other words, using slower processors to hide the delay in the network.) Another approach is to connect each processor only to certain neighbors. In this case, communication to a processor which is not a neighbor is expensive or impossible. And if the communication system is being used to access the data at a mesh point, all of the arguments about random memory access in a vector computer apply, with the cost of a non-local reference often being far more expensive than a local reference (local to the processor making the reference).

5.3. Load balancing

Load balancing is a catch-all term for any method used to keep all or most of the processors in a parallel computer busy, and is one of the keys to efficient use of parallel computers. There are two basic methods for accomplishing load balancing on p processors.

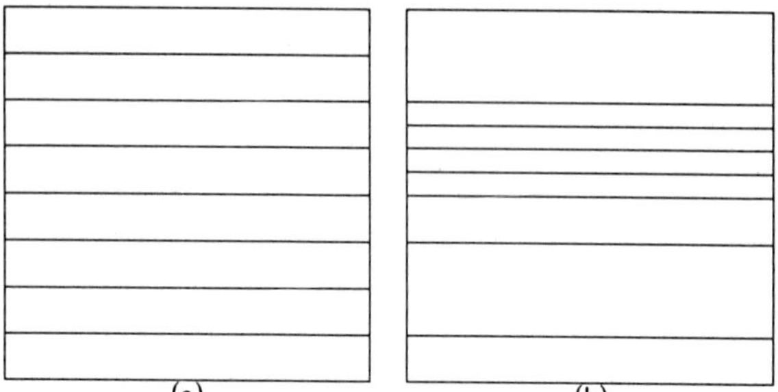

Figure 3: An example of load balancing. The division of the computational domain goes from the equal distribution in (a) to the unequal (but equal work) distribution in (b).

The first is to divide the work in to exactly p pieces, where each piece will take roughly the same amount of time. For a uniform grid, this is both the easiest and the most efficient because there is no overhead associated with the load balancing. For any adaptive computation, it becomes more difficult. For example, an $n \times n$ domain could be divided into p strips of size $n/p \times n$. As the computation progresses, the width of these strips could be adjusted to reflect the amount of work on each strip, including any refinements which may be present (c.f. Figure 3). The disadvantage with this method is that the interprocessor communication is now $2(\gamma + \beta n)$, since the boundary of the strip must be sent to processors on either side. Note that the communication does *not* depend on the number of processors, and hence may dominate for large numbers of processors. This is a 2-d version of the case discussed in Section 3, Equation (3.2).

The second approach to load balancing is to break the work into many more pieces than there are processors, and distribute the work evenly between processors. We will call a "piece of work" a *quantum*. An example quanta might be a $n/p \times n/p$ square (in 2-D) from the domain; there would then be $p^2 \gg p$ quanta. The difficulties with this approach are two-fold. First, each quantum must be *scheduled* or given to a processor. This will amount to an additional overhead. Second, it is impossible to maintain data locality as well as perfect load balancing; some quanta will be far from their neighbors. An additional problem is that since there are many more than p quanta, each quanta will have a smaller vector length than if there are only p pieces, and hence the "inner loops" will be smaller.

6. Parallel LUMR

In this section we describe an implementation of LUMR and the reasons for the various design choices. In the next section, we discuss some experiments performed with this code.

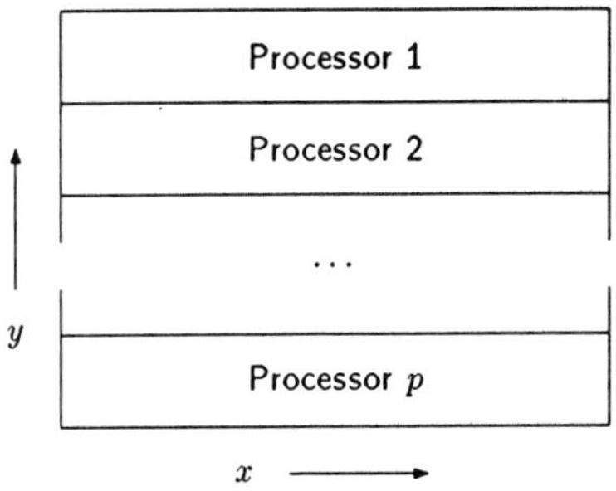

Figure 4: Partitioning of domain for a linear array of p processors.

To make this more concrete, we will consider only a loosely coupled parallel processor. We will be primarily concerned with a *bus* machine and a linear array. The bus machine consists of p processors, all communicating over the same IO bus. The linear array consists of p processors, connected in a line, with every processor connected to its left and right neighbors. The principle advantage of the linear array over the bus machine is that in the bus, all processors must share a fixed size resource (bytes/second on the bus). In the linear array, as each processor is added to the array, the rate at which data may be transfered between neighboring processors does not change.

The potential bottlenecks in a parallel version of Algorithm 1 lie in the steps which need to communicate between processors. These are the regridding step 2, the handling of the boundaries 3.1.2, and the update step 3.2. We can reduce each of these bottlenecks by taking advantage of the local or nearly local nature of solutions to PDEs and the local and uniform nature of the refinements to reduce the amount of interprocessor communication.

Note that the feedback between the grids means that we can't do the fine and coarse grids separately. This suggests that, for explicit methods at least, we partition the algorithm among many processors by keeping the fine grids as close to the coarse (parent) grids as possible, with closeness determined by the communication network.

These considerations lead to the following parallel LUMR algorithm. For each rectangle in the coarse grid, divide the grid into p (nearly) disjoint regions. These regions correspond to separate regions in the physical domain. Assign each region to a separate processor. The exact choice of division will depend on the architecture available. The rule to follow is that adjacent regions should be able to communicate quickly with each other. For the linear array and a 2-D problem, this suggests strips (see Figure 4); for a bus a different partitioning may be better (see Section 7). In a 3-D problem, slabs would be used instead of strips. We will concentrate on the strip partitioning.

$i \leftarrow 0.$

Repeat until done:

1. Integrate the coarse grid
2. *Regrid* if $i \bmod R_g = 0$
    2.1. Decide where to place the refinements. Each processor computes the data structure on its region; in doing so, it must receive from its neighbors the data structure for the fine grids at the region boundary.
    2.2. Initialize the new fine grids with the data in either the old fine grids or in the coarse (parent) grid. Must use the old fine grids if possible.
3. *Integrate*
    3.1. For R steps: Integrate each fine grid by integrating
        3.1.1. Interior of fine grid. Each processor does its own region.
        3.1.2. Boundary of fine grid. Each processor does its own region. In addition, it must get the fine/fine boundaries from the adjacent processors.
    3.2. Update the coarse grid with the new solution on the fine grid.
    $i \leftarrow i + 1$

Algorithm 2: Algorithm for Parallel LUMR.

Each processor is responsible for its assigned region in the domain. That is, a processor integrates the solution on its domain, and creates and integrates the solution on any refined grids it finds that it needs in its domain. The refined grids are *not* sent to other processors; by keeping them local to the processor which their parent grid lives on, we can significantly reduce data motion. Instead, each processor shares information along the overlap in the domain with it neighboring processor. Thus, the amount of data which must be moved is roughly the square root of the total data, or equivalently, the data to be moved is "one-dimensional" while the data being computed is "two-dimensional". This partitioning may change during the course of the solution process in order to distribute the load among the p processors. The algorithm is shown in Algorithm 2.

7. Estimate of performance

In this section we estimate the performance of parallel LUMR for two different divisions of the domain. More details of this analysis may be found in [15]. The speedup s is given by T_s/T_p, where T_s is the time for serial LUMR and T_p is the time for parallel

LUMR. We define $\beta(p)$ as proportional to the ratio of floating point speed and communication speed of the processors. β may depend on the number of processors communicating since there may be contention for communications resources. For slabs and either a bus machine or a linear array, it can be shown that

$$\frac{1}{s} = \frac{1}{p} + \beta(p).$$

Here we have ignored any communication startup times; except for large numbers of processors they don't have a significant impact on the results. From this, we can compute the speedup and the maximum speedup for both parallel architectures:

$\beta(p) = \beta$ (linear array)	$\beta(p) = \beta p$ (bus)
$\frac{1}{s} = \frac{1}{p} + \beta$	$\frac{1}{s} = \frac{1}{p} + \beta p$
$s = \frac{p}{1 + \beta p}$	$s = \frac{p}{1 + \beta p^2}$
$s_{\mathrm{max}} = \frac{1}{\beta}$	$s_{\mathrm{max}} = \frac{1}{2\sqrt{\beta}}$

Note that β must be less than 1 for the parallel algorithm to break even, thus the square root makes the bus architecture far less effective than the linear array architecture.

In these cases, the major bottleneck is in step 3.1.2, where the neighboring fine grid values must be sent to adjacent processors. By chosing a different partitioning, we can reduce this, though at a cost in added complexity for the communication links. For example, if we had a 2-D mesh connected array, we could cut up the domain into squares of side $1/\sqrt{p}$. This would essentially divide the term in step 3.1.2 by $\sqrt{p}$, which will improve these estimates. For example, for a mesh-connected (rather than linear) array, the speedup $s = p/(1 + \beta\sqrt{p})$, which is unbounded. However, the efficiency of processor usage is roughly $1/\beta\sqrt{p}$, and thus goes to 0 as the number of processors goes to infinity. For the bus architecture, the speedup would still be bounded, with $s = p/(1 + \beta\sqrt{p^3})$ and a maximum speed of $s = \frac{1}{3}\sqrt[2/3]{2/\beta}$. This speedup comes from reducing the absolute amount of data to be communicated by reducing the ratio of circumference to area for each of the domains to the minimum possible. Figures 5 and 6 show the speedup s as a function of β and p.

8. Experimental results

To test the above algorithm, we ran a parallel mesh refinement code on a 14 node Apollo ring. The ring consisted of 10 DN300s and 4 DN420s. The Apollo DN300 is a 68010 based workstation with no hardware floating point. The Apollo DN420 is a 68000 based workstation with hardware floating point and a hard disk. The Apollos are connected by a token passing network; at the user level, processes on different processors

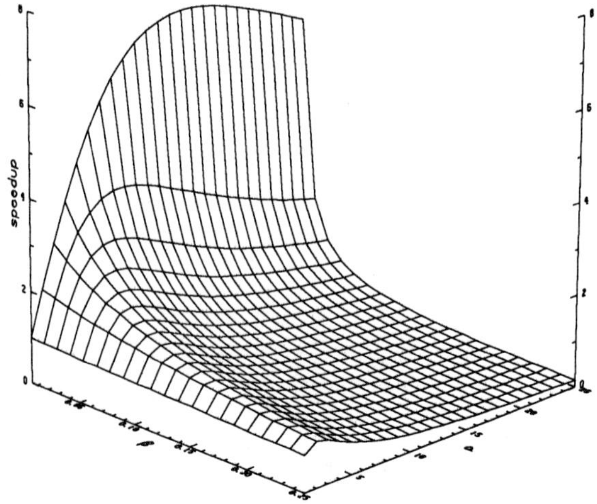

Figure 5: Speedup s as a function of β and p for a bus architecture. The minimum value of β is 0.005

communicate through mailboxes which live on a disk. The DN300s were used for the actual computations, with the DN420s being used to hold the actual mailboxes, monitor the computation, and act as partner to the DN300s. There is a fixed I/O bandwidth which is independent of the number of processors, so this configuration is similar to a bus architecture and will be compared with our results for a bus. More detailed results will be found in [15].

The sample problem is a variant of the revolving cone problem used in [3]. The region of refinement is concentrated around the cone; hence no static partitioning will make effective use of the available processors. In our test, we use Lax-Wendroff as the difference approximation, with inflow boundaries specified as $u = 0$ and outflow boundaries specified with first order extrapolation.

The processors were divided up as equal sized strips in y; for the two processor case, the initial domains were $-1 \leq y \leq h$ and $0 \leq y \leq 1$, where h is the space step size. The values of β in the above analysis was determined experimentally as $\beta = 0.0067p$.

Three sets of runs were made. The first run used no refinement and is a simple test of the parallel integration algorithm. The second test used fixed partitions. The third

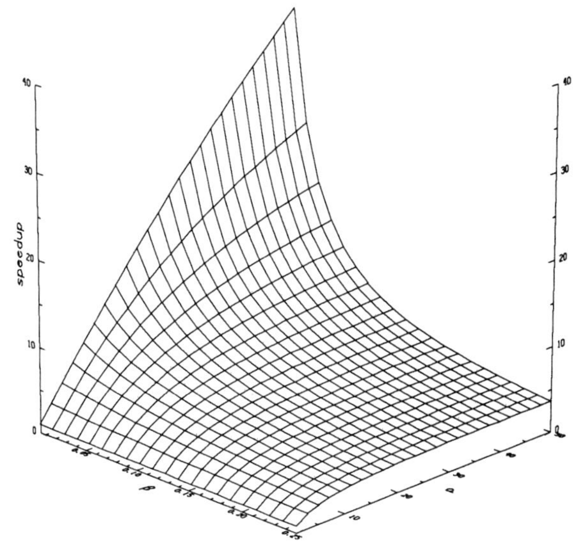

Figure 6: Speedup s as a function of β and p for a linear array architecture. The minimum value of β is 0.005.

test used dynamic load balancing to modify the partitioning in an attempt to more equally distribute the load. The results are presented graphically in Figure 7.

In Figure 8, we compare our experimental results with our predictions. We see that the estimated elapsed time is greater than we observed, but only by about 20%. The discrepancy is probably due to a combination of lack of load balancing and to items we have neglected, such as the computational time in handling the refinements (regriddings, updatings, data structure transfers). They clearly show that 10 Apollo DN300s are about all that could be used efficiently with this algorithm, because the interprocessor communication mechanism is too slow, and that our theory is an accurate model of the algorithm.

9. Conclusion

We have shown that a powerful adaptive technique can indeed be implemented on loosely coupled parallel processors, and that careful analysis, keeping highest order terms in computation and communication, can accurately predict the performance of such algorithms. Key to this however was designing the algorithm with the requirements of a parallel and vector computer in mind. Simply implementing some existing adaptive algorithm would probably have incurred enough extra interprocessor communication to make

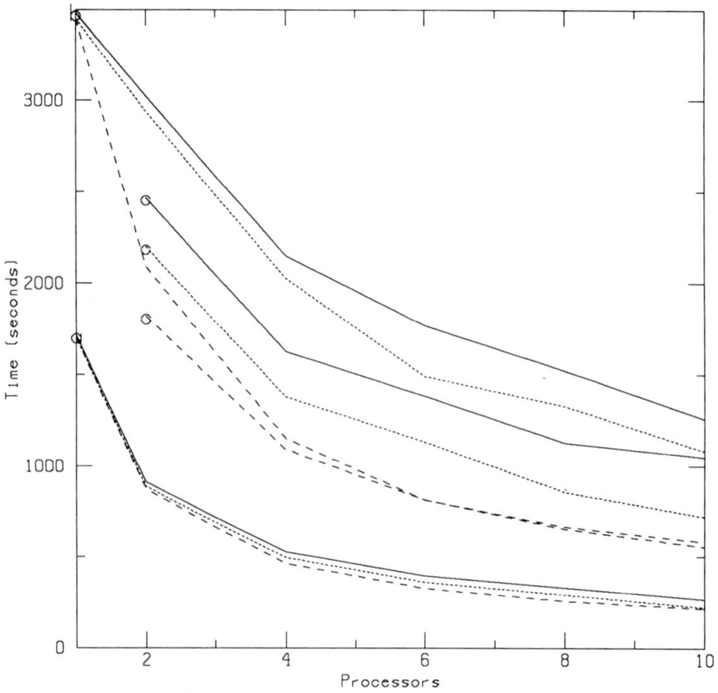

Figure 7: Timing results for experiments. Curves (a) are for no refinement, (b) refinement with no load balancing, and (c) refinement with dynamic load balancing. The solid lines are the total elapsed times, the dotted lines are the maximum cpu times and the dashed lines are the average cpu times.

it more expensive than a uniform mesh calculation. We have also tried to suggest how an analysis of the communication costs of different algorithms can be used to avoid bottlenecks in the program. In LUMR, a simple analysis was used to decide to partition the work by domain rather than by grid.

In addition, we showed one way in which load balancing may be done, and demonstrated its effectiveness in Figure 7. Other approaches to load balancing are necessary with different divisions of the problem domain. For example, if the domain is divided into cubes, then it is not possible to vary the sizes of the cubes without changing the communication structure, since when a cube becomes larger, it may share its boundary with more cubes than it originally did. Load balancing in this case will be a compromise between increased communication and increased processor utilization.

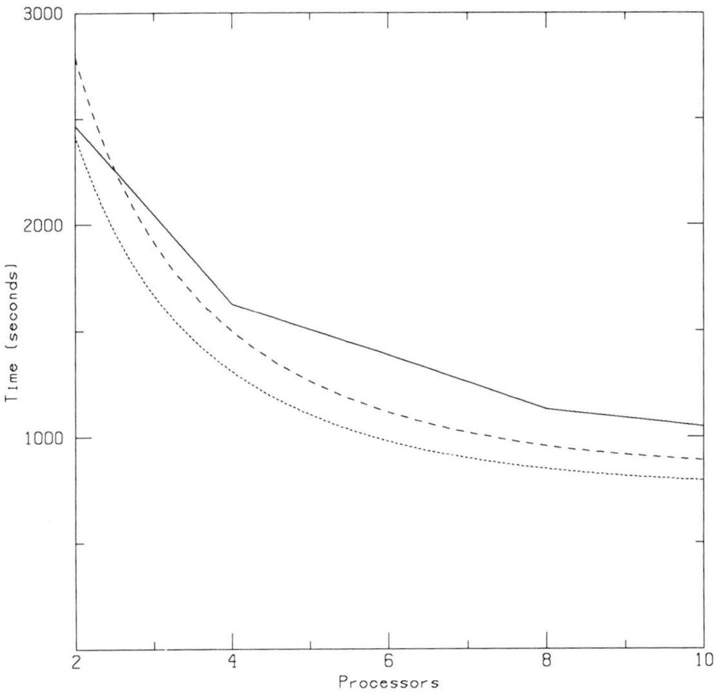

Figure 8: Comparison of predicted elapsed times with the observed times with load balancing. The solid line is the observed time. With f the fraction of the region which has been refined, the dashed line is the predicted time with $f = 0.25$, and the dotted line is the predicted time with $f = 0.20$.

This work was supported in part by Office of Naval Research Contract #N00014-82-K-0184, National Science Foundation Grant MCS-8106181, and Air Force Office of Scientific Research Contract AFOSR-84-0360.

References

[1] Ivo Babuska and Werner Rheinboldt, *A Posteriori error analysis of finite element solutions for one-dimensional problems,* SIAM Journal on Numerical Analysis, 18/3 June (1981), pp. 565–589.

[2] R. Bank, A Multi-Level Iterative Method for Nonlinear Elliptic Equations, *Elliptic Problem Solvers,* Academic Press, 1981, pp. 1–16.

[3] Marsha J. Berger and Joseph Oliger, *Adaptive mesh refinement for hyperbolic partial differential equations,* Technical Report ManuscriptNA-83-02, Stanford University, March 1983.

[4] Marsha J. Berger and Antony Jameson, *Automatic adaptive grid refinement for the Euler equations,* Technical Report DOE/ER/03077-202, Courant Mathematics and Computing Laboratory, New York University, October 1983.

[5] M. Bieterman and I. Babuska, *The finite element method for Parabolic Problems I,* Numerische Mathematik, 40 (1982), pp. 339–371.

[6] John H. Bolstad, *An adaptive finite difference method for hyperbolic systems in one space dimension,* Technical Report LBL–13287-rev, Lawrence Berkeley Laboratory, December 1982.

[7] A. Brandt, *Multi-Level Adaptive Solutions to Boundary Value Problems,* Mathematics of Computation, 31 (1977), pp. 333–390.

[8] S. Davis and J. Flaherty, *An Adaptive Finite Element Method for Initial-Boundary Value Problems for Partial Differential Equations,* SIAM Journal on Scientific and Statistical Computing, 3 (1982), pp. 6–27.

[9] Jack Dongarra and Stanley Eisenstat, *Squeezing the Most out of an Algorithm in CRAY FORTRAN,* ACM Transactions on Mathematical Software, 10/3 (1984), pp. 219–230.

[10] H. Dwyer, R. Kee, and B. Sanders, *Adaptive Grid Method for Problems in Fluid Mechanics and Heat Transfer,* AIAA Journal, 18 (1980), pp. 1205–1212.

[11] Phillip Ein-Dor, *Grosch's law re-revisited: CPU power and the cost of Computation,* Communications of the ACM, 28/2 February (1985), pp. 142–151.

[12] Dennis Gannon, *Self Adaptive Methods for Parabolic Partial Differential Equations,* Technical Report UIUCDCS-R-80-1020, Univ. of Illinois, 1980.

[13] William Gropp, *A test of mesh refinement for 2-d scalar hyperbolic problems,* SIAM Journal on Scientific and Statistical Computing, 1/2 June (1980), pp. 191–197.

[14] ———, *Local uniform mesh refinement for elliptic partial differential equations,* Technical Report YALE/DCS/RR-278, Yale University, Department of Computer Science, July 1983.

[15] ———, *Local uniform mesh refinement on loosely-coupled parallel processors,* Technical Report YALE/DCS/RR-352, Yale University, Department of Computer Science, December 1984.

[16] Alexandru Nicolau and Joseph Fisher. Using an oracle to measure parallelism in single instruction stream programs. *The 14th Annual Microprogramming Workshop,* ACM and IEEE Computer Society, 1981, pp. 171–182.

[17] J. Oliger. Approximate Methods for Atmospheric and Oceanographic Circulation Problems. *Lecture Notes in Physics, 91,* Springer-Verlag, 1979, pp. 171–184.

USING SUPERCOMPUTERS TO MODEL HEAT TRANSFER IN BIOMEDICAL APPLICATIONS

Linda J. Hayes

Abstract

This paper describes several approaches to converting numerical algorithms for use on supercomputers. The machine architecture which will be considered here is a vector computer. These approaches will be illustrated by considering two algorithms which arise in modelling heat transfer in biomedical applications. One algorithm is a vectorized matrix-vector multiply, and the other is a new, vectorized, overlapping-block iterative method for solving finite element equations.

1. Introduction

The motivating application of this work is the numerical solution of heat transfer problems in the biomedical field which arise in several situations. Human tissues are frozen for preservation and transplantation, and they are cooled in a surgical procedure for infants known as whole body cooling [4,14,15]. Human tissues are heated to enhance radiation and chemotherapy in certain cancer patients, and they are also heated in applications of lasers to human tissues. The majority of biomedical applications are characterized by having very irregular geometry and having a transient physical phenomenon which takes place in three spatial dimensions. If phase change occurs, then there is a nonlinear dependence of both the release of latent heat and the material properties upon the local temperature field. These aspects make accurate modelling of heat transfer in biomedical applications very difficult. The finite element method will be considered because of the very irregular geometry and nonhomogeneous thermal properties of the tissues being considered.

If one were to accurately model the human body or any complete organ in three dimensions, the number of elements or data points which would be required would far exceed the capacity of traditional, standard computers. In recent years, supercomputers have become widely available. The machines which are currently on the market include the CYBER 205, the CRAY XMP, the CRAY 2 and in the near future the ETA 10. These machines offer several orders of magnitude speed-up in computation times and a much larger storage capacity as compared to standard machines [9,10,11,19]. However, to gain the maximum benefit from these machines, one must be aware of the machines' architecture and must optimize computer codes for a particular machine. This paper discusses several approaches to vectorization and illustrates these approaches on two numerical algorithms which arise from solving the heat transfer equation for biomedical applications.

2. Biomedical Applications

Heat transfer equations from three specific applications will be considered. For simplicity, the grids considered here are two-dimensional and were taken through cross sections of the system of interest. This leads to relatively small systems of equations, but all of the conclusions readily extend to the full three-dimensional modelling. The first case which will be considered is modelling of a human heart during cryopreservation. At this time, this is not a clinically feasible procedure. However, this procedure was simulated to investigate the thermal gradients and resulting thermal stresses which would arise in the organ during such a procedure. In practice, cooling rates have a significant influence on the local cell survivability during cryopreservation, and one attempts to attain fairly uniform cooling rates throughout an organ. Figure 1 shows a cross section of the human heart including the muscle, the two chambers and the corresponding finite element grid. This is an extremely coarse grid which does resolve the tissue geometry.

The second application which will be discussed is the development of temperature fields in the human body when one is exposed to a cold environment. The long term application of this work is to design suits for deep space and for deep sea exploration which will protect a human from the cold environment and will not over-heat the human during periods of exercise. Figure 2 shows a cross section through the human arm which resolves the distinct muscle groups as well as the radius and ulna. Figure 2 also shows the corresponding finite element grid which again is very coarse but does resolve the internal geometry of the arm. Physical exercise is simulated by incorporating heat production in distinct muscle groups in the model.

The last application which will be considered arises from a medical procedure called whole body cooling which is used for premature infants. Some of these infants are so small that they cannot be attached to the life support systems for required surgery. Their entire body can be cooled with ice packs at the surface of the head and with a blood exchanger to cool their blood to 12-14°C. Blood flow stops and the heart ceases to beat. At this point, the physician has 30-60 minutes to operate in a blood free environment. After that time, the infants are rewarmed and the heart spontaneously begins beating. However, there is a significant number of these children who sustain brain damage. In an attempt to analyze the transient thermal fields in the brain, an experimental procedure is being done at the UT Health Science Center in Dallas by which macaque monkeys are instrumented and cooled using this same protocol [15]. The instrumentation on the monkey records temperatures near the surface of the skull. It is the feeling of the medical experts that brain damage occurs because of uneven heating and cooling at the base of the brain. For this reason, the finite element model shown in Figure 3 was developed which resolves the internal geometry of the monkey's head, and surface temperatures were correlated to experimental data.

These applications are very typical of biomedical applications. Each grid involves several different materials, each with unique thermal properties which may be temperature dependent. In vivo there may be heat production due to blood perfusion. Heat transfer in biomedical applications is governed by a simple transient diffusion equation in the body of interest

$$\rho C_p \frac{\partial T}{\partial t} - \nabla \cdot k \nabla T + \rho_b C_b \, \omega (T - T_a) = f(\underset{\sim}{x}, t) \qquad (1)$$

where T is temperature, t is time, $\underset{\sim}{x}$ is the spacial coordinate, ρ , ρ_b and C_p , C_b are densities and heat capacitances of the tissue and the blood, ω is a perfusion rate, T_a is a base arterial temperature, and $f(\underset{\sim}{x},t)$ is internal heat generation due to metabolism or an applied internal source such as occurs in laser or radio frequency heating of tissue. The initial temperature field is specified, and temperatures, fluxes or a convective resistance can be prescribed at the surface as a function of time [2]. Continuity of temperature and heat fluxes are imposed throughout the body.

The finite element method satisfies the governing equations in a weak or averaged sense over the elements [1]. The finite element modeling of these equations results in very large sparse matrices. The irregular grid structure gives rise to a very irregular sparsity structure in the matrix.

3. Vectorization

There are four basic approaches to converting a computer code from a conventional machine to a supercomputer. The first is simply to run the code as it is and hope that the large number-crunching ability of the supercomputer will give you a speed-up in computation time. In most cases, the computer code will in fact run faster on the supercomputer; however, experience has shown that one does not attain anywhere near the potential of the supercomputer if this approach is used. The second approach is to reorganize the computer code to facilitate vectorization. This might include a rearrangement of DO LOOPS to induce vectorization on longer vectors and to eliminate indirect addressing. This certainly will pay off in an increase in computational speed. One can also take advantage of special machine features, such as vector registers, to further increase the performance of the computer code. A third approach is to reexamine an existing algorithm and to create a new implementation of an existing algorithm which is optimized for a particular machine architecture. This approach will be illustrated here for a matrix-vector multiply. The fourth approach is to design a completely new algorithm for solving a particular problem which executes very quickly on the supercomputer. In many cases, this algorithm would not be competitive on a conventional computer. This will be illustrated by an element-by-element iterative solver for linear systems of equations.

The first approach to vectorization offers the least improvement in computation speed. The second approach offers increases in computational speed; however, this aspect of vectorization will not be addressed here because it is a very straightforward result of applying rules for vectorization. The third and fourth techniques will be discussed because they potentially offer more increase in computation speed and they will illustrate creative approaches to vectorization.

3.1 New Implementation of an Existing Algorithm: Matrix-Vector Multiply

The large irregular sparse matrices which arise from biomedical modelling are difficult to handle on vector computers. On traditional machines, sparse techniques, such as the Yale Sparse Matrix Package [3], have been developed and are very economical in terms of both computational time and storage. Compressed band algorithms are also attractive on traditional machines. On a vector machine, both the Yale Sparse storage form and the compressed storage form induce indirect addressing and vector operations of extremely short length [6] which degrades the machine performance. The algorithm presented here will be a new implementation of the standard matrix-vector multiplication for finite element matrices which was developed to be attractive on vector computers. This technique can be applied when a matrix, A , can be written as a multisplitting or sum of individual matrices.

$$A = \sum_{e=1}^{E} A_e \quad .$$

Multisplittings occur in a variety of applications [5,7,8,12,13,16, 18], and the individual matrices will be called 'element' matrices even when they arise from other applications.

The multiplication of a global coefficient matrix with a vector can be written as

$$Au = (\sum_e A_e)u = \sum_e A_e u_e \qquad (2)$$

$$= \sum_e b_e = b$$

This process can be divided into two parts. The first consists of the multiplication of the individual coefficient matrices A_e with their respective components of the vector u_e . The second part of the calculation consists of the assembly of the individual or element portions of the product $b = \sum b_e$. On a traditional sequential computer, one would perform the assembly or addition of the matrix first and then perform the multiplication, thus reducing the total computation costs. However, on a vector computer, the elementwise multiplication is done first in parallel and is followed by the assembly of the element vectors, b_e . This results in an increase in the number of arithmetic operations; however, since the calculations vectorize the total execution time is reduced.

In order to simplify the discussion, the algorithm will be described for a case where each of the element matrices of A_e is 4×4. This corresponds to a finite element grid consisting of bilinear finite elements. General geometry and mixed element types will be discussed at the end of this section. First, consider a single element matrix-vector multiply, $A_e u_e$. This multiplicaiton can be done in an inner product, outer product or diagonal form [9]. The algorithm described here uses a diagonal product form. One can write the multiplicaiton by diagonals as

$$\begin{bmatrix} A_1^e & A_5^e & A_9^e & A_{13}^e \\ A_{14}^e & A_2^e & A_6^e & A_{10}^e \\ A_{11}^e & A_{15}^e & A_3^e & A_7^e \\ A_8^e & A_{12}^e & A_{16}^e & A_4^e \end{bmatrix} \begin{bmatrix} u_1^e \\ u_2^e \\ u_3^e \\ u_4^e \end{bmatrix} =$$

$$\begin{bmatrix} A_1^e \\ A_2^e \\ A_3^e \\ A_4^e \end{bmatrix} \begin{bmatrix} u_1^e \\ u_2^e \\ u_3^e \\ A_4^e \end{bmatrix} + \begin{bmatrix} A_5^e \\ A_6^e \\ A_7^e \\ A_8^e \end{bmatrix} \begin{bmatrix} u_2^e \\ u_3^e \\ u_4^e \\ u_1^e \end{bmatrix} + \begin{bmatrix} A_9^e \\ A_{10}^e \\ A_{11}^e \\ A_{12}^e \end{bmatrix} \begin{bmatrix} u_3^e \\ u_4^e \\ u_1^e \\ u_2^e \end{bmatrix} \tag{3}$$

$$+ \begin{bmatrix} A_{13}^e \\ A_{14}^e \\ A_{15}^e \\ A_{16}^e \end{bmatrix} \begin{bmatrix} u_4^e \\ u_1^e \\ u_2^e \\ u_3^e \end{bmatrix} = \begin{bmatrix} b_1^e \\ b_2^e \\ b_3^e \\ b_4^e \end{bmatrix}$$

The individual element data structures can be stacked as shown in Figure 4. Conceptually, the A_1^e entries are multiplied with the u_1^e entries for all the elements in the grid, then the A_2^e entry is multiplied by the u_2^e entry for all the elements in the grid, etc. In this manner, multiplication (2) can be performed in four vector multiply operations and two matrix addition instruction whose lengths are four times the number of elements in the grid. (A divide-and-conquer scheme is used, so the first addition is twice this length.) Note that this calculation can be done by replicating the vector $\underset{\sim}{u}_1^e$ twice and by using pointers to initiate the second, third and fourth multiplication. One could perform all of the multiplication in one vector instruction of length sixteen times the number of elements; however, this would require replication of the vector $\underset{\sim}{u}^e$ four times which would increase the storage requirement. For reasonable sized finite element grids with fifty or more elements, the vectors are long enough so that the start-up time for the extra three-vector multiples can be ignored. Table 1 gives the operations counts for the vector multiplication and vector addition instructions in the multiplication process for a variety of types of finite elements. If several element types are used, then one performs the calculations as if all elements had the maximum number of nodes. The element matrices A_e corresponding to smaller elements are zero filled to the larger element size and operations are performed on these zero entries. The number of vector operations is independent of the number of elements in the finite element grid. It is also independent of the nodal point ordering in the grid. Increasing the number of elements simply increases the length of the vector operations which adds to their efficiency.

Once the elementwise multiplication has been performed, it is necessary to assemble or add the results into the final form, $b = \sum b_e$. In the finite element application, this corresponds to adding contributions to node i from all of the elements which contain that node. This operation vectorizes and involves one gather/scatter per product [8].

Comparisons in computation time will be given for a compressed band data structure and for the elementwise algorithm. Sparse Matrix algorithms and full band matrix multiplication are not included because the first does not vectorize and the second is not competitive for irregular finite element problems. Timing comparisons for a matrix-vector multiply where the matrix comes from a finite element grid will be given for several regular and irregular grids. All of the regular grids consist of triangular quadratic elements which are arranged as shown in Figure 5. The irregular finite element grids for the arm, heart and monkey head will be considered. Table 2 contains the computation time per product both for the elementwise algorithm and for the vectorized, compressed band algorithm. Both of these algorithms vectorize and both involve one gather and scatter per product. The elementwise algorithm requires approximately one-third of the time that is required with the compressed band storage algorithm. In order to illustrate the savings that can be realized in an algorithm which uses the elementwise matrix multiplication, timing data is given for the Jacobi-Conjugate Gradient algorithm (J-CG) for solving linear systems of equations. Two variations are presented. One uses the compressed band algorithm to perform all matrix-vector multiplications on the right-hand side, and the other uses the elementwise technique presented here. This data is shown in Table 3. The data in Table 3 was produced using the vectorized version of the ITPACK library which is available on the CYBER 205 at Colorado State. The version which uses the elementwise product requires 1/3-1/2 of the time required by using a compressed band technique. In the case of regular grids with quadratic elements approximately 30% of the entries in the compressed band structure storage are zeros, which leads to a loss of efficiency. The savings in execution time, using the elementwise technique, is greater for irregular grids, because there are even more zero entries in the compressed band data structure.

It is clear from the data in Table 3 that the element wise vectorized matrix multiplication can result in tremendous savings in algorithms where a significant portion of time involves a matrix multiplication; this illustrates the gains which can be achieved by implementing a standard algorithm in a new way on a vector computer.

3.2 Development of a New Vectorized Algorithm: Overlapping-Block Iteration

The following algorithm illustrates the fourth approach to vectorization of the solution to heat transfer problems in biomedical applications. The technique which is presented here is not competitive on a traditional architecture computer; however, on a vector computer, it is very competitive for irregular grids. This new algorithm is a variation of a block iterative method for solving linear systems of equations [9].

In a block method, the unknowns in the problem are decomposed into groups or blocks, and the matrix (1) is partitioned according to the groups of unknowns [17,18,20]. If there are p nodes in a block, this leads to the structure

$$A = \left[\begin{array}{c|c|c} B_{1,1} & B_{1,2} & \\ \hline & B_{2,2} & \\ \hline & & \\ \hline & & B_{E,E} \end{array}\right] \left[\begin{array}{c} x_{1_1} \\ x_{1_2} \\ \vdots \\ x_{1_p} \\ \hline x_{2_1} \\ \vdots \\ x_{2_p} \\ \hline \vdots \\ \hline x_E \end{array}\right] = \left[\begin{array}{c} b_{1_1} \\ b_{1_2} \\ \vdots \\ b_{1_p} \\ \hline b_{2_1} \\ \vdots \\ b_{2_p} \\ \hline \vdots \\ \hline b_E \end{array}\right] \tag{4}$$

The diagonal blocks $B_{p,p}$ are $N_p \times N_p$ matrices, x_{e_i} is the ith unknown in the eth group. Given an initial estimate for the solution vector $x^{(0)}$, one can define a block iterative method as

$$x^{(n+1)} = C^{-1}\, B\, x^{(n)} + C^{-1}\, b \tag{5}$$

where superscripts are used for iteration numbers. The matrix B is a block matrix of the form

$$B = \left[\begin{array}{c|c|c|c} 0 & B_{1,2} & B_{1,3} & \cdots \\ \hline B_{2,1} & 0 & \ddots & \cdots \\ \hline & \ddots & \ddots & B_{E-1,E} \\ \hline & & B_{E,E-1} & 0 \end{array}\right] \tag{6}$$

and C is the block diagonal matrix containing the block diagonal entries of A

$$C = \begin{bmatrix} B_{1,1} & 0 & 0 \\ 0 & B_{2,2} & 0 \\ & & \ddots & \\ 0 & 0 & B_{E,E} \end{bmatrix} \tag{7}$$

In order to carry out the iterative procedure (5), it is necessary to be able to solve equations corresponding to the diagonal of A. These equations have the form

$$\underset{\sim}{B}_{j,j}\,\underset{\sim}{x}_j = \underset{\sim}{b}_j \tag{8}$$

where $\underset{\sim}{x}_j$ and $\underset{\sim}{b}_j$ are the unknowns and right-hand side from the jth block. If each block contains only one unknown, then this would reduce to a standard point method. However, the advantage of using block methods is that they will converge faster than point methods. The increase in the rate of convergence is directly proportional to the size of the matrix $B_{j,j}$ that must be inverted [7]. If there were only one block which contained all of the unknowns, C would coincide with the matrix A and the solution would be obtained directly in one step. If the blocks overlap and an unknown appears in more than one block at the end of each iteration, each block will have a new value for that unknown and $X_i^{(n+1)}$ is obtained by averaging the values obtained from each of the blocks which contain node i.

In this application, the unknowns will be grouped according to nodes on each element in the finite element grid. Since a given node will be in several elements, it will appear in several of the groups or blocks of unknowns, and the blocks will overlap. Ostrowski [18] has proved convergence for overlapping block techniques. On a traditional computer, the gain in convergence rate for this overlapping block method is offset by the increase in computations associated with having a particular node appear in several elements. On a vector machine, the elementwise operations vectorize, and the overlapping block method is attractive.

The basic overlapping block iterative method (5) can be accelerated using the conjugate gradient technique [8]. The conjugate gradient acceleration of this overlapping block method can be written

$$X^{n+1} = X^n + \lambda_n p_n \tag{9}$$

$$r^n = b - Ax^n \quad \text{(residual)} \tag{10}$$

$$\delta^n = X^{n+1} - X^n = [D]^{-1} \sum_{e=1}^{E} C_e^{-1} (F^e - (Ax^n)^e) \quad \text{pseudo-residual} \tag{11}$$

$([D]_{ii} = NE_i\delta_{ij})$ NE_i = # elements which contain node i

$$p^n = \begin{cases} \delta^0 & n = 0 \\ \delta^n + \alpha_n p^{n-1} & n = 1,2,\ldots.. \end{cases} \tag{12}$$

$$\alpha_n = \frac{(\delta^n, r^n)}{(\delta^{n-1}, r^{n-1})} \qquad n = 1,2,\ldots.. \tag{13}$$

$$\lambda_n = \frac{(\delta^n, r^n)}{(p^n, Ap^n)} \qquad n = 0,1,2,\ldots \tag{14}$$

The matrix-vector multiplies which appear in equations (11) and (14) were done using the elementwise data structure presented previously.

Timing comparisons for the overlapping block method were made on the CYBER 205 at Colorado State University. Comparisons were made for both regular grids and irregular grids. The regular grids had a uniform mesh of tensor product quadratic elements on a rectangular region. The irregular grids resulted from the three biomedical applications. The elementwise matrix-vector multiply presented here was used for all right-hand side calculations.

The timing comparisons for the regular grid are shown in Table 4 and the timing comparisons for the irregular grids are shown in Table 5. The band solver was a vectorized algorithm available in the MAJEV Library of the CYBER 205. J-CG ITPACK is the vectorized version of the point Jacobi conjugate gradient which uses the compressed band storage scheme. J-CG elem uses the vectorized element-by-element matrix multiplication technique which we developed for right-hand side calculations. Overlap.Block is the overlapping block iterative method described here. Several observations can be made. The band solver which was used in this study was a very efficiently vectorized version of the algorithm. It was particularly fast in solving tri-diagonal systems. However, once the problem size exceeds a few hundred unknowns, iterative methods solve the linear system of equations faster than the band solver. The vectorized element-by-element matrix-vector multiply improved the performance of the point Jacobi conjugate gradient method. For this reason, the overlapping block method should be compared to the Jacobi conjugate gradient method, both of which use the vectorized element-by-element right-hand side calculations. In the case of the irregular grids, even for a few number of unknowns (N = 148), the overlapping block iterative method has definite advantages over standard equation solving techniques. For the very regular type grids, the overlapping block method is competitive with standard iterative methods but shows no special advantages. The storage requirements for several standard equation solving techniques are given in Table 6. The vectorized ITPACK software uses the Purdue Sparse Matrix Storage Pattern, which requires less storage than the vectorized band solver. Of these two vectorizable algorithms, the overlapping block method requires far less storage. In fact, it is comparable to the storage required by the non-vectorizable Yale Sparse Matrix routines which are considered to have minimal storage requirements for sparse matrices.

4. Conclusion

When one works on a supercomputer, one can no longer simply count arithmetic operations to evaluate numerical algorithms. One must take into consideration the particular machine architecture and how well the numerical algorithm can take advantage of the machine architecture. Two practical approaches to vectorization have been demonstrated. One is to change the implementation of a current algorithm to enhance vectorization and the other is to develop new vectorized algorithms. The standard implementation of a matrix-vector multiply was modified to produce a very fast, vectorizable algorithm for finite element applications. A new iterative algorithm was developed which is very attractive for vector computers and for irregular finite element grids. This algorithm would not have been competitive on a scalar computer.

Acknowledgement: This work was supported in part by a grant from Control Data Corporation and the University Research Institute at The University of Texas at Austin.

References

1. E. B. Becker, G. F. Carey and J. T. Oden, Finite Elements: An Introduction, Vol. I, Prentice Hall, 1981.

2. R. C. Eberhardt and A. S. Shitzer, Eds., Heat Transfer in Medicine and Biology, Plenum Press, Nwe York, 1985.

3. S. C.Eisenstat, M. C. Gursky, M. H. Schultz and A. H. Sherman, "Yale Sparse Matrix Package I: The Symmetric Codes," Intl. J. Numer. Meth. Engin., 18, 1145-1151 (1982).

4. J. Farrant, "General Observations on Cell Preservation," in Low Temperature Preservation in Medicine and Biology, M. J. Ashwood-Smith and J. Farrant, Eds., Univ. Park Press, Baltimore, p. 1 (1980).

5. B. N. Jiang and G. F. Carey, "Subcritical Flow Computation Using Element-by-Element Conjugate Gradient Method," Proceedings: Fifth International Symposium on Finite Elements and Flow Problems, Austin, Texas, January 1984.

6. D. R. Kincaid, T. C. Oppe, J. R. Respess and D. M. Young, ITPACKV2C User's Guide, Center for Numer. Anal. Rpt., CNA-192, The University of Texas at Austin, February 1984.

7. L. J. Hayes, "A Vectorized Matrix-Vector Multiply and Overlapping Block Iterative Method," Proceedings: Super Computer Applications Symposium, Purdue, R. Numerich, Ed., Plenum Publishing Co., 1984.

8. L. J. Hayes and Ph. Devloo, "An Element-by-Element Block Iterative Method for Large Non-Linear Problems," ASME-WAM, New Orleans, Innovative Methods for Nonlinear Behavior, W. K. Liu, K. C. Park and T. Belytschko, eds., Pineridge Press, 1985.

9. L. J. Hayes, "Programming Supercomputers," Chapter 22 in Handbook of Numerical Heat Transfer, M. C. Minkowicz and E. M. Sparrow, Eds., John Wiley & Sons, 1986.

10. R. Hockney and C. Jesshope, Parallel Computers: Architecture, Programming and Algorithms, Adam Hilger, Ltd., Bristol, 1981.

11. K. Hwang and F. A. Briggs, Computer Architecture and Parallel Processing, McGraw-Hill, 1984.

12. J. R. Hughes, M. Levit and J. Winget, "Element-by-Element Implicit Algorithms for Heat Conduction,' J. of Engin. Mech., 109, No. 2, 576-585 (1983).

13. J. R. Hughes, M. Levit and J. Winget, "Element-by-Element Implicit Solution Algorithm for Problems of Structural and Solid Mechanics," Comp. Meth. in Appl. Mech. and Engin., 36, 241-254 (1983).

14. A. P. Mazur, "Cryobiology: The Freezing of Biological Systems," Science, 168, 939-949 (1970).

15. R. W. Olsen, R. C. Eberhardt, L. J. Hayes, et al., "Influence of Extracerebral Temperature on Cerebral Temperature Distribution During Induction of Deep Hypothermia with Subsequent Circulatory Arrest," 1984 ASME/WAM, 1984 Advances in Bioengineering, New Orleans.

16. D. P. O'Leary and E. B. White, "Multi-Splitting of Matrices and Parallel Solutions of Linear Systems," Computer Science Technical Report 1362, University of Maryland, Dec. 1983.

17. A. M. Ostrowski, "Iterative Solution of Linear Systems of Functional Equations," J. of Math. Anal. and Appl., 2, 351-369 (1961).

18. A. M. Ostrowski, "On the Linear Iteration Procedures for Symmetric Matrices," National Bureau of Standards Report No. 1844, August 1952, p. 23, 1968.

19. T. M. Shih, L. J. Hayes, M. C. Minkowicz, et al., "Parallel Computations in Heat Transfer," Int'l. J. of Heat Transfer, in press, 1985.

20. D. M. Young, Iterative Solutions of Large Linear Systems, Academic Press, New York (1971).

TABLE 1

Operation Counts for the Multiplication Process
(E = # Elements in the Grid)

Order of A_3	# of Vector Multiplications	# of Vector Additions	Length of Multiplies	Length of First Add	Length of Second Add	Length of Third Add
Linear Elements						
3 x 3	3	2	3 * E	3 * E	3 * E	
4 x 4	4	2	4 * E	8 * E	4 * E	
Quadratic Elements						
6 x 6	6	3	6 * E	18 * E	12 * E	6 * E
8 x 8	8	3	8 * E	32 * E	16 * E	8 * E
9 x 9	9	4	9 * E	36 * E	18 * E	9 * E

TABLE 2

Comparison of Computation Time per Product for a Matrix-Vector Multiply

	Elementwise (msec/product)	Compressed Band (msec/product)	SPEEDUP
REGULAR GRIDS:			
200 Elements (441 Unknowns)	0.3	8.0	63%
800 Elements (1,681 Unknowns)	1.0	2.4	58%
5408 Elements (11,025 Unknowns)	6.5	14.0	54%
IRREGULAR GRIDS:			
ARM 51 Elements (148 Unknowns)	0.16	0.46	65%
HEART 76 Elements (223 Unknowns)	0.20	0.60	66%
MONKEY HEAD 118 Elements (351 Unknowns)	0.30	0.90	66%

TABLE 3

Comparisons of Computation Times for the Jacobi-Conjugate Gradient Algorithm

		Total Time (Sec)		
	Number of Iterations	Compressed Band	Elementwise	SPEEDUP
REGULAR GRIDS:				
200 Elements (441 Unknowns)	33	0.045	0.015	67%
800 Elements (1,681 Unknowns)	68	0.208	0.102	49%
5408 Elements (11,025 Unknowns)	176	2.60	1.70	35%
IRREGULAR GRIDS:				
ARM (148 Unknowns)	77	0.058	0.017	71%
HEART (223 Unknowns)	128	0.117	0.037	68%
MONKEY HEAD (351 Unknowns)	249	0.280	0.100	64%

TABLE 4

CYBER 205 Timing Comparisons for Regular Grids

Method	Time/Iteration (Sec)	# Iterations	Total Time (Sec)
CASE A: 200 ELEM	441 Unknowns		
Bandsolver			0.053
J-CG ITPACK	0.00136	33	0.045
J-CG ELEM	0.00044	33	0.015
Overlap.Block	0.0007	36	0.025
CASE B: 800 ELEM	1681 Unknowns		
Bandsolver			0.433
J-CG ITPACK	0.0031	68	0.203
J-CG ELEM	0.0015	68	0.102
Overlap.Block	0.0024	60	0.144
CASE C: 5408 ELEM	11,025 Unknowns		
Bandsolver			--
J-CG ITPACK	0.015	176	2.60
J-CG ELEM	0.0096	176	1.70
Overlap.Block	0.015	114	1.71

TABLE 5

CYBER 205 Timing Comparisons for Irregular Grids

Method	Time/Iteration (Sec)	# Iterations	Total Time (Sec)
CASE I: Arm Cross-Section		148 Unknowns	
Bandsolver			0.021
J-CG ITPACK	0.00075	77	0.058
J-CG ELEM	0.000225	77	0.017
Overlap.Block	0.000374	36	0.013
CASE II: Heart		223 Unknowns	
Bandsolver			0.046
J-CG ITPACK	0.00091	128	0.117
J-CG ELEM	0.00029	128	0.037
Overlap.Block	0.00048	52	0.025
CASE III: Monkey Head		351 Unknowns	
Bandsolver			0.089
J-CG ITPACK	0.0011	249	0.280
J-CG ELEM	0.00041	249	0.100
Overlap.Block	0.00067	95	0.064

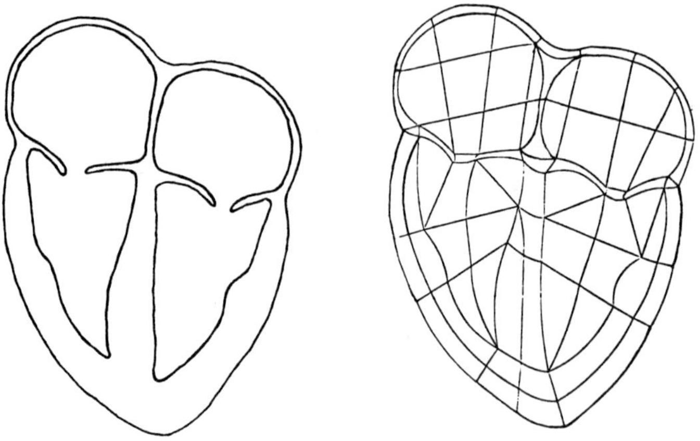

FIGURE 1 Human Heart and Finite Element Grid

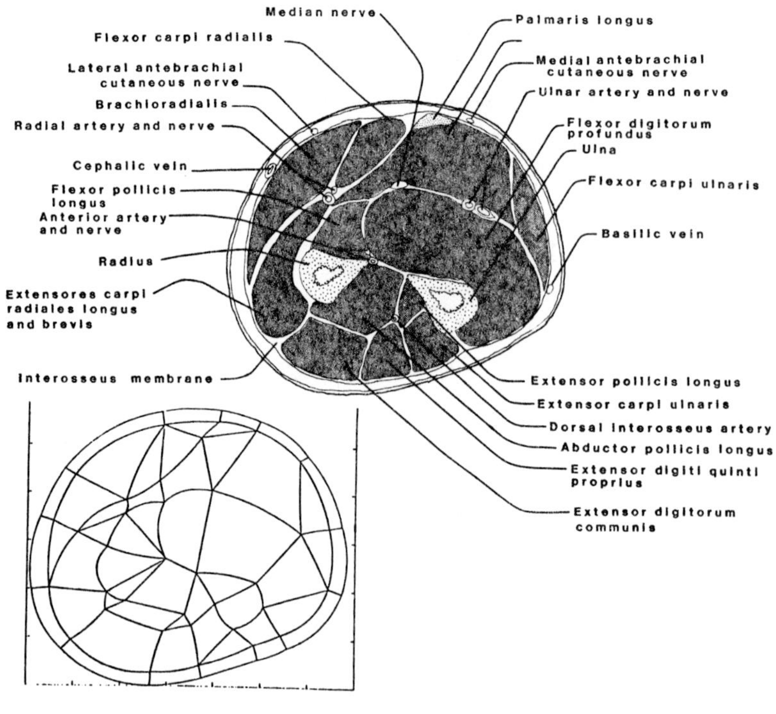

FIGURE 2 Human Forearm and Finite Element Grid

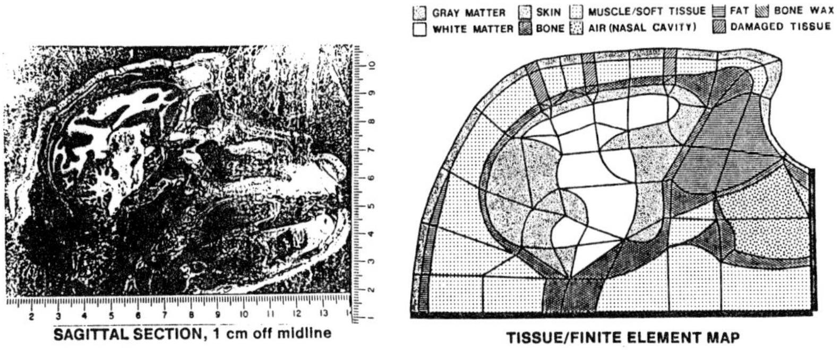

FIGURE 3 Maque Monkey Head and Finite Element Grid

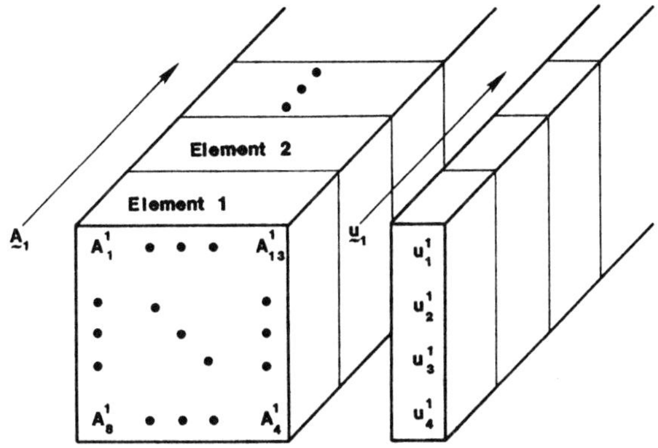

FIGURE 4 Stacked Element Matrices

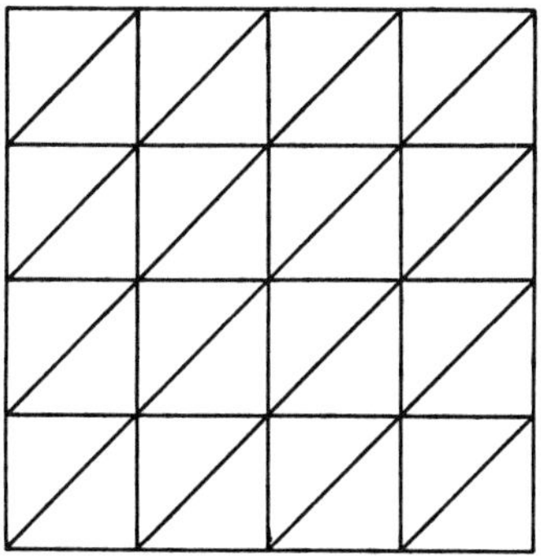

FIGURE 5 Regular Finite Element Grid

SPEAKERS

Prof. U. Ascher
Department of Computer Science
The University of British Columbia
2075 Westbrook Mall
Vancouver, B.C.
Canada V6T 1W5

Prof. R.E. Bank
University California San Diego
La Jolla, CA 92093
USA

Prof. P. Deuflhard
Konrad-Zuse-Zentrum fuer
Informationstechnik Berlin (ZIB)
Heilbronner Strasse 10
D-1000 Berlin 31
Fed. Rep. Germany

Dr. I.S. Duff
Computer Science and System Division
AERE Harwell
Oxford OX11 ORA
United Kingdom

Prof. B. Engquist
Department of Mathematics
University of California, Los Angeles
Los Angeles, CA 90024
USA

Prof. V. Friedrich
Technische Hochschule
DDR-90 Karl-Marx-Stadt
German Democratic Republic

Prof. Fu Hong-Yuan
Institute of Applied Physics and
Computational Mathematics
P.O. Box 8009, Beijing
The People's Republic of China

Dr. W.D. Gropp
Box 2158 Yale Station
Department of Computer Science
Yale University
New Haven, CT 06520
USA

Prof. Linda J. Hayes
Aerospace Engineering
Mechanics Department
The University of Texas at Austin
Austin, Texas 78712-1085
USA

Dr. F.K. Hebeker
Fachbereich Mathematik und Informatik
Universitaet Paderborn
Warburger Strasse 100
D-4790 Paderborn
Fed. Rep. Germany

Dr. H. Jarausch
RWTH Aachen
Institut fuer Geometrie
und Praktische Mathematik
Templergraben 55
D-5100 Aachen
Fed. Rep. Germany

Dipl.-Math. H. Kruse
Westfaelische Wilhelms-Universitaet Muenster
Institut fuer Numerische und Instrumentelle
Mathematik
Einsteinstrasse 62
D-4400 Muenster
Fed. Rep. Germany

Dr. P. Kraemer-Eis
Institut fuer Angewandte Mathematik
der Universitaet Bonn
Wegelerstrasse 6
5300 Bonn
Fed. Rep. Germany

Dr. P. Knabner
Universitaet Augsburg
Mathematisches Institut
Memmingerstrasse 6
D-8900 Augsburg
Fed. Rep. Germany

Prof. Dr. A. Louis
Institut fuer Mathematik
Technische Universitaet Berlin
Strasse des 17. Juni 136
D-1000 Berlin 12
Fed. Rep. Germany

Dipl.-Math. U. Nowak
Konrad-Zuse-Zentrum fuer
Informationstechnik Berlin (ZIB)
Heilbronner Strasse 10
D-1000 Berlin 31
Fed. Rep. Germany

Prof. A. Rizzi
Box 11021
S-16111 Bromma
Sweden

Prof. M. Smooke
Department of Computer Science
Yale University
10 Hillhouse Avenue
P.O. Box 2158 Yale Station
New Haven, CT 06520
USA

Prof. Uwe H. Suhl
FB10 WE6 Wirtschaftsinformatik
Freie Universitaet Berlin
Garystrasse 21
D-1000 Berlin 33
Fed. Rep. Germany

Prof. H.J. Wacker
Johannes Kepler Universitaet
Altenberger Strasse 69
A-4045 Linz
Austria

Dr. J. Warnatz
Institut fuer Physikalische Chemie
Im Neuenheimer Feld 253
D-6900 Heidelberg
Fed. Rep. Germany

Prof. H. Yserentant
Universitaet Dortmund
Fachbereich Mathematik
Postfach 500 500
Vogelpothweg
D-4600 Dortmund 50
Fed. Rep. Germany